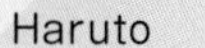

Haruto Akari Sara Yu

Study with your Friends!

How do we learn mathematics?

Based on the problem you find in your daily life or what you have learned, let's come up with a purpose.

1

The first problem of the lesson is written. On the left side, what you are going to learn from now on through the problem is written.

When you see the problem and think that you "want to think", "want to represent", "want to know", and "want to explore", that will be your "purpose" of your learning. You can find the purpose not only at the beginning of the lesson but in various timings and settings.

You can check your understanding and try more using what you have learned.

①

Let's try this problem first.

The starting point

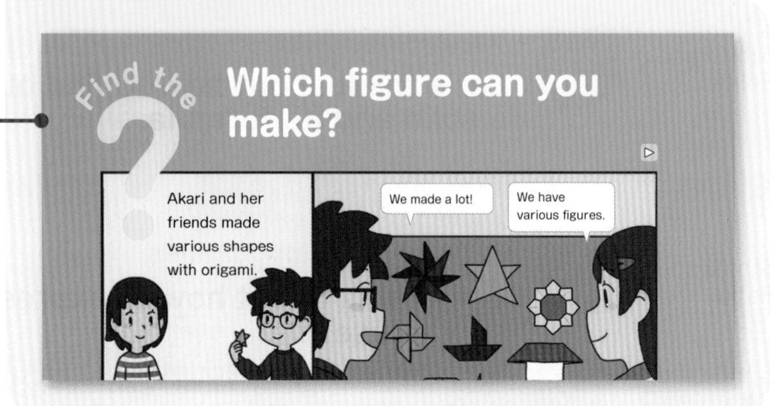

Find the
Which figure can you make?
Akari and her friends made various shapes with origami.
We made a lot!
We have various figures.

What you have learned today

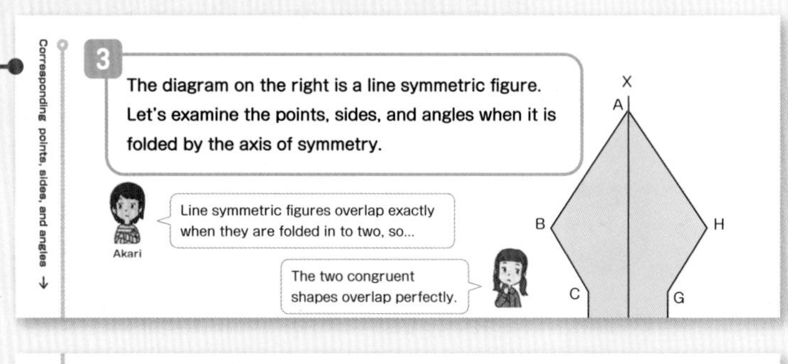

Corresponding points, sides, and angles
3
The diagram on the right is a line symmetric figure. Let's examine the points, sides, and angles when it is folded by the axis of symmetry.
Line symmetric figures overlap exactly when they are folded in to two, so...
Akari
The two congruent shapes overlap perfectly.
X
A
B
H
C
G

Want to explore

? Purpose When a line symmetric figure is folded by the axis of symmetry, what are the sizes of corresponding sides and angles?

1 The diagram below is a line symmetric figure that has straight line XY as the axis of symmetry. Let's answer the following questions.

2 Let's calculate the following.

① $\frac{2}{5} \times 2$ ② $\frac{4}{9} \times 2$ ③ $\frac{7}{11} \times 4$ ④ $\frac{6}{7} \times 5$

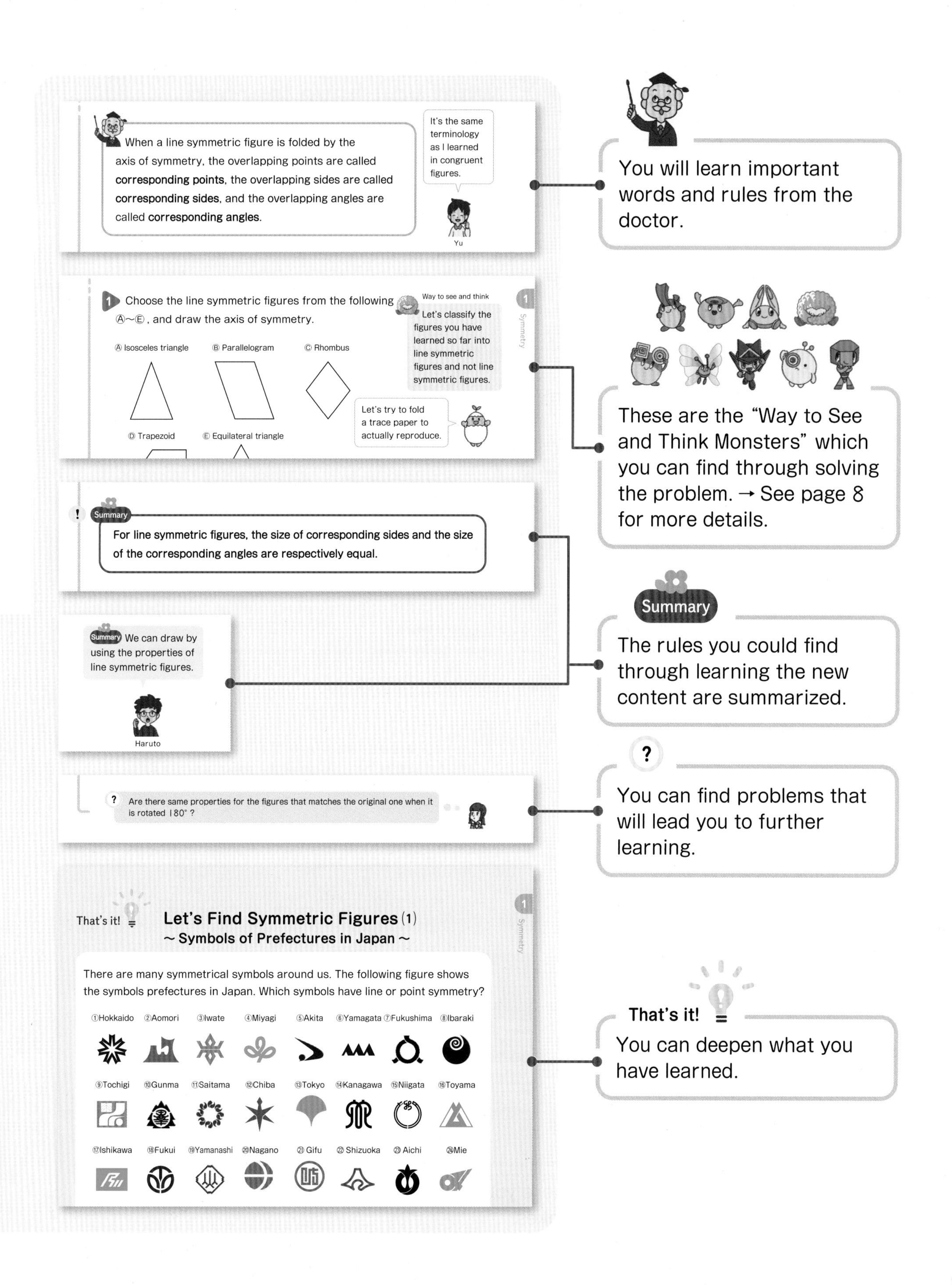
When a line symmetric figure is folded by the axis of symmetry, the overlapping points are called corresponding points, the overlapping sides are called corresponding sides, and the overlapping angles are called corresponding angles.
It's the same terminology as I learned in congruent figures.
Yu
You will learn important words and rules from the doctor.
1 Choose the line symmetric figures from the following Ⓐ~Ⓔ, and draw the axis of symmetry.
Ⓐ Isosceles triangle
Ⓑ Parallelogram
Ⓒ Rhombus
Ⓓ Trapezoid
Ⓔ Equilateral triangle
Way to see and think
Let's classify the figures you have learned so far into line symmetric figures and not line symmetric figures.
Let's try to fold a trace paper to actually reproduce.
Symmetry
These are the "Way to See and Think Monsters" which you can find through solving the problem. → See page 8 for more details.
Summary
For line symmetric figures, the size of corresponding sides and the size of the corresponding angles are respectively equal.
Summary We can draw by using the properties of line symmetric figures.
Haruto
Summary
The rules you could find through learning the new content are summarized.
? Are there same properties for the figures that matches the original one when it is rotated 180°?
?
You can find problems that will lead you to further learning.
That's it!
Let's Find Symmetric Figures (1)
~ Symbols of Prefectures in Japan ~
There are many symmetrical symbols around us. The following figure shows the symbols prefectures in Japan. Which symbols have line or point symmetry?
①Hokkaido ②Aomori ③Iwate ④Miyagi ⑤Akita ⑥Yamagata ⑦Fukushima ⑧Ibaraki
⑨Tochigi ⑩Gunma ⑪Saitama ⑫Chiba ⑬Tokyo ⑭Kanagawa ⑮Niigata ⑯Toyama
⑰Ishikawa ⑱Fukui ⑲Yamanashi ⑳Nagano ㉑Gifu ㉒Shizuoka ㉓Aichi ㉔Mie
That's it!
You can deepen what you have learned.

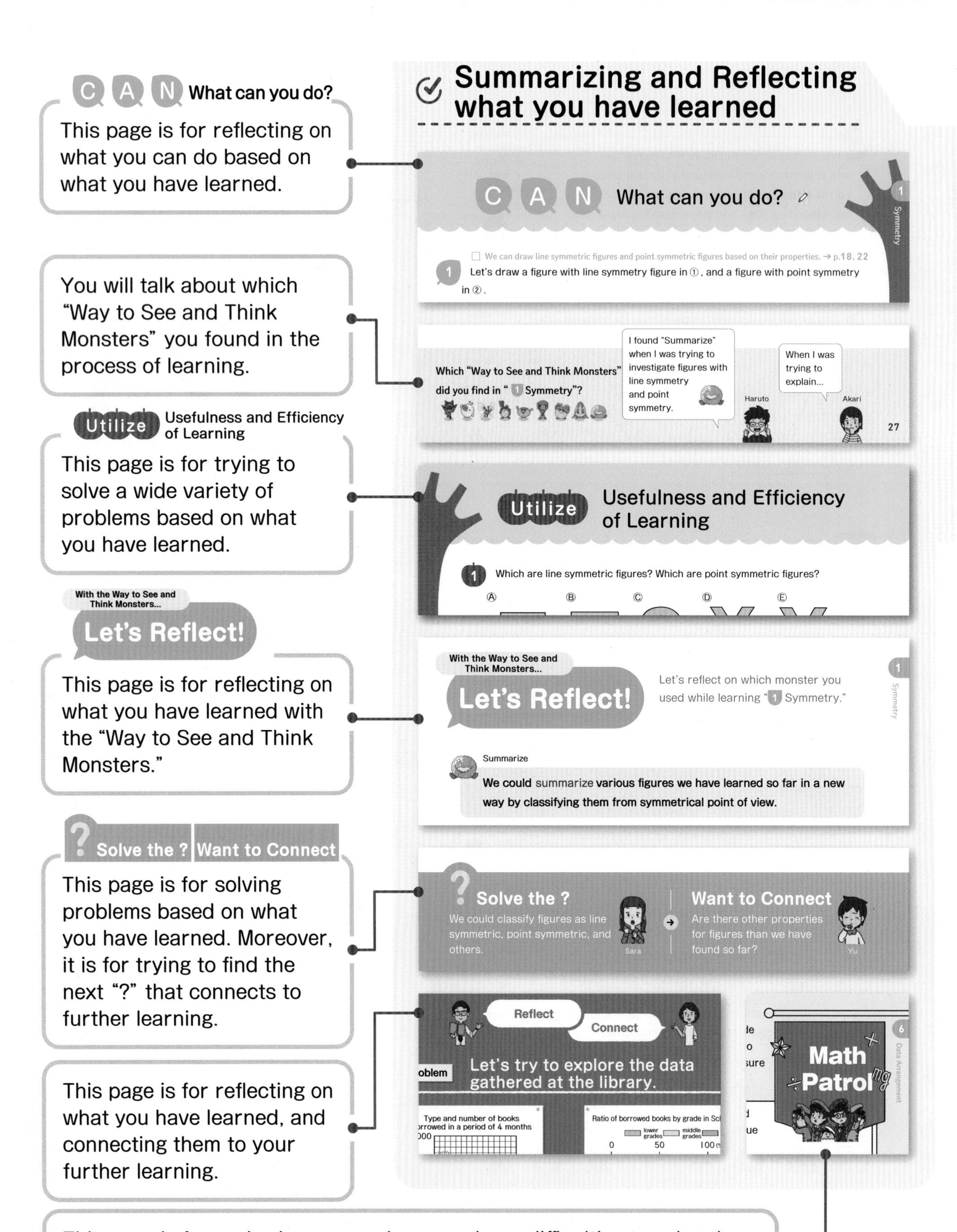
Summarizing and Reflecting what you have learned
C A N What can you do?
This page is for reflecting on what you can do based on what you have learned.
C A N What can you do?
Symmetry
We can draw line symmetric figures and point symmetric figures based on their properties. → p.18, 22
1 Let's draw a figure with line symmetry figure in ①, and a figure with point symmetry in ②.
You will talk about which "Way to See and Think Monsters" you found in the process of learning.
Which "Way to See and Think Monsters" did you find in "1 Symmetry"?
I found "Summarize" when I was trying to investigate figures with line symmetry and point symmetry.
Haruto
When I was trying to explain...
Akari
27
Utilize Usefulness and Efficiency of Learning
This page is for trying to solve a wide variety of problems based on what you have learned.
Utilize Usefulness and Efficiency of Learning
1 Which are line symmetric figures? Which are point symmetric figures?
Ⓐ Ⓑ Ⓒ Ⓓ Ⓔ
With the Way to See and Think Monsters...
Let's Reflect!
This page is for reflecting on what you have learned with the "Way to See and Think Monsters."
With the Way to See and Think Monsters...
Let's Reflect!
Let's reflect on which monster you used while learning "1 Symmetry."
Summarize
We could summarize various figures we have learned so far in a new way by classifying them from symmetrical point of view.
Solve the ? Want to Connect
This page is for solving problems based on what you have learned. Moreover, it is for trying to find the next "?" that connects to further learning.
Solve the ?
We could classify figures as line symmetric, point symmetric, and others.
Sara
Want to Connect
Are there other properties for figures than we have found so far?
Yu
This page is for reflecting on what you have learned, and connecting them to your further learning.
Reflect
Connect
Let's try to explore the data gathered at the library.
Type and number of books borrowed in a period of 4 months
Ratio of borrowed books by grade in Sc
lower grades
middle grades
0
50
100
Math Patrol
Data Arrangement
This page is for reviewing areas where you have difficulties to solve the problem, or are likely to make mistakes.

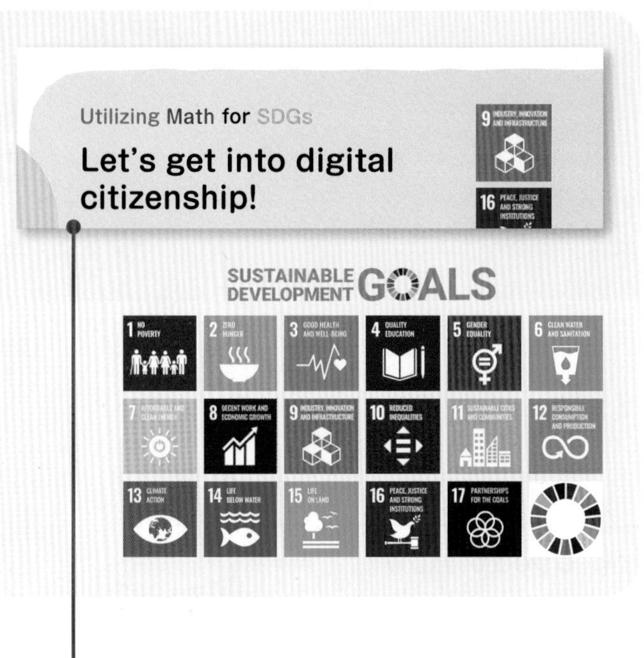

Utilizing Math for SDGs

The Sustainable Development Goals (SDGs) are a set of goals that we aim to achieve in order to create a world where we can live a life of safety and security. This page will help you think about what you can do for society and the world through math.

About the QR Code

Some of the pages include the QR code which is shown on the right.

▷ … You can learn how to draw a diagram and how to calculate by watching a movie.

… You can learn by actually moving and operating the contents.

… You can learn by reflecting on what you have learned previously in your previous grades.

… You can utilize it to know the solution to the problems that you couldn't find out the answer, or to try various problems.

… You can deepen your learning by actually looking at the materials including the website.

Dear Teachers and Parents

This textbook has been compiled in the hope that children will enjoy learning through acquiring mathematical knowledge and skills.
The unit pages are carefully written to ensure that students can understand the content they are expected to master at that grade level. In addition, the "More Math!" section at the end of the book is designed to ensure that each student has mastered the content of the main text, and is intended to be handled selectively according to the actual conditions and interests of each child. We hope that this textbook will help children develop an interest in mathematics and become more motivated to learn.

The sections marked with this symbol deal with content that is not presented in the Courses of Study for that grade level, thus do not have to be studied uniformly by all children.

QR codes are used to connect to Internet content by launching a QR code-reading application on a smartphone or tablet and reading the code with a camera.
The QR Code can be used to access content on the Internet.

https://r6.gakuto-plus.jp/s601

Note: This book is an English translation of a Japanese mathematics textbook. The only language used in the contents on the Internet is Japanese.

Infectious Disease Control
In this textbook, pictures of activities and illustrations of characters do not show children wearing masks, etc., in order to cultivate children's rich spirit of communicating and learning from each other. Please be careful to avoid infectious diseases when conducting classes.

Becoming a Writing Master

The notebook can be used effectively.

- To organize your own thoughts and ideas
- To summarize what you have learned in class
- To reflect on what you have learned previously Let's all try to become notebook masters.

Write today's date. → January 25th

Write the problem of the day that you must solve. → Problem

Let's think about how to find out the volume of the figure shown on the right.

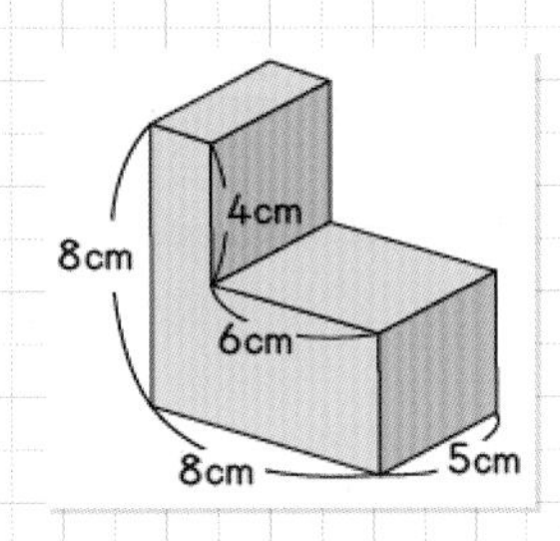

Let's write down what you thought while thinking about the solution of the problem as "purpose". → 〈Purpose〉 How can we find out the volume of a figure that is not a cuboid nor a cube?

Write your ideas or what you found about the problem. → ○ My idea

I divided the figure into 2 cuboids and calculated.

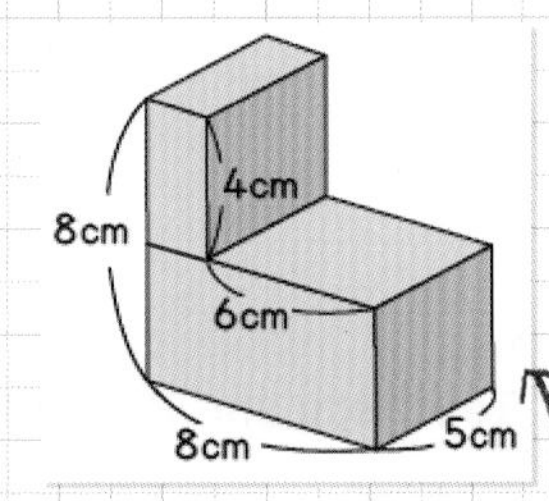

~~$5 \times 2 \times 4 = 40$~~
~~$5 \times 8 \times 5 = 200$~~
~~$40 + 200 = 240$~~
~~Answer: 240 cm^3~~

Here, I made a mistake with 5 cm.

Tips for Writing 1

$5 \times 8 \times 4 = 160$
$40 + 160 = 200$
Answer: 200 cm^3

Tips for Writing ❶

When you make a mistake, don't erase it so that it will be easier to understand when you look back at your notebook later.

Tips for Writing ❷

By finding the "Way to See and Think Monsters," it will connect you to what you have learned previously.

Tips for Writing ❸

By writing down what you would like to try more, it will lead you to further learning.

○ Sara's idea

Subtract the small cuboid from the big cuboid.

$5 \times 8 \times 8 = 320$

$5 \times 6 \times 4 = 120$

$320 - 120 = 200$

Answer: $200\ cm^3$

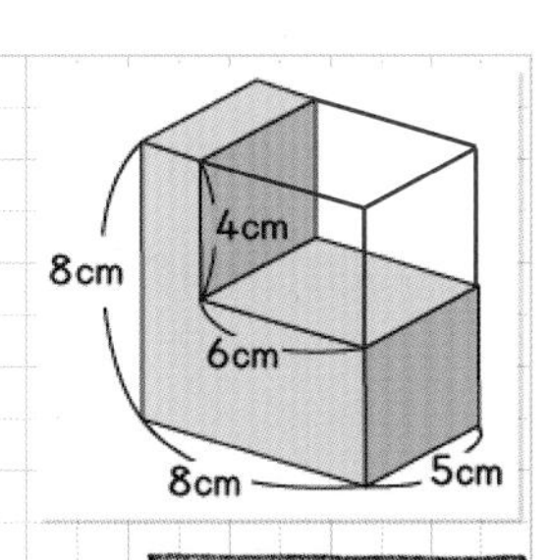

where did I find it before?

This is the same as in the case of area.

Same way

Write the classmate's ideas you consider good.

<Summary>

Even when the figure is not a cuboid or a cube, we can find out the volume by dividing the figure or subtracting the dented part.

Summarize what you have learned today.

<Reflection>

I found out that the volume of a figure can be found out by changing the unknown figure to a figure such as cuboids.

<What I want to do next>

I want to find out the volume of various figures.

Reflect on your class, and write down the following;

- What you learned.
- What you found out.
- What you can do now.
- What you don't know yet.

While learning mathematics...

Based on what I have learned previously...

Why does this happen?

There seems to be a rule.

You may be in situations like above. In such case, let's try to find the "Way to See and Think Monsters" on page 9. The monsters found there will help you solve the mathematics problems. By learning together with your friend and by finding more "Way to See and Think Monsters," you can enjoy learning and deepening mathematics.

What can we do at these situations?

I think I can use 2 different monsters at the same time...

→ You may find 2 or 3 monsters at the same time.

I came up with the way of thinking which I can't find on page 9.

→ There may be other monsters than the monsters on page 9. Let's find some new monsters by yourselves.

Now let's open to page 9 and reflect on the monsters you found in the 5th grade. They surely will help your mathematics learning in the 6th grade!

Representing ways of thinking in mathematics

Way to See and Think Monsters

Unit

If you set the unit...

Once you have decided one unit, you can represent how many using the unit.

Summarize

If you try to summarize..

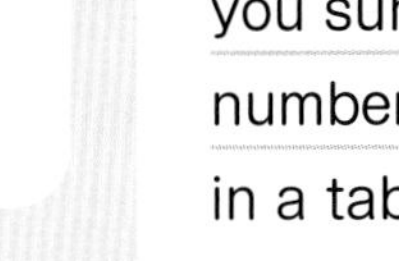

It makes it easier to understand if you summarize the numbers or summarize in a table or a graph.

Other Way

If you represent in other ways...

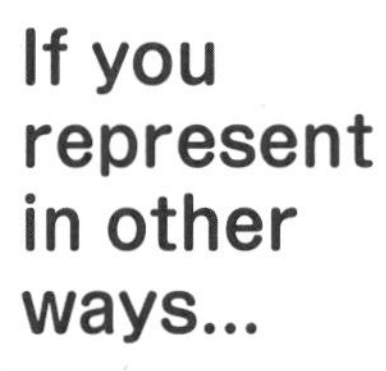

If you represent in other something depending on your purpose, it is easier to understand.

Align

If you try to align...

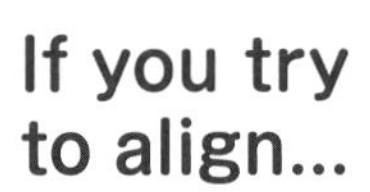

You can compare if you align the number place and align the unit.

Change

If you try to change the number or the figure...

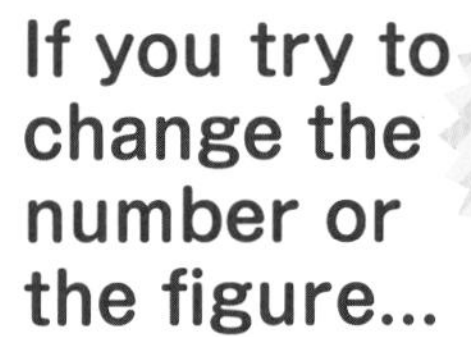

If you try to change the problem a little, you can understand the problem better or find a new problem.

Divide

If you try to divide...

Decomposing numbers by place value and dividing figures makes it easier to think about problems.

Why

You wonder why?

Why does this happen? If you communicate the reasons in order, it will be easier to understand for others.

Rule

Is there a rule?

By examining, you can find rules and think using rules.

Same Way

Can you do it in a similar way?

If you find something the same or similar to what you have learned, you can understand.

Ways to think learned in the 5th grade

Numbers and Calculations

Unit

Setting the unit for each place value, and thinking how many you have.

38.05 = 10 × [3] + 1 × [8] + 0.1 × [0] + 0.01 × [5]

Summarize

Categorizing whole numbers based on the remainder when divided by 2.

Whole Number	
Even numbers	Odd numbers
0, 2, 4, 6, …	1, 3, 5, 7, …

Rule / Same Way

Calculations with decimal numbers can be done using rules, in the same way as in whole numbers.

2.1 × 2.3 = [4.83]
(10 times ↓, 10 times ↓, ↑ $\frac{1}{100}$)
21 × 23 = [483]

5.76 ÷ 3.2 = [1.8]
(10 times ↓, 10 times ↓)
57.6 ÷ 32 = [1.8]

Other Way

When the quotient is represented in the fractions, the quotient can be obtained for any kind of division.

$$● \div ▲ = \frac{●}{▲}$$

Align

Even if fractions, decimals, and whole numbers are mixed, they can be compared by aligning them to fractions or decimals.

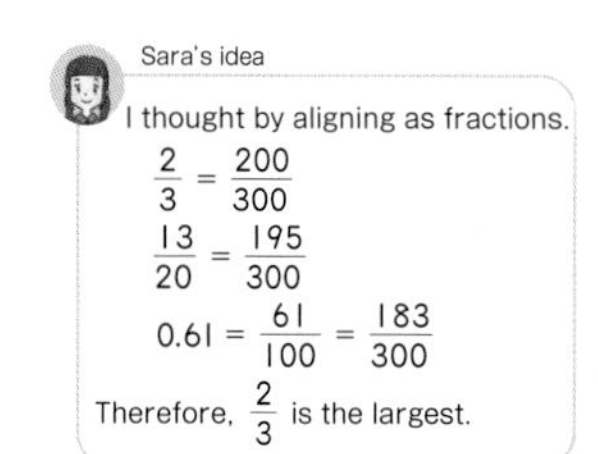
Sara's idea

I thought by aligning as fractions.

$\frac{2}{3} = \frac{200}{300}$

$\frac{13}{20} = \frac{195}{300}$

$0.61 = \frac{61}{100} = \frac{183}{300}$

Therefore, $\frac{2}{3}$ is the largest.

Align

Fractions with different denominators can be compared by aligning the denominators.

$$\frac{5}{6} = \frac{5 \times 4}{6 \times 4} = \frac{20}{24}$$

$$\frac{7}{8} = \frac{7 \times 3}{8 \times 3} = \frac{21}{24}$$

Figure

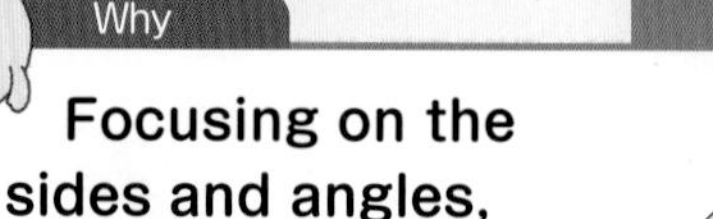

Why

Focusing on the sides and angles, you can explain why the two figures are congruent.

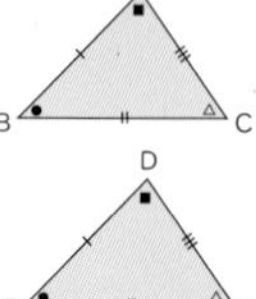

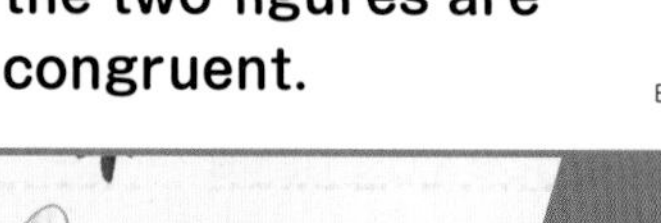

Divide / Rule

Finding out the rules while confirming with various methods.

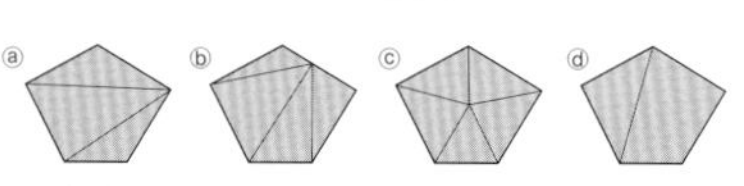

Change

Transform the figure you want to find the area of into the figure you have studied before.

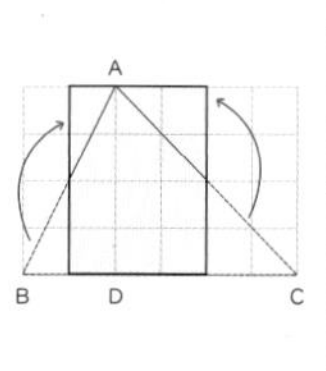

Rule / Same Way

Understanding the rule between the circumference and diameter.
Pi = Circumference ÷ Diameter
Pi is always the same, and it is 3.14159….

	4 cm diameter	8 cm diameter	12 cm diameter
Circumference (cm)	12.6	25.1	37.7
Diameter (cm)	4	8	12

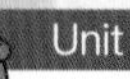

Unit

The same as in length and area, we can set one unit to represent volume as a number.

1 cm, 1 cm, 1 cm

Rule

Find out the rule of the two quantities if one becomes double, triple,..., then the other also becomes double, triple,...

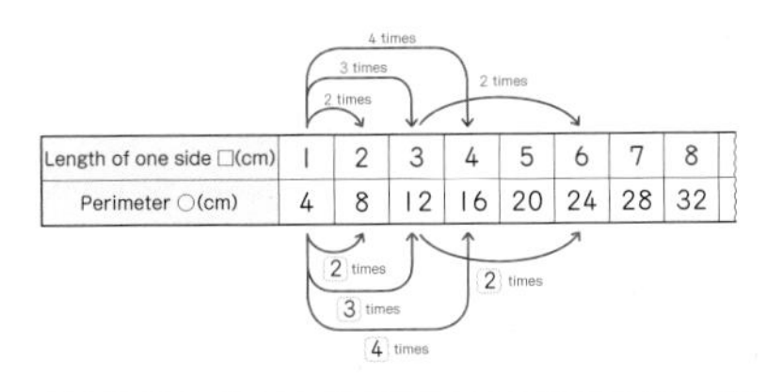

Align

We can compare two related quantities by aligning to either one of them, such as the size and the number of people for the degree of how crowded it is, or the time and the distance for the speed.

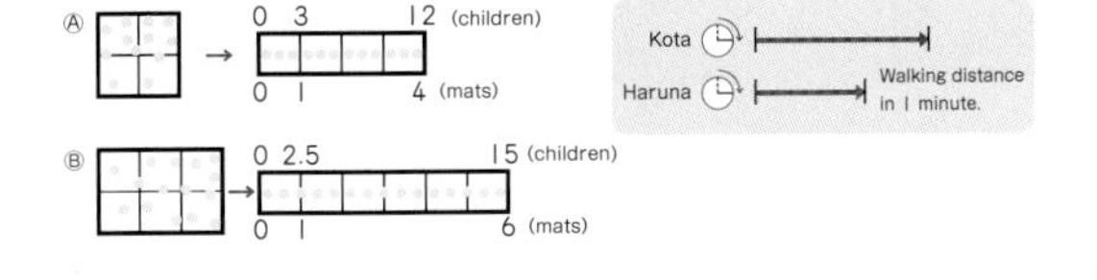

Unit

When there are two quantities, the relationship between the two can be expressed using proportions by setting either quantity as "1".

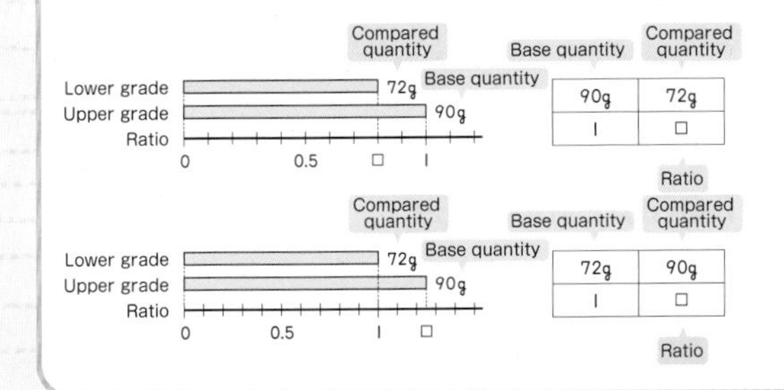

Other Way

According to what you want to know, decide the type of graph and represent using it.

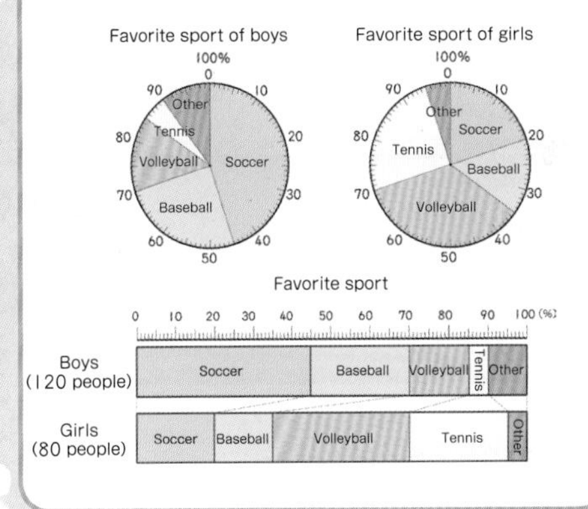

Why

Identify the problem, decide how to collect and organize data, and can explain the conclusion.

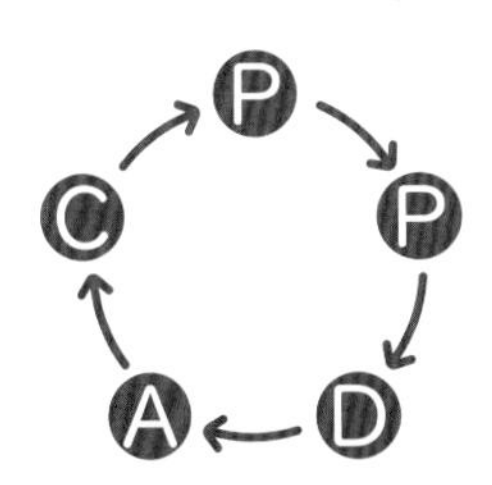

Find the ?

Which figure can you make?

Akari and her friends made various shapes with origami.
1
We made a lot!
We have various figures.
2
Can we classify them into groups?
Windmill
Ship
House
Sun flower
Crane
Shuriken
Star
Happo Shuriken
Samurai helmet
Dolphin
What should I pay attention to?
3

What are the properties of each figure?

Symmetry

Let's explore the shapes that overlap through folding or rotating.

1 Figures with line symmetry

Classifying figures →

1 Akari classified the figures as follows. Let's answer the following questions.

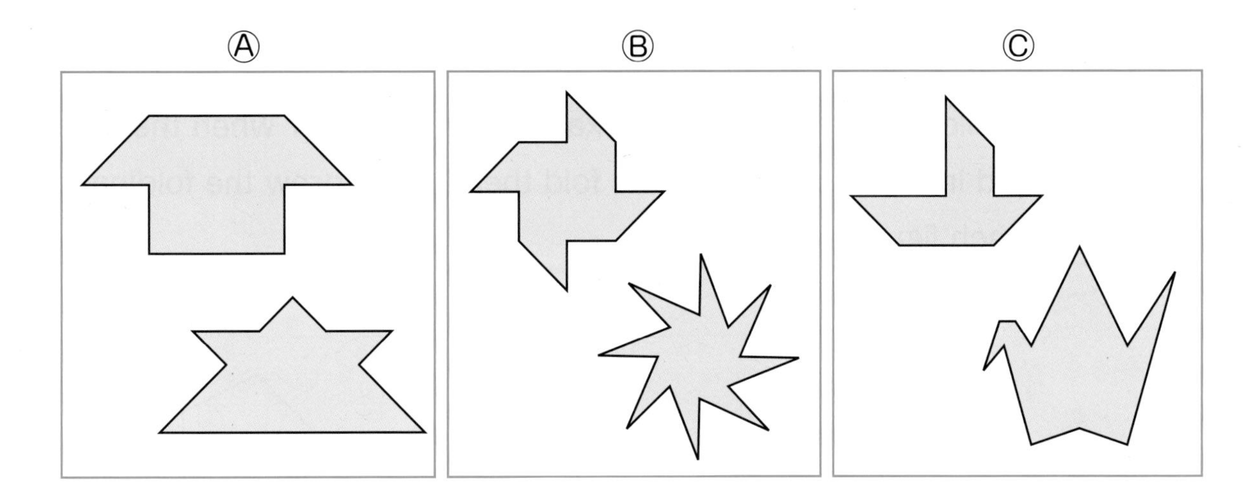

\ Want to explore /

? Purpose From what point of view are the figures classified?

1. What kind of group can Ⓐ, Ⓑ, Ⓒ be described?
2. Which of the groups Ⓐ, Ⓑ, Ⓒ do the following ⓐ, ⓑ, and ⓒ belong to?

ⓐ Dolphin

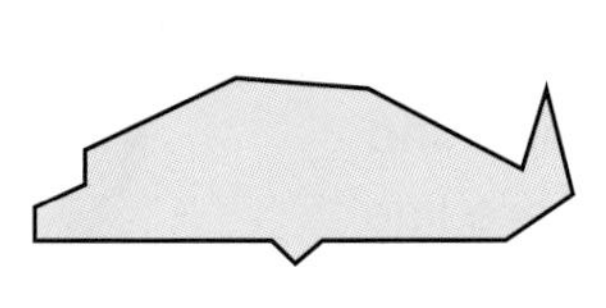

ⓑ Star

ⓒ Shuriken

Let's examine by cutting the figures on page 257.

Some shapes overlap exactly with the other part when folded in two, or when it is rotated 180°. ▷

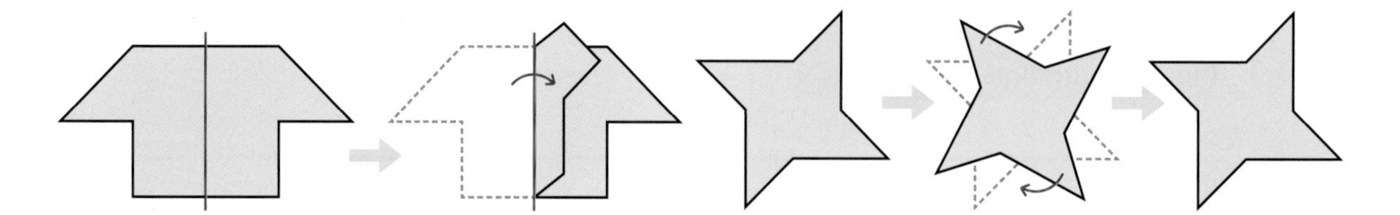

? What kind of figures can be overlapped exactly when folded in two?

2

The following figures overlap exactly with the other when they are folded in two. How should you fold them? Let's draw the folding line on each figure below.

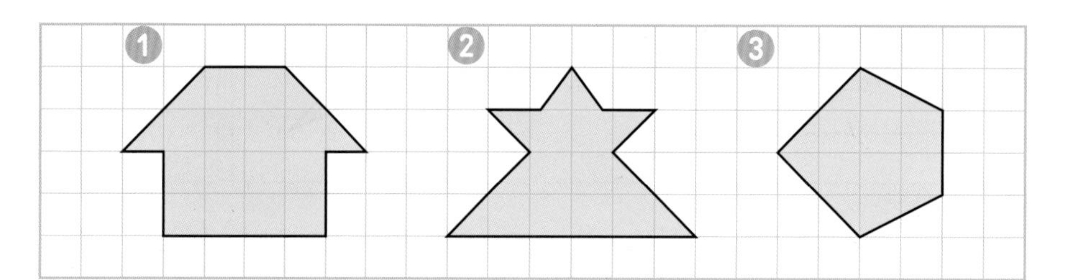

A figure has **line symmetry** when it can be folded along a straight line and the two halves overlap exactly. The folding line is called the **axis of symmetry** or **line of symmetry**.

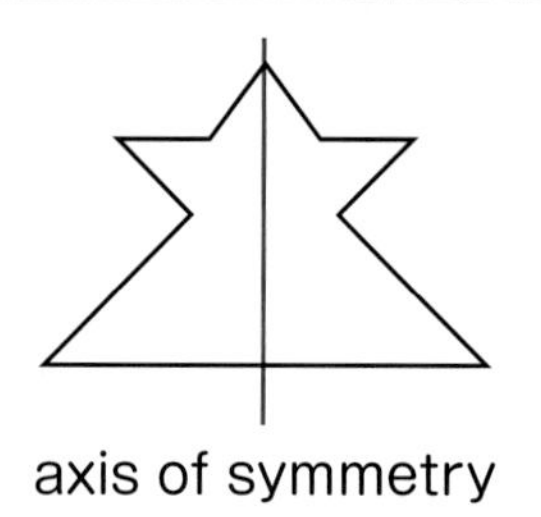

Two congruent figures overlapped perfectly.

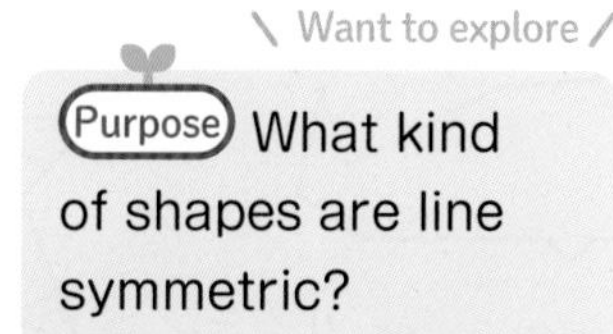

Want to explore
Purpose What kind of shapes are line symmetric?

1 Choose the line symmetric figures from the following Ⓐ～Ⓔ, and draw the axis of symmetry.

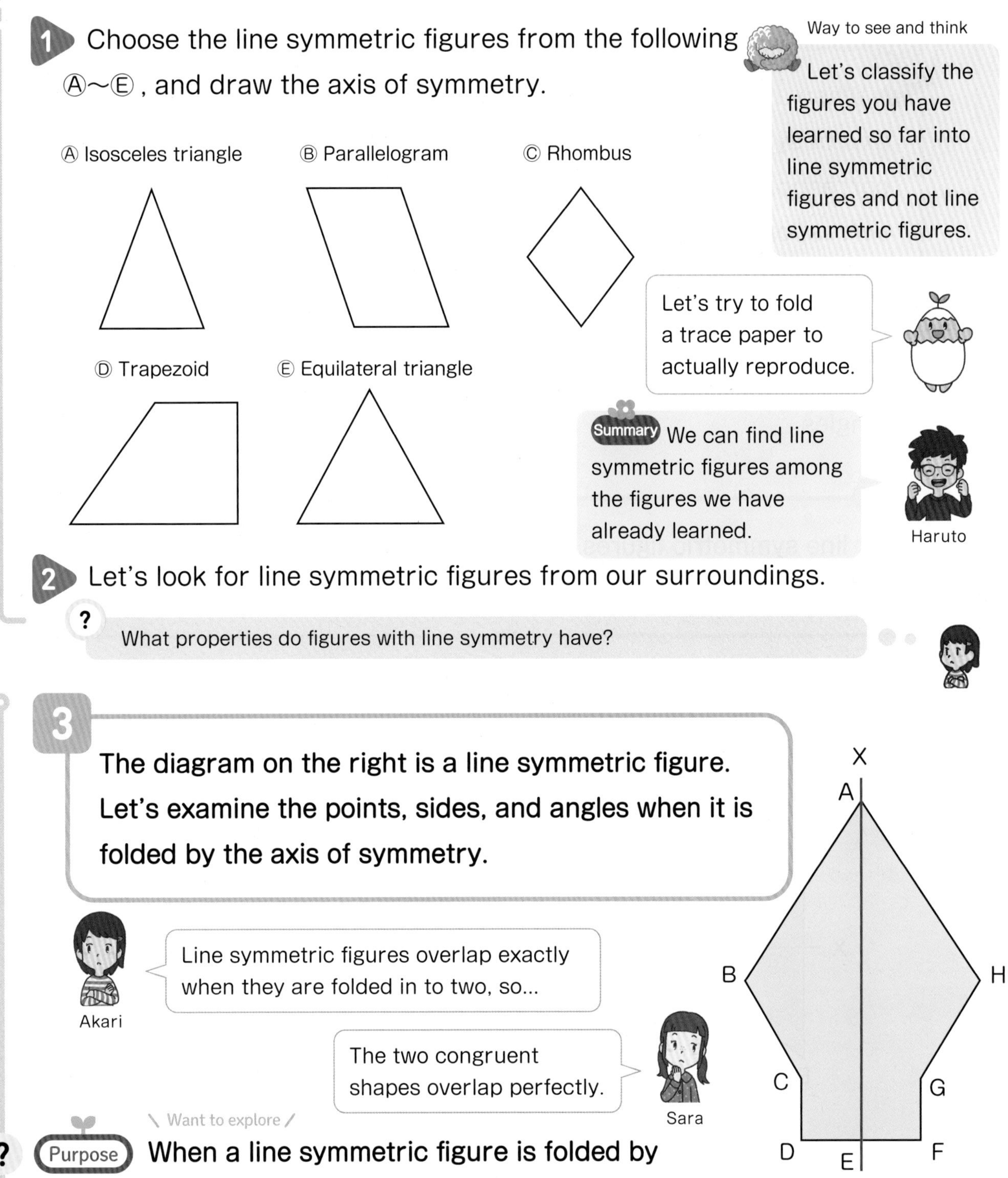
Ⓐ Isosceles triangle
Ⓑ Parallelogram
Ⓒ Rhombus
Ⓓ Trapezoid
Ⓔ Equilateral triangle
Way to see and think
Let's classify the figures you have learned so far into line symmetric figures and not line symmetric figures.
Let's try to fold a trace paper to actually reproduce.
Summary We can find line symmetric figures among the figures we have already learned.
Haruto
Line symmetric figures overlap exactly when they are folded in to two, so...
Akari
The two congruent shapes overlap perfectly.
Sara
X
A
B
H
C
G
D
E
F
Y
axis of symmetry

2 Let's look for line symmetric figures from our surroundings.

? What properties do figures with line symmetry have?

Corresponding points, sides, and angles ↓

3 **The diagram on the right is a line symmetric figure. Let's examine the points, sides, and angles when it is folded by the axis of symmetry.**

＼ Want to explore ／

? Purpose **When a line symmetric figure is folded by the axis of symmetry, what are the sizes of corresponding sides and angles?**

① Which are the respective overlapping points for points B and point C?

② Which are the respective overlapping sides for side AB and side CD?

③ Which are the respective overlapping angles for angle B and angle F?

When a line symmetric figure is folded by the axis of symmetry, the overlapping points are called **corresponding points**, the overlapping sides are called **corresponding sides**, and the overlapping angles are called **corresponding angles**.

It's the same terminology as I learned in congruent figures.
Yu

④ Let's find out the relationship between the corresponding sides and angles.

Summary

For line symmetric figures, the size of corresponding sides and the size of the corresponding angles are respectively equal.

1 The diagram below is a line symmetric figure that has straight line XY as the axis of symmetry. Let's answer the following questions.

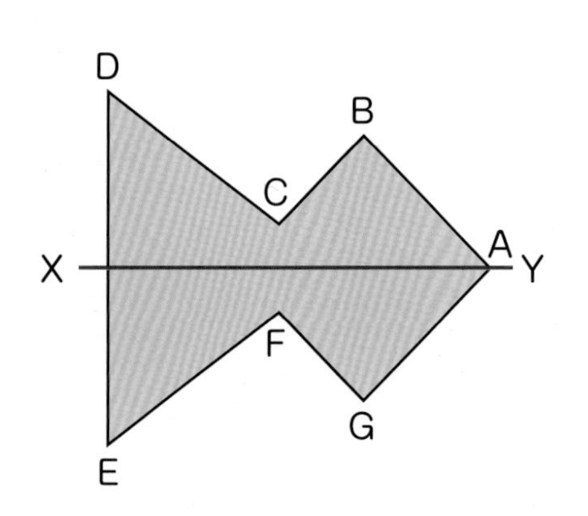
D
B
C
X
A
Y
F
G
E

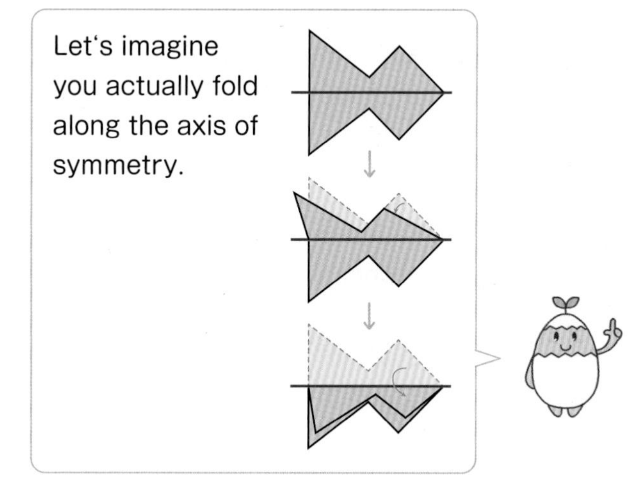
Let's imagine you actually fold along the axis of symmetry.

① Which are the respective corresponding points for points D, C, and B?

② Which are the respective corresponding sides for sides AB, BC, and CD?

③ Which are the respective corresponding angles for angles B, C, and D?

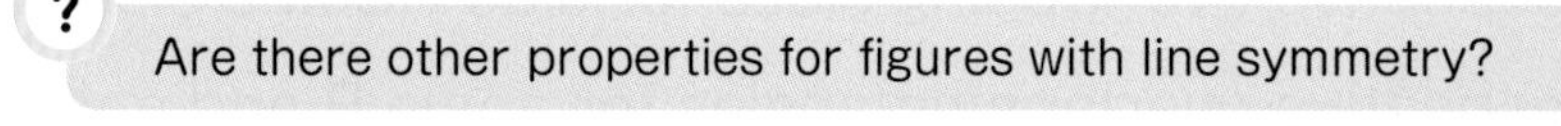
Are there other properties for figures with line symmetry?

4

Let's examine the following things about the line symmetric figure on the right.

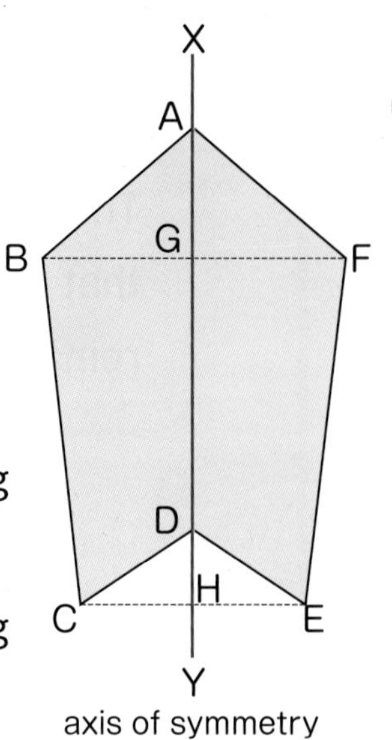

\ Want to explore /

?  Are there other properties for symmetric figures?

1. How does straight line BF, connecting the corresponding points B and F, intersect with the axis of symmetry?
2. How does straight line CE, connecting the corresponding points C and E, intersect with the axis of symmetry?
3. Let's compare the length of straight lines BG and FG. Let's compare the length of straight lines CH and EH.

! **Summary**

For line symmetric figures, the intersection between the straight line that connects two corresponding points and the axis of symmetry is perpendicular. The length from the axis of symmetry to each of the two corresponding points is equal.

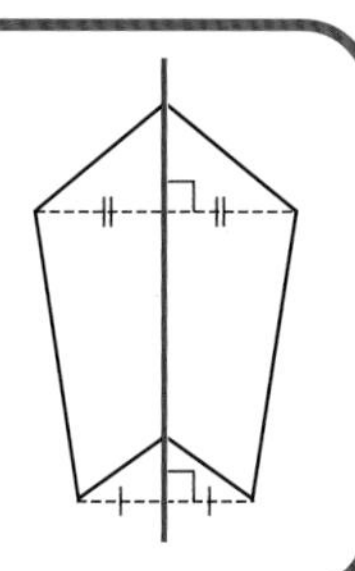

1 The diagram on the right is a line symmetric figure that has straight line XY as the axis of symmetry. Let's answer the following questions.

① How does line CE intersect with the axis of symmetry?

② The length of line BJ is 25 mm. What is the length, in millimeters, is line FJ?

③ Let's draw in the diagram the corresponding point M to point L.

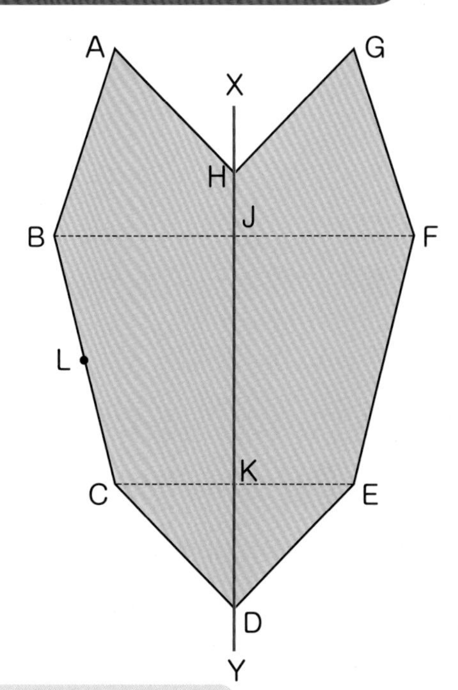

? Can we draw a figure with line symmetry based on the properties we have found so far?

5 **The following diagram represents half of a line symmetric figure that has the straight line XY as the axis of symmetry. Let's draw the remaining half to complete the figures.**

①

X

Y

\ Want to try /

(Purpose) How can we draw a line symmetric figure?

Way to see and think

Can you explain this using the fact that the lengths of the corresponding sides are equal and that the length from the axis of symmetry to the two corresponding points are equal?

②

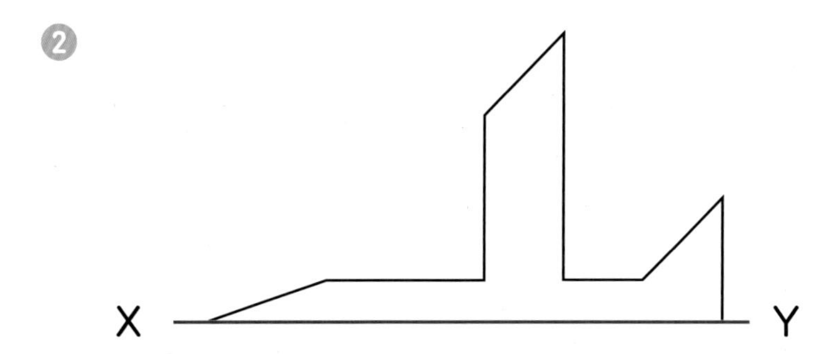

Summary We can draw by using the properties of line symmetric figures.

1 Write down the properties of line symmetry that you used to complete figure ② in your notebook, and explain them to your friends.

? Are there same properties for the figures that matches the original one when it is rotated 180° ?

1 In the following diagram, the figure matches the original figure when they are rotated 180° around point " · ". Let's confirm.

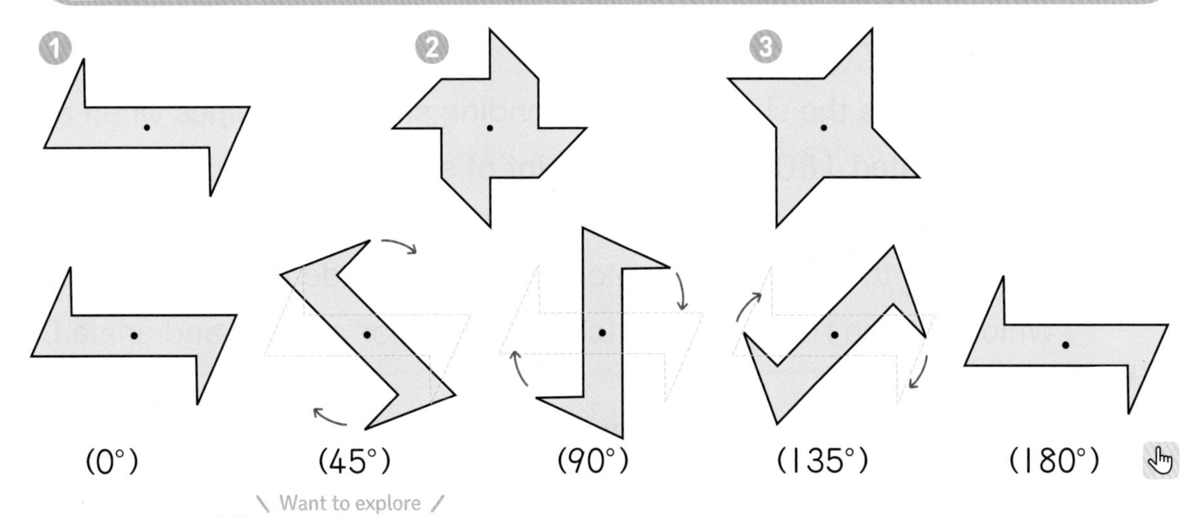

\ Want to explore /

Purpose Let's examine the figures classified on page 13 by cutting out the figures on page 257 and rotating them 180 degrees.

A figure has **point symmetry** if the figure matches the original one when it is rotated 180° with respect to a center point. The center point is called **point of symmetry** or **center of symmetry**.

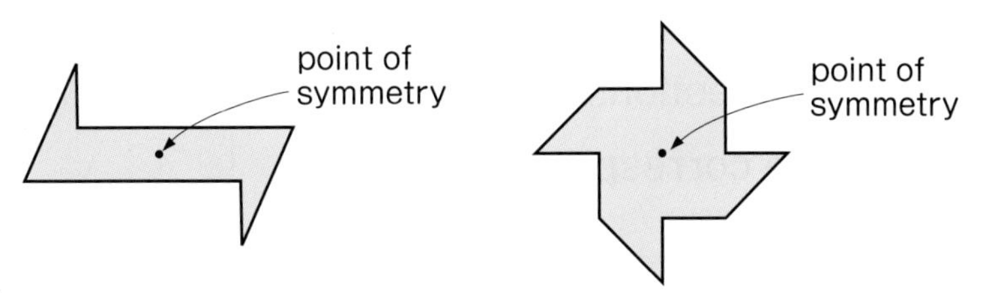

The figure of sunflower on page 12 is both line symmetry and point symmetry.

1 Let's look for point symmetric figures from our surroundings.

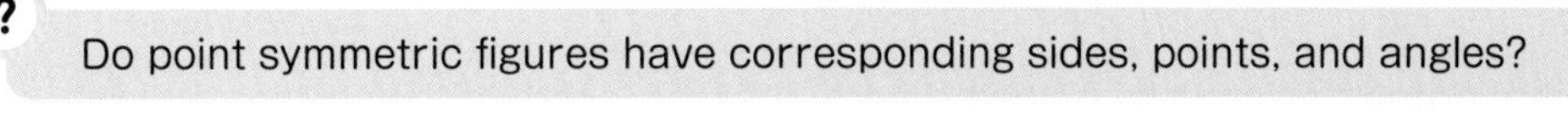

?
Do point symmetric figures have corresponding sides, points, and angles?

2

The diagram on the right is a point symmetric figure considering point O as the point of symmetry. Let's reproduce the figure in a trace paper and explore the points, sides, and angles when the figure is rotated 180° around the point of symmetry.

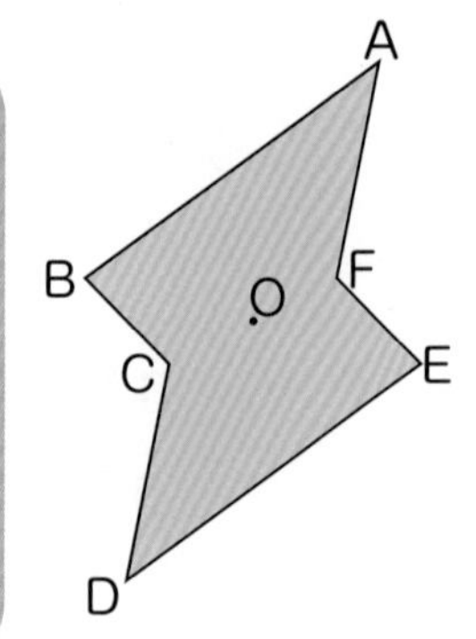

\ Want to explore /

What are the sizes of corresponding sides and angles when a figure is rotated 180° around its point of symmetry?

1. Which are the respective matching points for point B and point C?
2. Which are the respective matching sides for side AB and side BC?
3. Which are the respective matching angles for angle B and angle D?

When a point symmetric figure is rotated 180° around the point of symmetry, the matching points are called **corresponding points**, the matching sides are called **corresponding sides**, and the matching angles are called **corresponding angles**.

4. Let's examine the sizes of corresponding sides and angles.

Summary

In point symmetric figures, the length of the corresponding sides and the size of the corresponding angles are respectively equal.

1 The diagram on the right is a point symmetric figure considering point O as the point of symmetry. Let's answer the following questions.

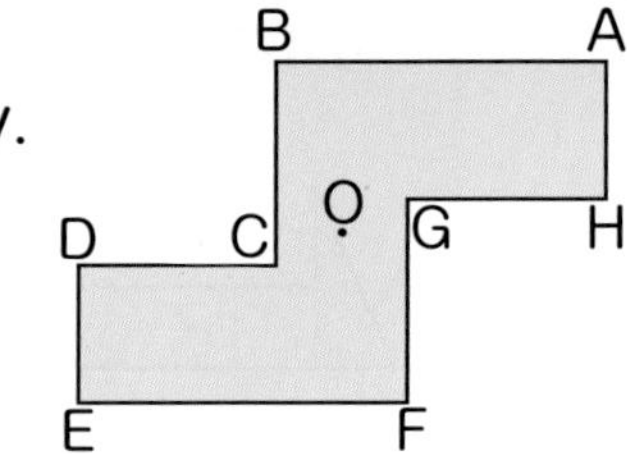

① Which are the respective corresponding points for points A, B, and C ?

② Which are the respective corresponding sides for sides AB, BC, and CD ?

③ Which are the respective corresponding angles for angles A, D, and F ?

? Are there properties other than the property that the length of the corresponding lines and the size of the corresponding angles are respectively equal?

3 Let's examine the following about the point symmetric figure on the right.

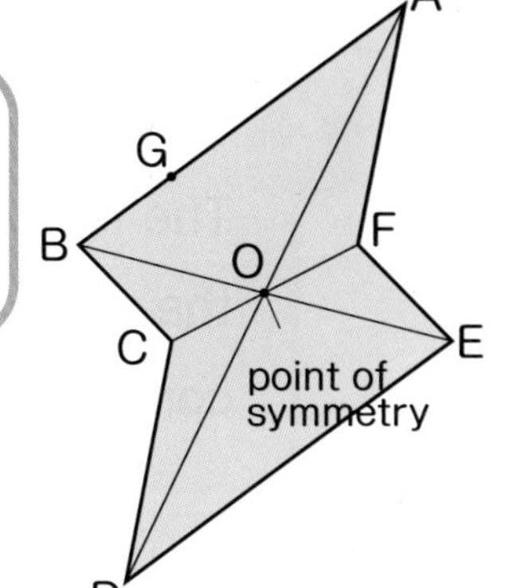

\ Want to explore /

Purpose What other properties do figures with point symmetry have?

1. The straight lines AD, BE, and CF are connecting corresponding points. Where do these straight lines intersect?
2. Let's draw in the figure the corresponding point H to point G on side AB.
3. Let's compare the length of straight lines AO and DO. Let's compare the length of straight lines GO and HO.

Let's draw on a trace paper and actually rotate the figure.

Summary

For point symmetric figures, the straight line that connects two corresponding points always passes through the point of symmetry. The length from the point of symmetry to each of the corresponding points is equal.

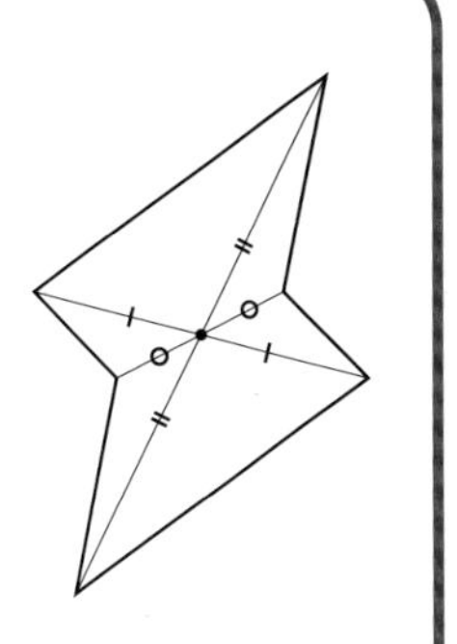

Way to see and think

The point of symmetry is on the line connecting the two corresponding points.

1 The diagram on the right is a point symmetric figure. Let's answer the following questions.

① Let's draw the point of symmetry. Then, explain how you found the point of symmetry.

② Let's draw in the figure the corresponding point B to point A.

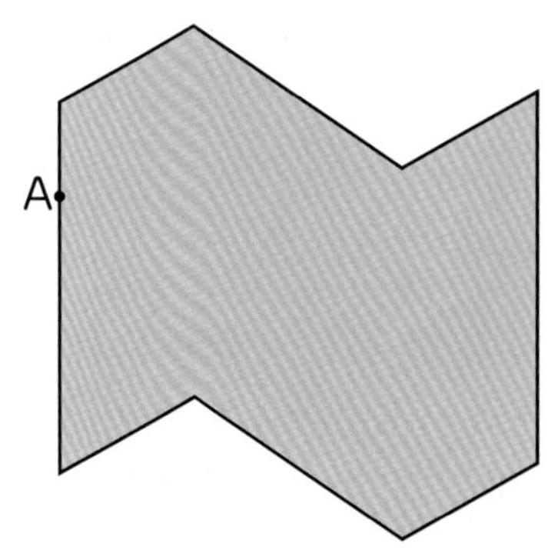

? Can we draw a figure with point symmetry based on the properties we have found so far?

4

The following figure shows half of a point symmetric figure that has the point O as the point of symmetry. Let's draw the remaining half to complete the figure.

①

O

②

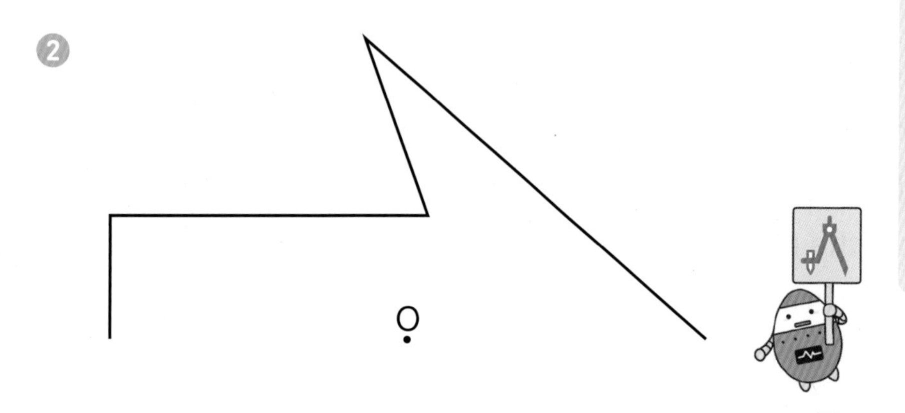

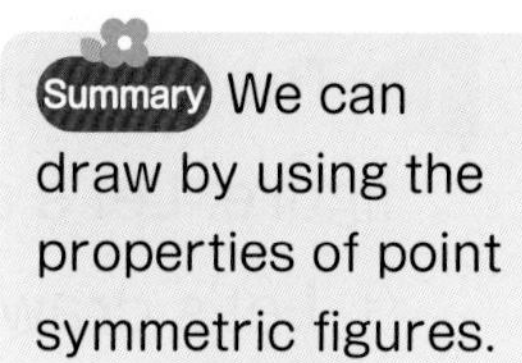

\ Want to try /

Purpose How can we draw point symmetric figures?

Akari

Way to see and think

Can you explain this using the fact that the lengths of the corresponding sides are equal and that the lengths from the point of symmetry to the two corresponding points are equal?

Summary We can draw by using the properties of point symmetric figures.

Haruto

1 Write down the properties of point symmetry that you used to complete figure ② in your notebook, and explain them to your friends.

? Are some of the figures we have learned so far line symmetry or point symmetry?

That's it!

Let's Find Symmetric Figures (1)
~ Symbols of Prefectures in Japan ~

There are many symmetrical symbols around us. The following figures shows the symbols prefectures in Japan. Which symbols have line or point symmetry?

①Hokkaido ②Aomori ③Iwate ④Miyagi ⑤Akita ⑥Yamagata ⑦Fukushima ⑧Ibaraki

⑨Tochigi ⑩Gunma ⑪Saitama ⑫Chiba ⑬Tokyo ⑭Kanagawa ⑮Niigata ⑯Toyama

⑰Ishikawa ⑱Fukui ⑲Yamanashi ⑳Nagano ㉑ Gifu ㉒ Shizuoka ㉓ Aichi ㉔Mie

 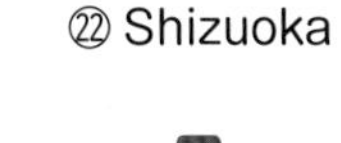              

㉕Shiga ㉖Kyoto ㉗Osaka ㉘Hyogo ㉙Nara ㉚Wakayama ㉛Tottori ㉜Shimane

               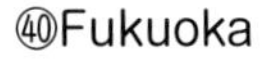

㉝Okayama ㉞Hiroshima ㉟Yamaguchi ㊱Tokushima ㊲Kagawa ㊳Ehime ㊴Kochi ㊵Fukuoka

㊶Saga ㊷Nagasaki ㊸Kumamoto ㊹Oita ㊺Miyazaki ㊻Kagoshima ㊼Okinawa

How about the symbols of the prefecture and the municipality you live in?

Yu

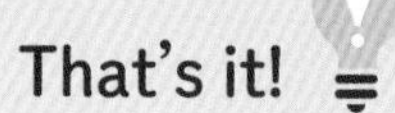

Let's Find Symmetric Figures (2)

There are many symmetrical symbols around us.
Look for shapes you have seen in town.

Traffic Signs in Japan

Ⓐ  Ⓑ Ⓒ  Ⓓ Ⓔ Ⓕ  Ⓖ

Ⓗ  Ⓘ Ⓙ Ⓚ Ⓛ  Ⓜ  Ⓝ

      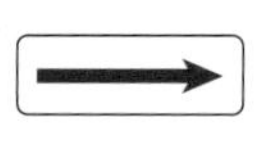

Map symbols

ⓐ  ⓑ  ⓒ ⓓ  ⓔ ⓕ ⓖ ⓗ

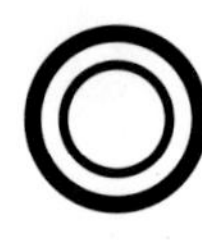       

ⓘ ⓙ ⓚ ⓛ

Pictograms were used in the Tokyo Olympics held in 2021.

Akari

Pictograms

ⒶToilets 

ⒷMen 

ⒸWomen 

ⒹNursery 

ⒺEscalator 

ⒻTrash box 

ⒼCollection facility for the recycling products

ⒽTelephone

ⒿStairs 

ⓀBus 

ⓁCoin lockers 

ⓂTrain 

1 Let's explore the things shown below about the following quadrilaterals.

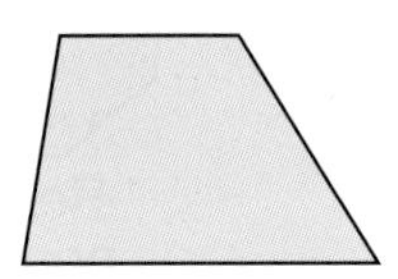
Trapezoid

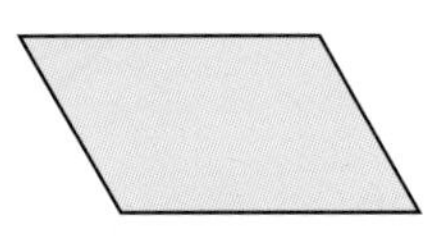
Parallelogram

Rectangle

Square

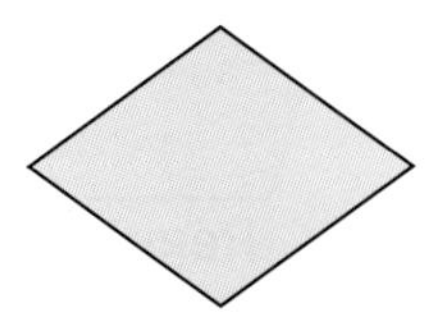
Rhombus

① Let's classify the quadrilaterals to line symmetric and point symmetric figures, and write ◯ or × in the table. As for line symmetric figures, how many axes of symmetry does each quadrilateral have?

\ Want to explore /

Purpose Let's find out if any of the various quadrilaterals have line symmetric or point symmetric figures.

Sara

	Line symmetry	Number of axes of symmetry	Point symmetry
Trapezoid			
Parallelogram			
Rectangle			
Square			
Rhombus			

② Let's draw the point of symmetry in the figures of point symmetry.

③ Let's make a presentation of what you notice from the table.

1 Let's explore the things below about the following triangles.

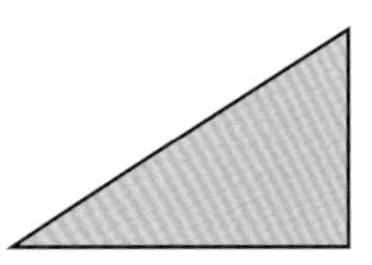
Right triangle

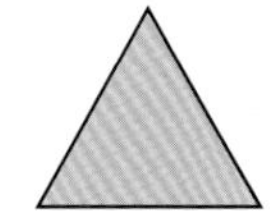
Equilateral triangle

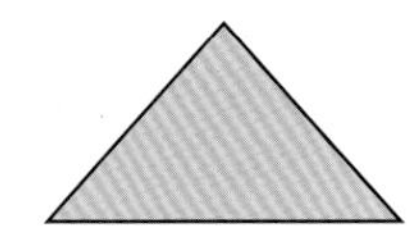
Isosceles triangle

① Which are line symmetric triangles? How many axes of symmetry does each triangle have?

② Is there any point symmetric triangle?

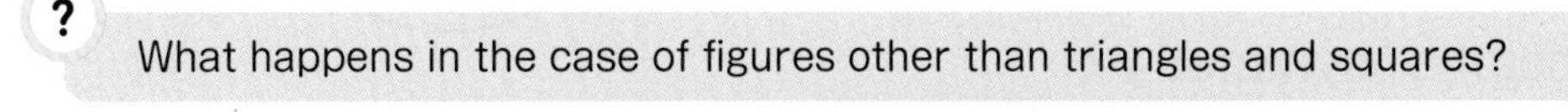

?
What happens in the case of figures other than triangles and squares?

2 Let's explore the things shown below about the following regular polygons.

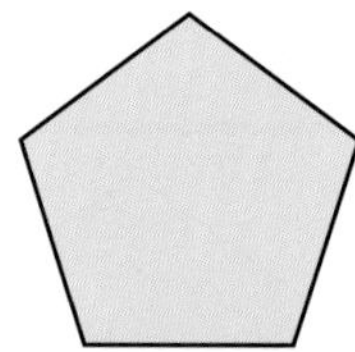

Regular pentagon

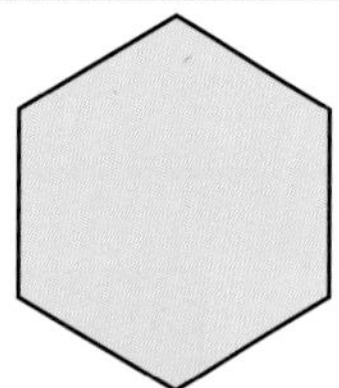
Regular hexagon

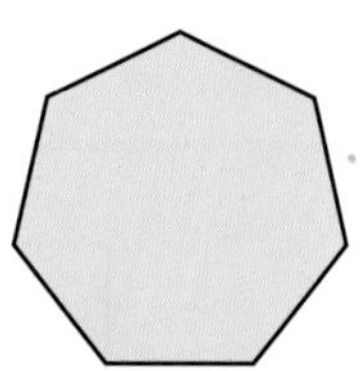
Regular heptagon

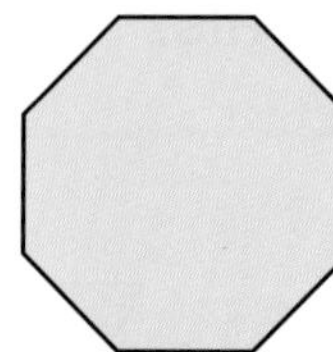
Regular octagon

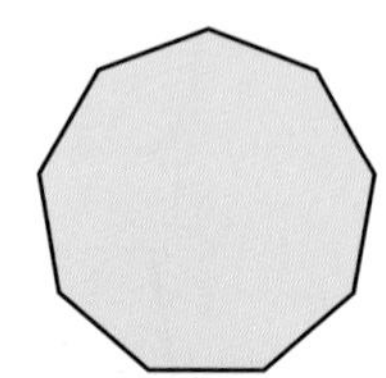
Regular nonagon

\ Want to explore /

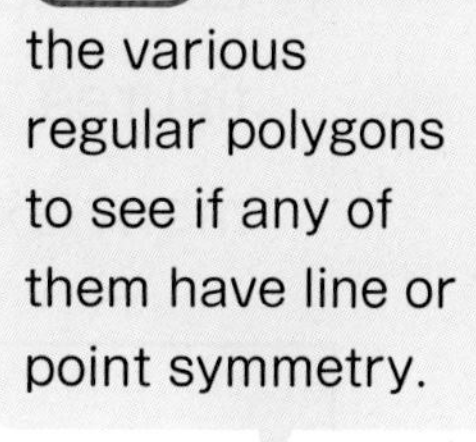
Purpose Examine the various regular polygons to see if any of them have line or point symmetry.

Yu

❶ Let's classify the regular polygons to line symmetric and point symmetric figures, and write ○ or × in the table. As for line symmetric figures, how many axes of symmetry does each have?

	Line symmetry	Number of axes of symmetry	Point symmetry
Regular pentagon			
Regular hexagon			
Regular heptagon			
Regular octagon			
Regular nonagon			

Way to see and think

We can find out the rules by putting the data on the table.

❷ Let's draw the point of symmetry in the figures of point symmetry.

❸ Let's write down what you found out so far and discuss with your friends.

Let's add an equilateral triangle and square in the table as regular polygons.

1 Let's explore the circle.

① Is the circle a line symmetric figure? How many axes of symmetry does it have?

② Is circle a point symmetric figure? Where is the point of symmetry?

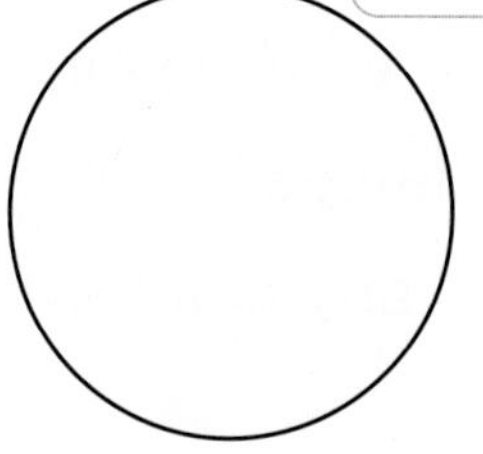

CAN What can you do?

□ We can draw line symmetric figures and point symmetric figures based on their properties. → p.18, 22

1 Let's draw a figure with line symmetry in ①, and a figure with point symmetry in ②.

① Line AB is the axis of symmetry.

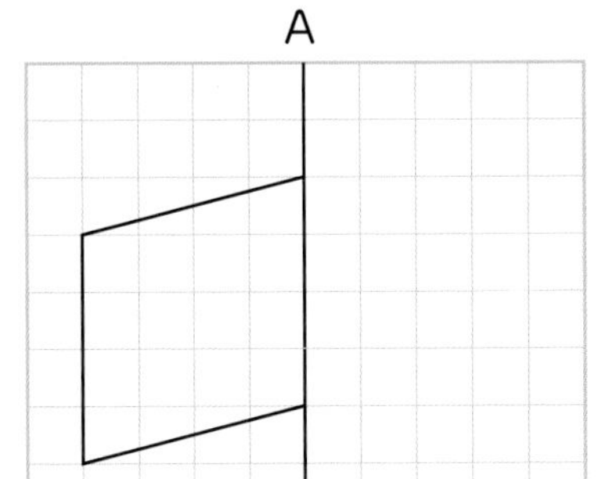

② Point O is the point of symmetry.

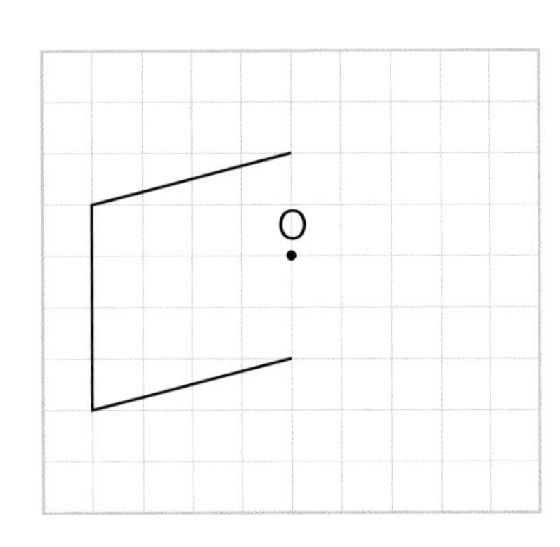

□ We can classify and organize figures in a symmetrical way. → p.25

2 Let's summarize the properties of the following quadrilaterals in the table below.

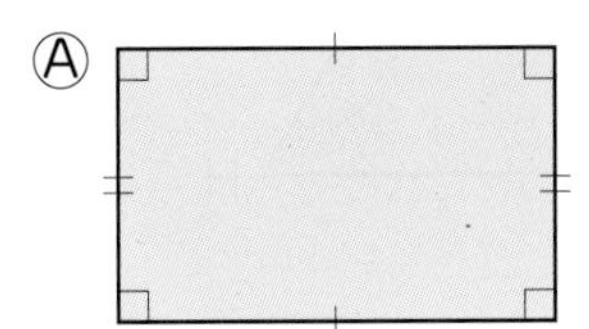

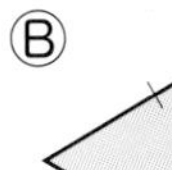

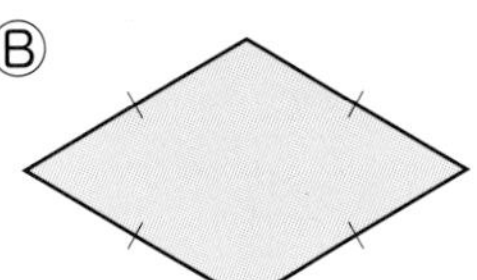

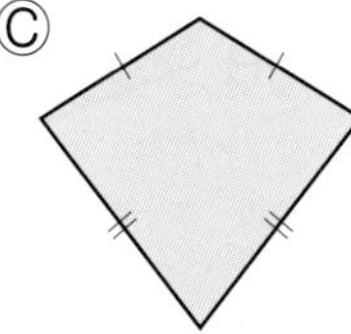

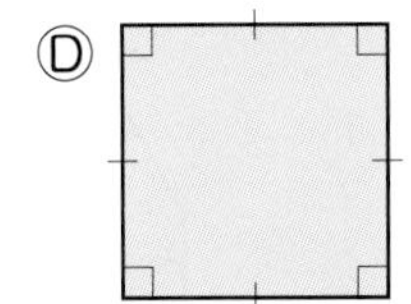

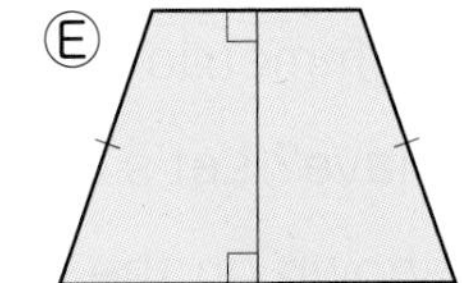

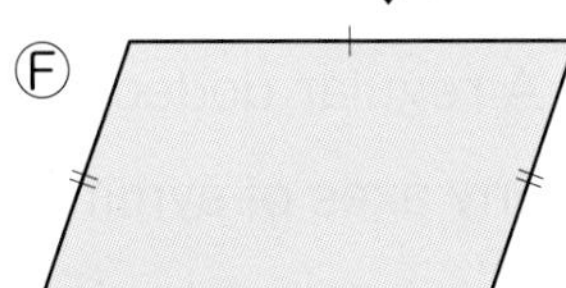

	Ⓐ	Ⓑ	Ⓒ	Ⓓ	Ⓔ	Ⓕ
Line symmetric figures	○					
Number of axes of symmetry	2					
Point symmetric figures	○					

Supplementary Problems → p.232

Which "Way to See and Think Monsters" did you find in "1 Symmetry"?

I found "Summarize" when I was trying to investigate figures with line symmetry and point symmetry.

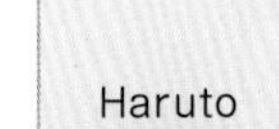

When I was trying to explain...

Usefulness and Efficiency of Learning

1 Which are line symmetric figures? Which are point symmetric figures?

Ⓐ Ⓑ Ⓒ Ⓓ Ⓔ

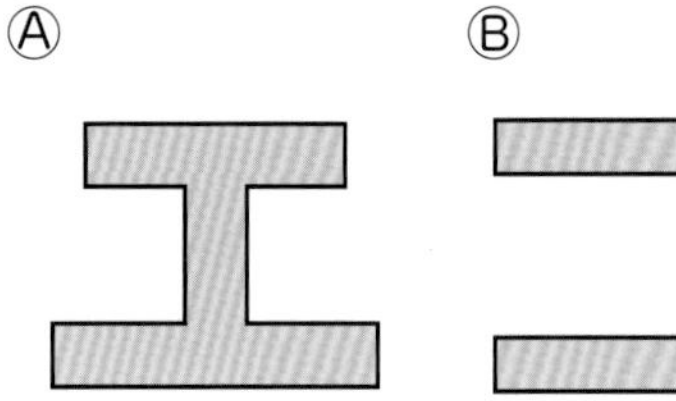

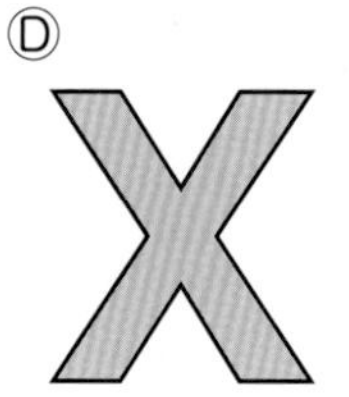

2 ① is a line symmetric figure. Let's draw the axis of symmetry. ② is a point symmetric figure. Let's draw the point of symmetry.

①

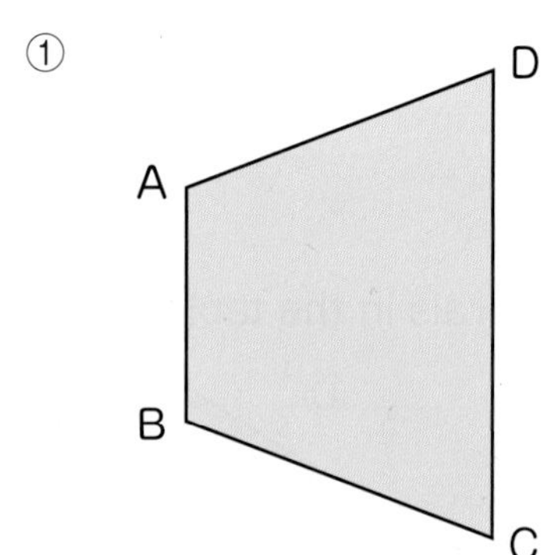

②

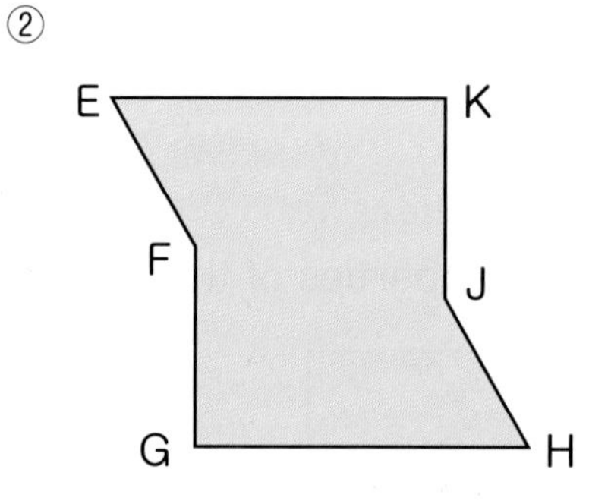

3 A regular dodecagon is a line symmetric figure. How many axes of symmetry does it have? Let's draw all the lines of symmetry in the figure shown on the right.

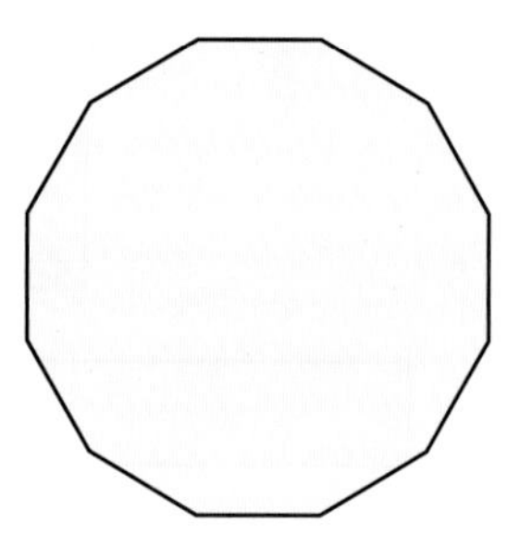

4 A square is a line symmetric figure. Let's divide it to two congruent figures by one line.

(example)

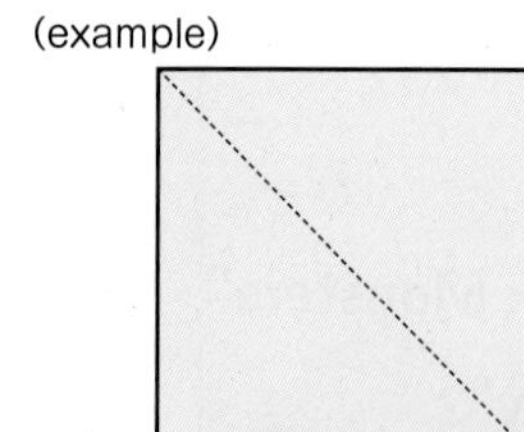

(example)

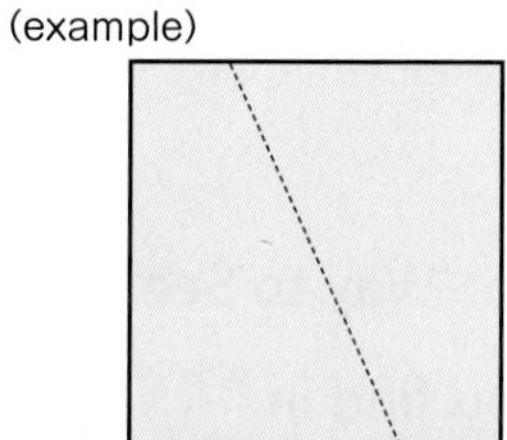

With the Way to See and Think Monsters...

Let's Reflect!

Let's reflect on which monster you used while learning "1 Symmetry."

Summarize

We could summarize various figures we have learned so far in a new way by classifying them from symmetrical point of view.

① The following table shows how the various figures we have studied so far can be regarded as line symmetric or point symmetric figures. Let's discuss what you have noticed.

	Line symmetry	Number of axes of symmetry	Point symmetry
Trapezoid	×	0	×
Parallelogram	×	0	○
Rectangle	○	2	○
Square	○	4	○
Rhombus	○	2	○

	Line symmetry	Number of axes of symmetry	Point symmetry
Regular pentagon	○	5	×
Regular hexagon	○	6	○
Regular heptagon	○	7	×
Regular octagon	○	8	○
Regular nonagon	○	9	×

Among the quadrilaterals, ___, ___, and ___ has both properties of line symmetry and point symmetry.

Sara

Regular polygons can always be classifies as ___ figures.

Haruto

Among the regular polygons, those who have properties of both line symmetry and point symmetry have ___ vertices.

Akari

? Solve the ?

We could classify figures as line symmetric, point symmetric, and others.

Sara

Want to Connect

Are there other properties for figures than we have found so far?

Yu

Find the ? What should I do when there is an unknown quantity?

Can you represent various situations around you in a math equation?

2

Mathematical Letter and Equation

Let's try to use letters to represent quantities and relationships in math equations.

1 Various quantities and math equations

1 I bought 6 buns. Let's think about how to find out the total cost.

Yu: If the bun is 50 yen each, we can find out by 50×6. If the bun is 60 yen each, we can find out by 60×6.

Akari: But we don't know the cost for one bun...

\ Want to represent /

? Purpose When there are unknown quantities, how should we represent math expressions?

	Quantity for 1 unit		How many units
If the bun is 50 yen·········	50	×	6
If the bun is 60 yen·········	60	×	6
⋮		⋮	
If the bun is □ yen·········	□	×	6

In mathematics, numbers and quantities can be represented using mathematical letters such as x and a other than □ and ○. ▷

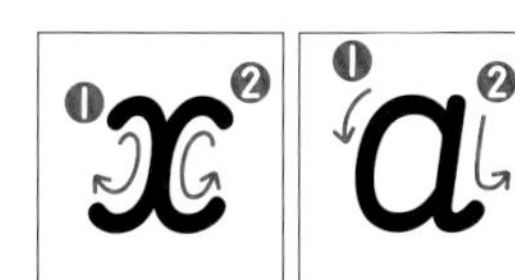

By using mathematical letters, "the price when buying 6 buns that cost a yen" can be represented as $a \times 6$ yen.

$a \times 6$ represents the total cost for the buns.

1 I bought a ribbon that costs 80 yen per meter. Let's answer the following questions.

① Let's write math expreesions that represent the prices when buying 2 m and 3 m of this ribbon, respectively.

② Let's write a math expression that represents the price for buying x m of this ribbon using x.

Summary

When there are unknown quantities, each quantity can be represented by x or a in the math expression.

? Can we represent other situations of shopping into a math expression?

2

Shun went to a vegetable shop. The cost of the vegetables were the following: one carrot for a yen, one tomato for 50 yen, and one radish for 120 yen. Let's think about how to find out the price.

The price of the carrot changes every day.

Since we don't know the current price, it is set as a yen.

Want to represent

Purpose **Can we represent in a math expression to show the way of buying?**

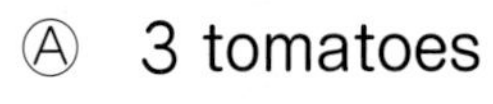

1 Let's write the math expression to find out the cost.

Ⓐ 3 tomatoes

□ × □

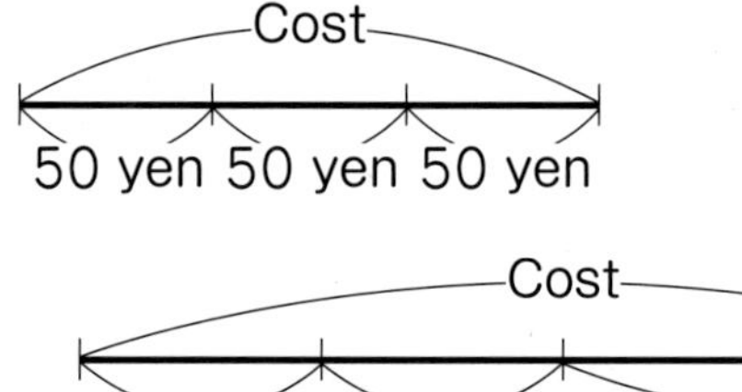

Ⓑ 2 tomatoes and 1 radish

□ + □

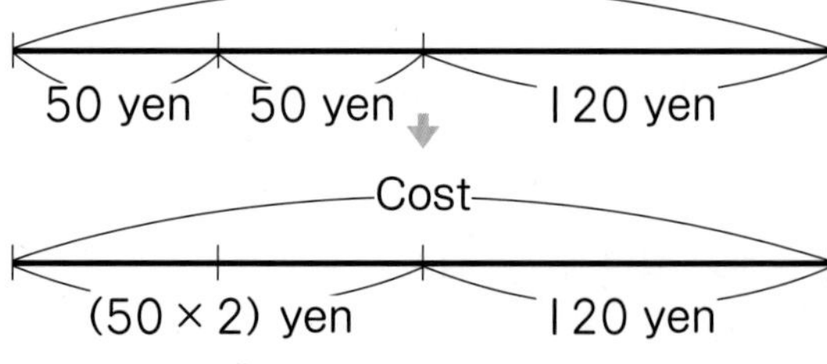

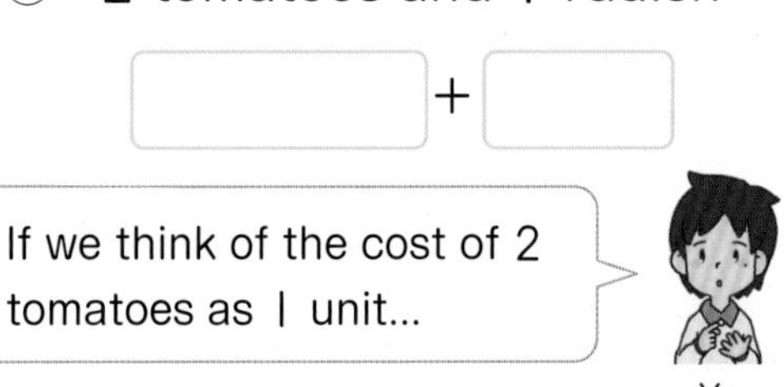

If we think of the cost of 2 tomatoes as 1 unit...
Yu

Math expression that represents cost →

Ⓒ 5 carrots

☐ × ☐

Ⓓ 3 carrots and 2 tomatoes

☐ + ☐

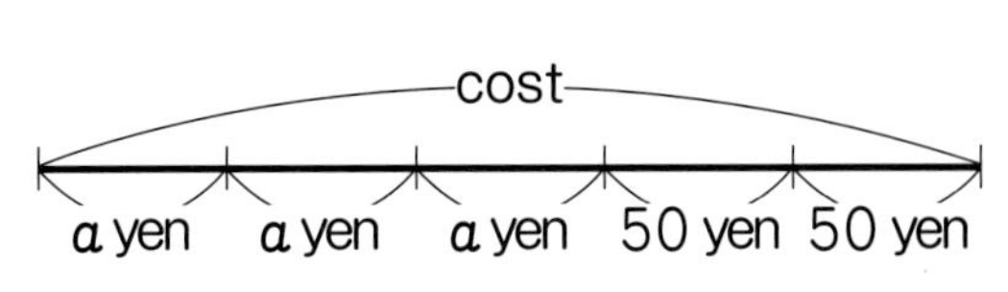

\ Want to think /

Purpose Can we know the way of buying looking at the math expression to find out the cost?

Akari

❷ What does the following math expression Ⓔ~Ⓖ represent?

Ⓔ $a + 50$ → 1 carrot and ☐

Ⓕ $a \times 7$ → ☐

Ⓖ $a \times 3 + 120 \times 2$ → ☐ and ☐

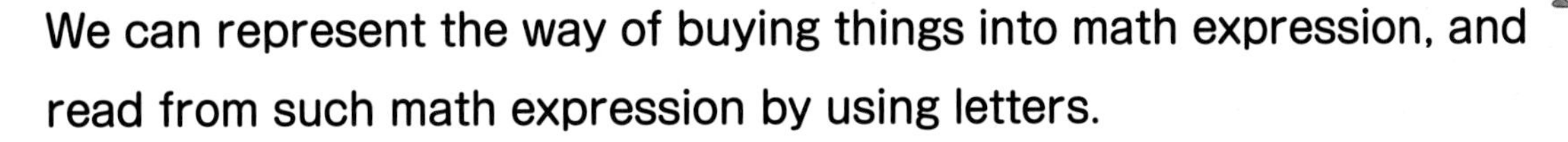

Summary Way to see and think

We can represent the way of buying things into math expression, and read from such math expression by using letters.

1 1 bottle of juice costs 150 yen and one bottle of tea costs a yen at a supermarket. Let's represent the following situations into a math expression.

① 1 bottle of juice and 3 bottles of tea

② 4 bottles of juice and 8 bottles of tea

2 I went shopping to a stationery store. What is set as x? Let's look at the diagram on the right and explain what cost the following math expressions represent.

① $70 \times x$

② $70 \times x + 200 \times 4$

? Are there other situations that can be represented by using letters?

1

There are equilateral triangles with side lengths of 1 cm, 2 cm, 3 cm, ... Let's calculate the surrounding length.

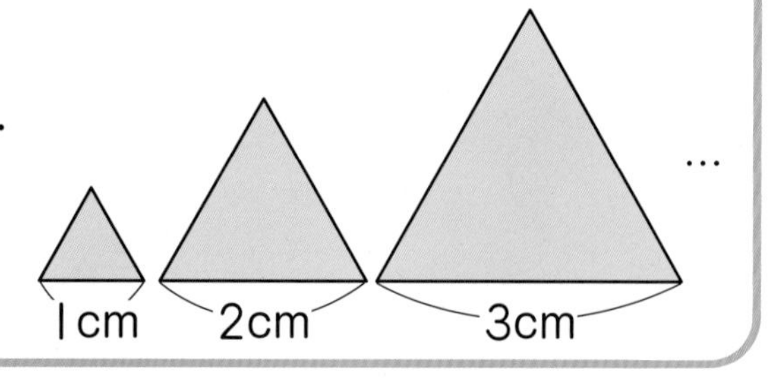

❶ Let's write a math equation to find out the surrounding length of each equilateral triangle.

	Length of one side		Number of sides		Surrounding length
Case for 1 cm···	1	×	3	=	3
Case for 2 cm···	□	×	3	=	□
Case for 3 cm···	□	×	3	=	□

❷ Let's write a math equation to find out the surrounding length of the equilateral triangle with which the length of one side is 5 cm and 8 cm respectively.

Haruto

As the length of the side changes, the length of the surroundings will change.

We represented the changing quantities into math equations in 5th grade.

Sara

? **Purpose** ＼ Want to represent ／ **As for the relationship of two quantities changing together, how should we represent math equations?**

❸ Among the math equations made in ❶ and ❷, what number always stays the same?

❹ Let's represent with a math equation the relationship between □and ○, when □ cm is the length of one side and ○ cm is the surrounding length.

Case for □ cm··· [ ]

⑤ Let's write a math equation considering x cm as the length of one side and y cm as the surrounding length.

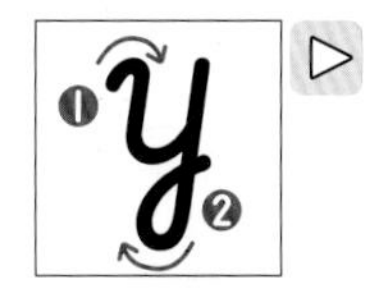

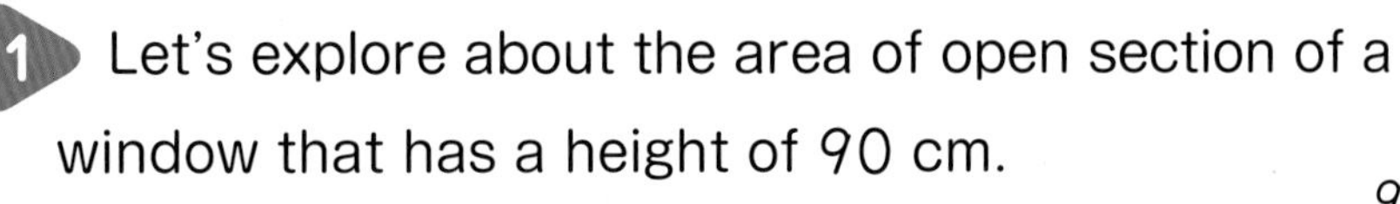

$x \times \square = y$

Way to see and think

We can represent □ as x and ○ as y.

1 Let's explore about the area of open section of a window that has a height of 90 cm.

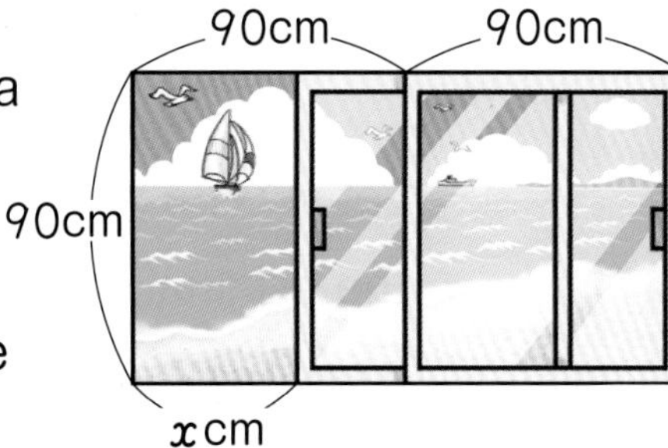

① Let's write a math equation to find out the area when the width of the open section of the window is 5 cm, 10 cm, and 12.5 cm.

The width of the open section is 5 cm… $90 \times 5 = 450$

The width of the open section is 10 cm… $90 \times \square = \square$

The width of the open section is 12.5 cm… $\square \times \square = \square$

② Let's write a math equation to find the area of the open section y cm^2 when the width of the open section is x cm.

$\square \times \square = y$

Way to see and think

Since the area of the open section is a rectangle with a length of 90 cm, the formula for area can be used.

Summary

The relationship of two quantities that change together can be represented in math equations by using x, y, etc.

2 The length of the circumference can be found as diameter × 3.14.

① Let's write a math equation to find out the length of the circumference y cm when the radius is x cm.

② Let's find out y when x is 2.

If you put a number in either x or y in a math equation using the letters, the other number can be found out by calculation.

Haruto

? We could find out the number that fits □ in a math equation using □. Can we also find out in the math equation using the letters?

1

There were certain number of origami sheets. When 7 sheets were added, the number of sheets became 35. Let's answer the following questions.

If we set the unknown quantity as x, we can represent the math problem into math equation.

Sara

❶ Let's write a math equation considering that the original number of sheets is x and the total number is 35.

[ ] $= 35$

Total number, x sheets, 7 sheets

❷ How many sheets of origami were there in the beginning?

The number that is applied to x is the number of origami papers in the beginning.

Akari

\ Want to think /

? **Purpose** **How can we find the number that applies x?**

❸ Yu considered the following. Let's explain Yu's idea.

Yu's idea

$x + 7 = 35$
$x = 35 - 7$
$x = 28$

35 sheets, x sheets, 7 sheets

It's easier to see if the equality sign is vertically aligned.

! **Summary**

In the case of an addition equation such as $x + 7 = 35$, the number that applies for x can be found using subtraction as the inverse operation of addition.

1 Let's find out the number that applies for x.

① $x + 4 = 22$　② $38 + x = 54$

③ $x - 6 = 15$　④ $x - 2.7 = 1.8$

? Can we express in a math equation or find out the number that fits the letter in other cases?

2

Let's find out the height of a parallelogram with an area of 18 cm^2 and a base of 5 cm.

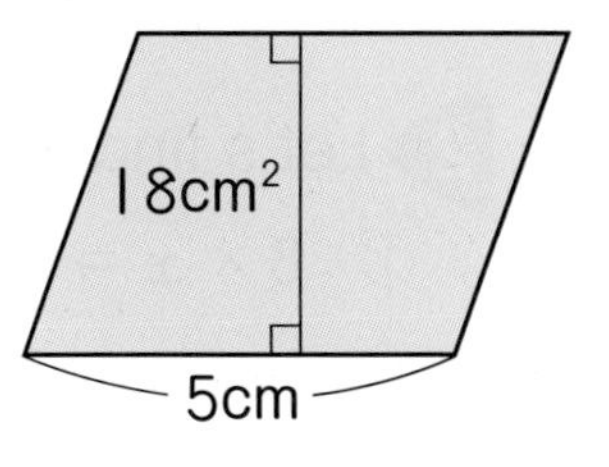

Akari: The math equation to find out the area of a parallelogram is a multiplication.

Haruto: Since the height is unknown, let's consider it as x cm.

Way to see and think
The area of a parallelogram can be found as base × height.

❶ Let's write a math equation to find out the area, considering the height as x cm. $\square = 18$

\ Want to think /

? Purpose **How can we find out the number that applies for x ?**

❷ Based on the math equation from ❶ , let's find out the height of the parallelogram.

$\square \times x = 18$

$x = 18 \div \square$

$x = \square$

x represents not only whole numbers but also decimal numbers.

! Summary

In the case of a multiplication equation such as $5 \times x = 18$, the number that applies for x can be found using division as the inverse operation of multiplication.

1 After connecting three tapes with the same length, the total length became 2 m. Let's find out the length of one section.

① Let's write a math equation to find the total length, considering x m as the length of one section.

$\square = 2$

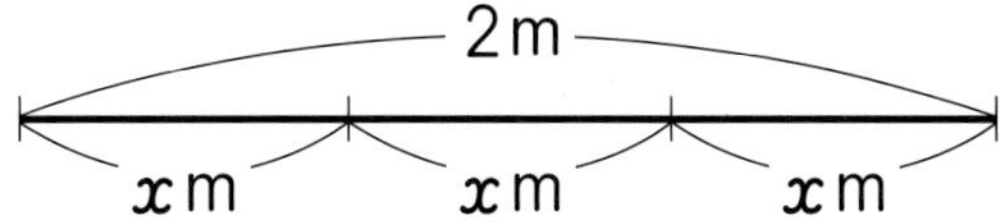

② What is the length of one section of the tape in meters (m)?

x represents not only whole numbers or decimal numbers but also fractions.

2 Let's find out the numbers that applies for x.

① $8 \times x = 20$ ② $7 \times x = 5$ ③ $x \div 4 = 8$ ④ $x \div 6 = 3$

? Can we find the number that satisfies the letter in a math equation where addition and multiplication are mixed?

3

There are 2 boxes and 3 chocolates. Each box has the same number of chocolates. Let's answer the following questions.

❶ Let's write a math equation to find out the total number of chocolates, considering that the number of chocolates inside one box is x and the total number of chocolates is y.

$\square \times \square + \square = \square$

❷ If all the chocolates including the individual ones are taken out of the boxes, the total number of the chocolates is 29. Let's represent this situation in a math equation.

 $= 29$

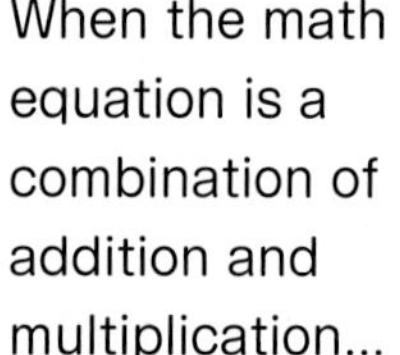

When the math equation is a combination of addition and multiplication...

Sara

\ Want to explore /

? Purpose **How can we find out the number that applies the letter?**

❸ How many chocolates were inside one box? Let's place 10, 11, 12... as x and explore the total number of chocolates using the following table.

Way to see and think

The same as for □ or ○, there is a way to find the number that applies for x by placing various numbers as x.

x	10	11	12					
$x \times 2$	20							
$x \times 2 + 3$	23							

1 There are 8 bundles and 3 sheets of colored paper. Let's answer the following questions.

① Let's write a math equation to represent the total number of sheets of paper, considering the number of sheets in one bundle as x.

② There were 115 sheets of paper in total. How many sheets of paper were gathered in one bundle? Let's place 10, 11, 12,... as x and explore the total number of sheets of colored paper.

When exploring various numbers as x, let's find an estimate answer using numbers like 10 or 20 that are easy to calculate.

Way to see and think

Summary

We can find out the number that satisfies the letter by applying various numbers.

2 Let's find out the number that applies for x by placing 8, 9, 10,... as x in the following math equations.

① $x \times 3 + 4 = 37$　　② $x \times 8 - 5 = 67$

? I can understand the idea when it is represented in a math equation, but does the math equation change when the idea is different?

ADVANCED
Junior High School

That's it! What is the number that applies for x?

In **1** above, we can find out the number x that satisfies $x \times 8 + 3 = 115$ as follows.

$$x \times 8 + 3 = 115$$
$$x \times 8 = 115 - 3$$
$$x \times 8 = 112$$
$$x = 112 \div 8$$
$$x = 14$$

If we subtract 3 sheets from the total number, that will be the number for 8 bundles.

If we divide the number of sheets for 8 bundles with 8, that will be the number of sheets for 1 bundle.

Way to see and think

If we assume $x \times 8$ as one unknown quantity, it becomes an addition equation. Therefore, you can use subtraction as the inverse operation of addition.

Yu

I can see that 3 is subtracted from both $x \times 8 + 3$ and 115.

It seems that there are some calculation rules.

Akari

Let's try to find the number x that satisfies the math equation in ② of **3** using the finding method above.

4 Reading the math equation

We want to find out the area of the flower bed shown on the right. Let's answer the following questions.

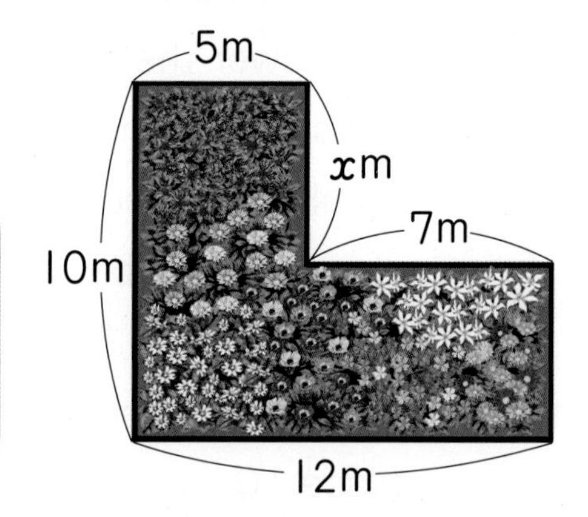

① Haruto wrote the following math expression to find out the area of the flower bed.

$10 \times 5 + (10 - x) \times 7$

Let's explain Haruto's idea using a diagram.

He first calculates 10×5, so...
Sara

What length does $10 - x$ represent?
Yu

\ Want to think /

? Purpose **Can we find out the way of thinking by observing the math expression?**

② Sara and Yu wrote the math expressions as follows.

Let's choose which of the diagram Ⓐ and Ⓑ below represents their math expression.

Sara's math expression $10 \times 12 - x \times 7$

Yu's math expression $x \times 5 + (10 - x) \times 12$

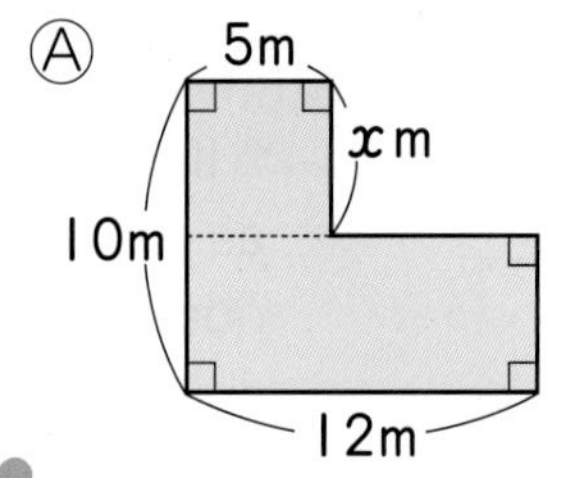

Ⓑ 7m, xm, 10m, 12m

Way to see and think

In mathematics, it is important to represent a diagram by math expressions, but it is also important to represent a math expression with diagrams.

! Summary

We can find out the way of thinking by observing the math expression. If the way of thinking differs, the math expression may differ.

CAN What can you do?

☐ We can represent in math equations by using mathematical letters. → pp.31 ~ 33

1 Let's represent in math equations by using x.

① There are x sheets of origami. If 30 sheets were added, the number becomes 44 sheets.

② There were 15 chocolates. Since you ate x chocolates, 6 chocolates were left.

③ The area of a rectangle with a length of 8 cm and a width of x cm is 48.8 cm^2.

④ If x kg of flour is divided into 9 bags with equal amounts, each bag weighs 0.6 kg.

☐ We can find out the number that satisfies the math equation. → pp.36 ~ 38

2 Let's find the number that applies to x.

① $x + 8 = 12$ ② $7 + x = 13$

③ $x - 9 = 11$ ④ $x - 19 = 13$

⑤ $4 \times x = 28$ ⑥ $x \div 7 = 8$

⑦ $x - 3.5 = 7$ ⑧ $x \times 3 = 4.2$

☐ We can read and understand the meaning of a math expression. → p.40

3 There is a bottle with x L of juice inside it. Which of the following scenes Ⓐ~Ⓓ represent the math expressions ①~④?

① $x + 6$ ② $x - 6$ ③ $x \times 6$ ④ $x \div 6$

Ⓐ The amount of juice per person when x L is divided equally among 6 people.

Ⓑ The amount of juice when x L is combined with 6 L.

Ⓒ The total amount of juice when there are 6 bottles with x L of juice per bottle.

Ⓓ The remaining amount of juice when 6 L were drunk from x L.

Supplementary Problems → p.233

Which "Way to See and Think Monsters" did you find in "2 Mathematical Letter and Equation"?

I found "Other Way" when I was trying to represent my idea with math expressions and diagrams.

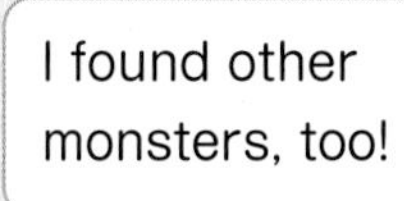

Sara

I found other monsters, too!

Haruto

Utilize Usefulness and Efficiency of Learning

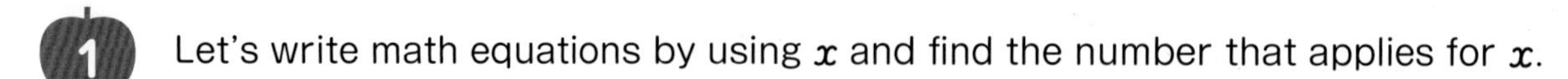

1 Let's write math equations by using x and find the number that applies for x.

① A bundle of envelopes costs x yen, and bundles of envelopes cost 720 yen.

② The cost of 1 notebook is x yen, and the cost of 5 notebooks is 650 yen.

③ There are 20 marbles. If x marbles were added, the number became 52 marbles.

④ There is a ribbon with a length of x cm. When 50 cm were used, the remaining length was 60 cm.

2 Let's explore the number that applies for x by placing 6, 7, 8, ⋯ as x in the following math equations.

① $x \times 4 + 7 = 39$　　　② $x \times 5 - 9 = 36$

3 Which of the following scenes Ⓐ~Ⓓ represent the math expressions ①~④?

① $x + 30$　② $x \times 30$　③ $x \div 30$　④ $x - 30$

Ⓐ If x candies are distributed equally among 30 people, how many candies will each person receive?

Ⓑ If an x m ribbon is connected with a 30 m ribbon, what is the total length of the ribbon in meters (m)?

Ⓒ A rectangle with an area of 30 cm^2 has been subtracted from a square with an area of x cm^2? What is the area for the remaining part in square centimeters (cm^2)?

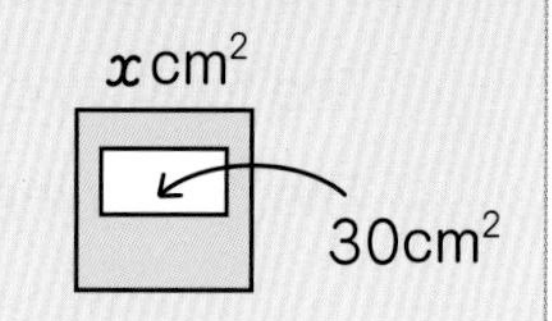

Ⓓ What is the area of the parallelogram that has a base of x cm and a height of 30 cm? Answer in square centimeters (cm^2).

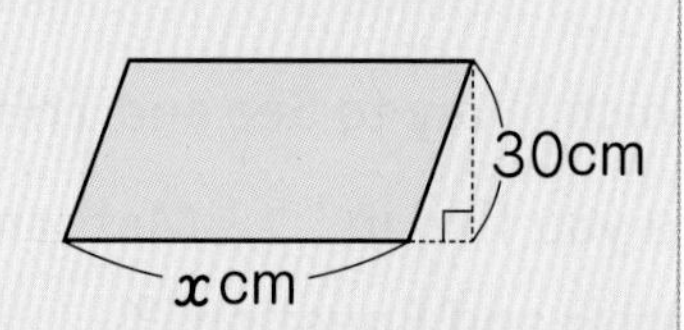

With the Way to See and Think Monsters...

Let's Reflect!

Let's reflect on which monster you used while learning " 2 Mathematical Letter and Equation."

Other Way

When explaining my ideas, I could make it easier for others to understand by representing in math equations **using letters and** representing in diagrams.

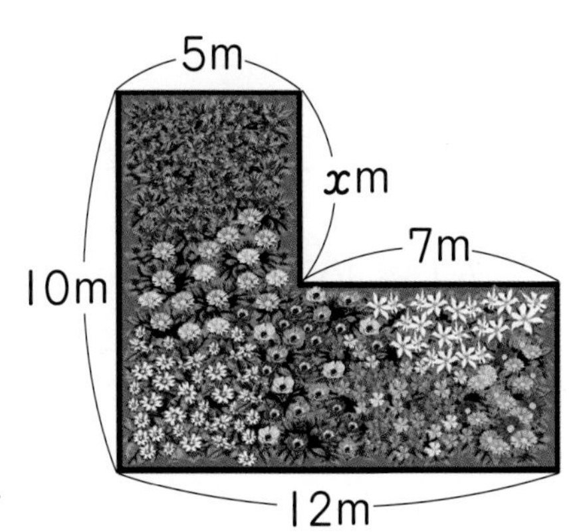

① In what way did you think about the math expression to find out the area of the flower bed shown on the right?

Based on the diagram below, I represented as $10 \times 5 + (10 - x) \times 7$.

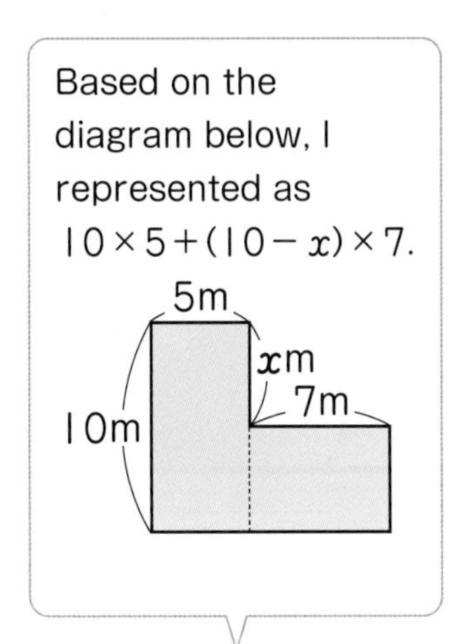

Haruto

Based on the diagram below, I represented as $10 \times 12 - x \times 7$.

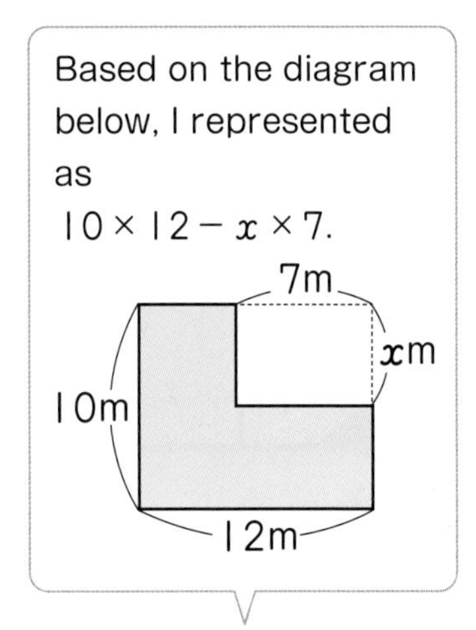

Sara

Based on the diagram below, I represented as $x \times 5 + (10 - x) \times 12$.

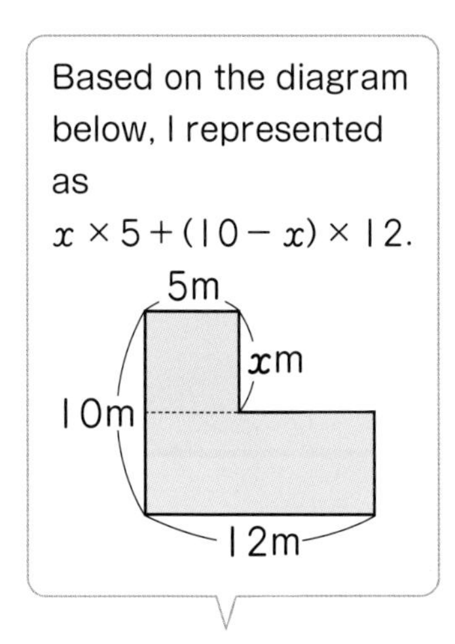

Yu

? Solve the ?

We could use letters to express ideas in math equations. We could also find the numbers that fit those letters.

Yu

Want to Connect

What other relationships or rules can we represent by a math equation using letters?

Akari

Find the ?

How much can we paint?

The fence is being painted with green paint. If x m² can be painted per 1 dL of paint, how many square meters can be painted with 3 dL of this paint?
Let's think about this problem.
I don't know the area that can be painted per 1 dL.
Does any number apply?
1
Painted area
Amount of paint
4
□ (m²)
1
3 (dL)
4 m² | □m² | 1 dL | 3 dL
If x is 4, we can find out by 4 × 3.
2
Painted area
Amount of paint
0.4
□ (m²)
1
3 (dL)
0.4 m² | □m² | 1 dL | 3 dL
When x is a decimal number such as 0.4, we can find out in the same way by 0.4 x 3.
3
What happens if x is a fraction?
We considered how many 1s for whole numbers and how many 0.1s for decimals.
If x is $\frac{4}{5}$ ...
Can we find out the answer by $\frac{4}{5}+\frac{4}{5}+\frac{4}{5}$?
4

Want to think

Purpose How can we multiply when the multiplicand is a fraction?

3 Multiplication and Division of Fractions and Whole Numbers

Let's think about the meaning of multiplication and division and how to calculate.

【Procedure for problem solving】

Read the problem.
↓
Represent it with a table or a diagram.
↓
Write a math expression.
↓
Think about how to calculate.
↓
Try to confirm.
↓
Think for a better method.
↓
Try to confirm.

1 The case: fraction × whole number

1 The fence is being painted with green paint. If $\frac{4}{5}$ m^2 can be painted per deciliter of paint, how many square meters can be painted with 3 dL?

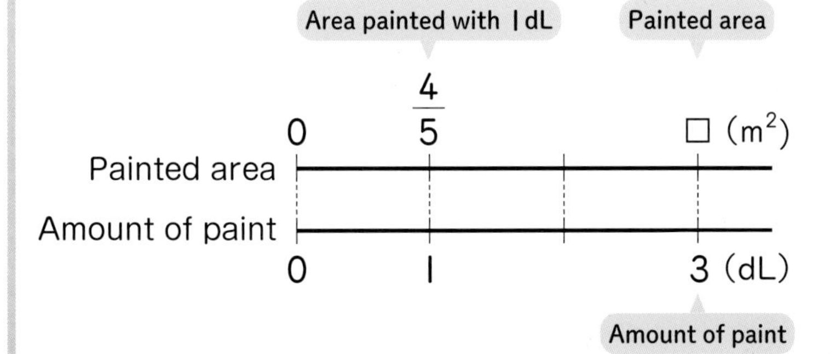

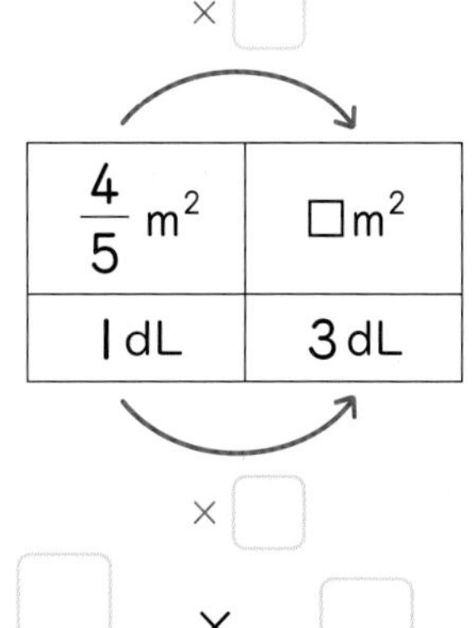

Since you know "measurement per unit quantity" and "how many units," you can use a multiplication expression.

① Let's write a math expression.

□ × □

Area painted with 1 dL　Amount of paint

② Let's think about how to calculate.

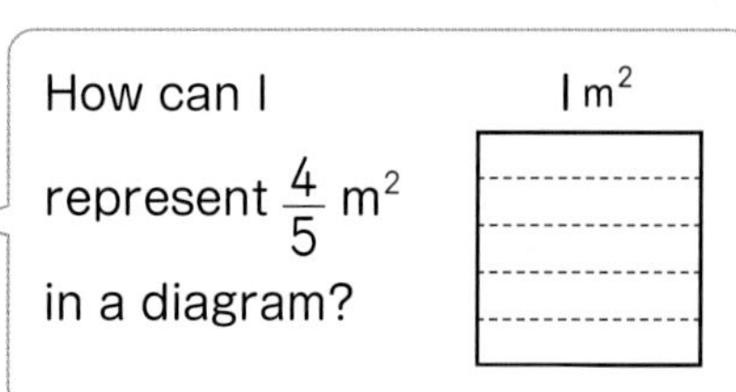

How can I represent $\frac{4}{5}$ m² in a diagram?
1 m²

Since $\frac{4}{5}$ has four sets of $\frac{1}{5}$, so...

Haruto's idea

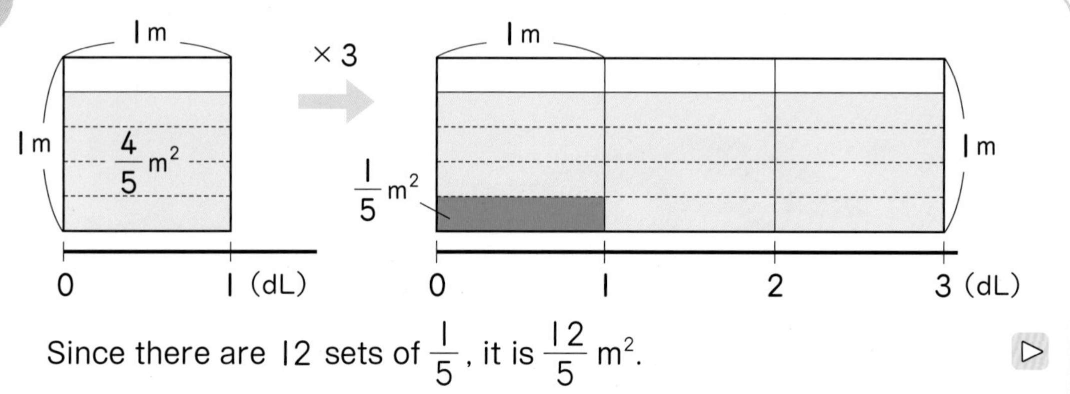

Since there are 12 sets of $\frac{1}{5}$, it is $\frac{12}{5}$ m².

Sara's idea

If I consider $\frac{1}{5}$ as one unit, $\frac{4}{5}$ has four sets of $\frac{1}{5}$. $\frac{4}{5} \times 3$ is three sets of $\frac{4}{5}$, so it becomes (4×3) sets of $\frac{1}{5}$.

$$\frac{4}{5} \times 3 = \frac{4 \times 3}{5} = \frac{12}{5}$$

Way to see and think

They are thinking about "how many sets of $\frac{1}{5}$?" in both cases.

When comparing two people's ideas, we can explore the following:

- What are their ideas?
- What are the differences?
- What are the similarities?

③ Let's try to compare the calculation methods of Haruto and Sara.

Summary

When you multiply an improper fraction or proper fraction by a whole number, leave the denominator as it is and multiply the numerator by the whole number.

$$\frac{b}{a} \times c = \frac{b \times c}{a}$$

? Can we calculate in the same way for other fractions?

2

$\frac{2}{9} \times 3$ was calculated as shown below. Let's explain how each child calculated.

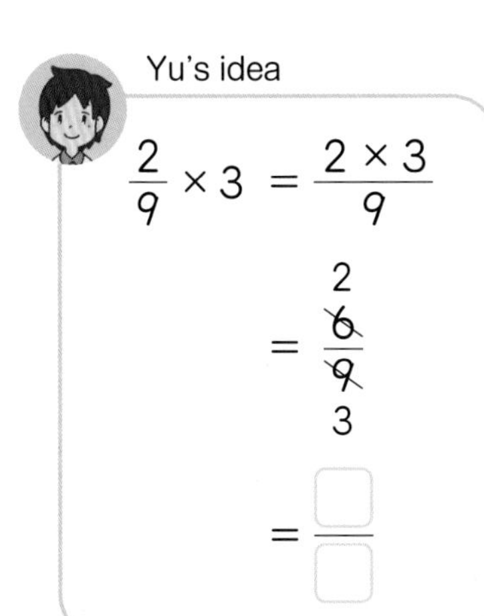

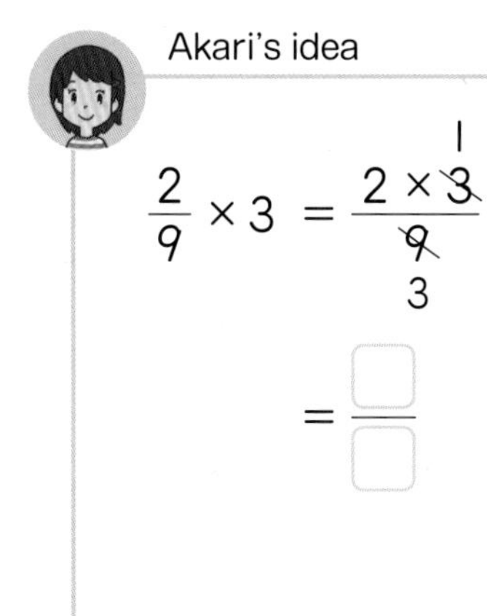

\ Want to know /

Purpose What if I can reduce in the middle of the calculation?

Summary It is easier to calculate if fractions can be reduced in the middle of the calculation.

1 Let's calculate $\frac{10}{9} \times 3$ based on Yu and Akari's idea in **1**.

2 Let's calculate the following.

① $\frac{2}{5} \times 2$	② $\frac{4}{9} \times 2$	③ $\frac{7}{11} \times 4$	④ $\frac{6}{7} \times 5$
⑤ $\frac{11}{8} \times 3$	⑥ $\frac{7}{6} \times 5$	⑦ $\frac{9}{4} \times 7$	⑧ $\frac{11}{5} \times 2$
⑨ $\frac{3}{8} \times 2$	⑩ $\frac{5}{18} \times 3$	⑪ $\frac{9}{16} \times 8$	⑫ $\frac{7}{12} \times 6$
⑬ $\frac{7}{6} \times 4$	⑭ $\frac{13}{10} \times 25$	⑮ $\frac{3}{2} \times 4$	⑯ $\frac{4}{3} \times 6$

What should I do if the multiplicand is a mixed fraction?

3

4 pieces of tape will be made, each with a length of $1\frac{2}{5}$ m. What is the total length, in meters, of the tape that will be needed?

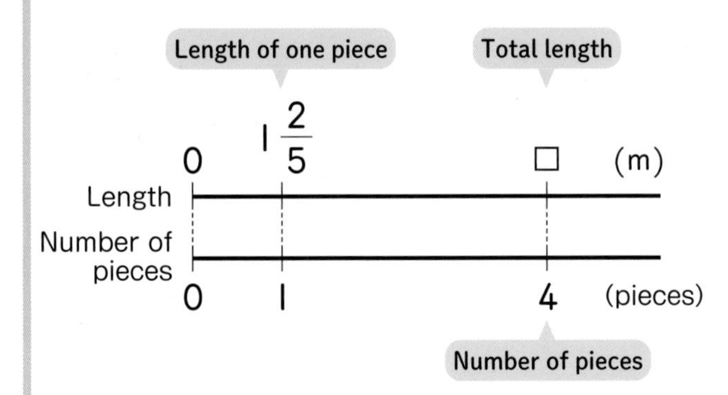

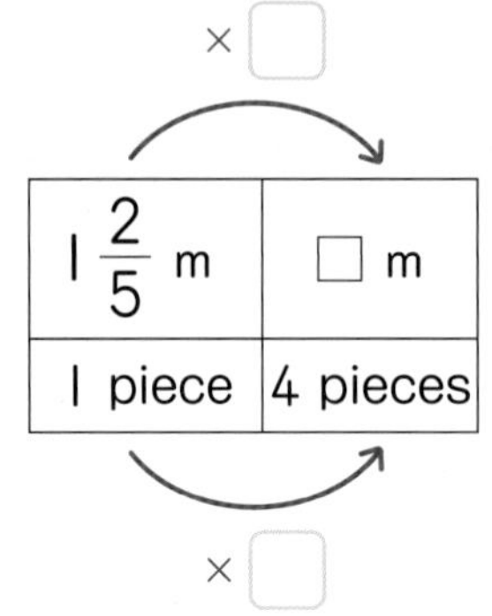

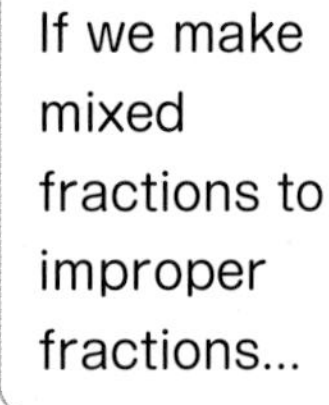

If you think about decomposing $1\frac{2}{5}$ into a whole number and a fraction, the approximate answer is easy to understand.

① Let's write a math expression.

② What is the approximate length that is needed?

③ Let's think about how to calculate.

If we decompose mixed fraction to a whole number and fraction...
Haruto

If we make mixed fractions to improper fractions...
Sara

\ Want to think /

? Purpose What should we do to multiply a mixed fraction by a whole number?

Haruto's idea

Calculate by decomposing $1\frac{2}{5}$ into 1 and $\frac{2}{5}$.

$1\frac{2}{5} \times 4$ ⟨ $1 \times 4 = \square$; $\frac{2}{5} \times 4 = \square$

$\square + \square = \square\frac{\square}{\square}$

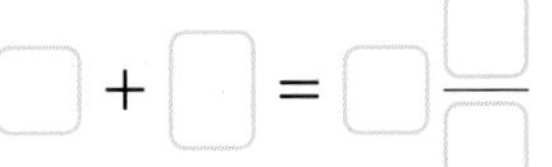

Sara's idea

Calculate by changing $1\frac{2}{5}$ into an improper fraction.

$1\frac{2}{5} \times 4 = \frac{7}{5} \times 4$

$= \square$

$= \square\frac{\square}{\square}$

Decompose mixed fractions into whole numbers and fractions or make them improper fractions.

④ Let's try to compare the methods from Haruto and Sara. Let's present the good points of their ideas.

As for Haruto's idea, it is easy to estimate the answer.
If you change to an improper fraction, you can use the previous method.

1 $2\frac{3}{10}$ dL of soda is used to make one fruit punch.
How many deciliters of soda will be needed to make four of these fruit punches?
Let's use the ideas of Haruto and Sara on the previous page respectively.

Summary — Way to see and think

When you multiply a mixed fraction by a whole number, if the mixed fraction is changed into an improper fraction then it can be solved as we have learned before.

2 What is the area, in square meters, of a rectangular flower bed with a length of $4\frac{2}{3}$ m and a width of 6 m?

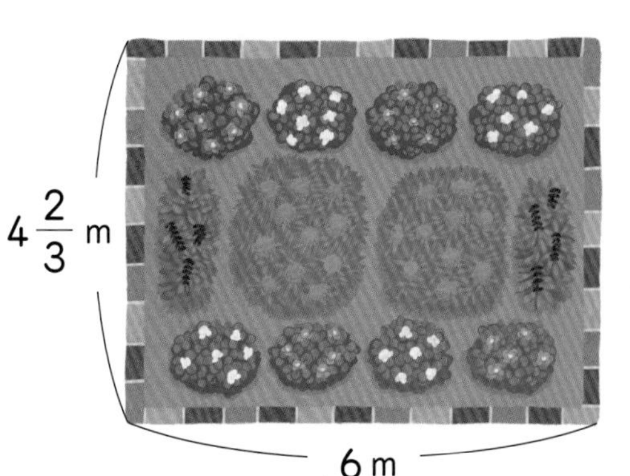

3 Let's calculate the following.

① $1\frac{3}{7} \times 2$　② $2\frac{2}{3} \times 2$　③ $1\frac{5}{8} \times 2$　④ $2\frac{5}{6} \times 12$

Can we do fraction ÷ whole number like in multiplication?

2 The case: fraction ÷ whole number

1 2 dL of blue paint is used to paint a $\frac{4}{5}$ m^2 wall. What is the area, in square meters, that can be painted per 1 dL of this paint?

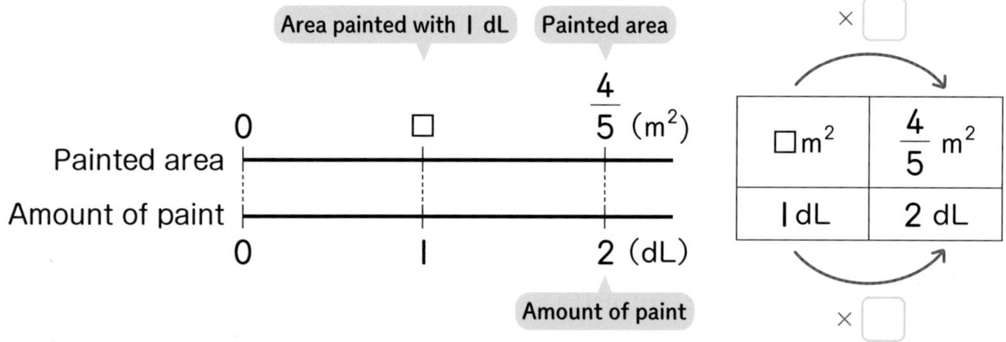

① Let's write a math expression.

② Let's think about how to calculate.

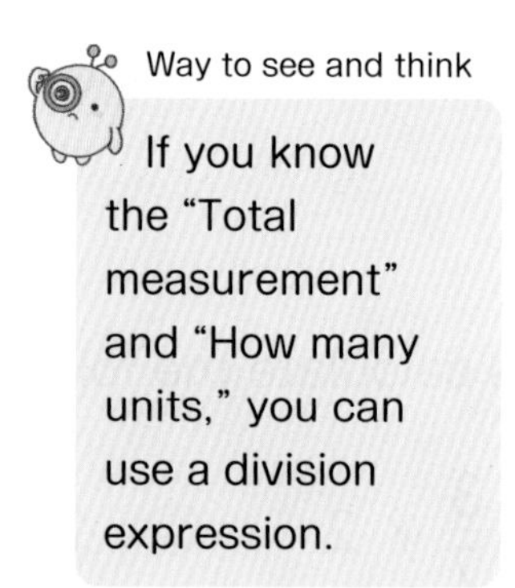

Way to see and think
If you know the "Total measurement" and "How many units," you can use a division expression.

\ Want to think /

? Purpose **What should we do to divide a fraction by a whole number?**

③ Let's explain the ideas of Sara, Akari, and Haruto.

Sara's idea

Considering the same as the multiplication of fractions, as for the case of division, divide the numerator.

$$\frac{4}{5} \div 2 = \frac{4 \div 2}{5} = \frac{\square}{\square}$$

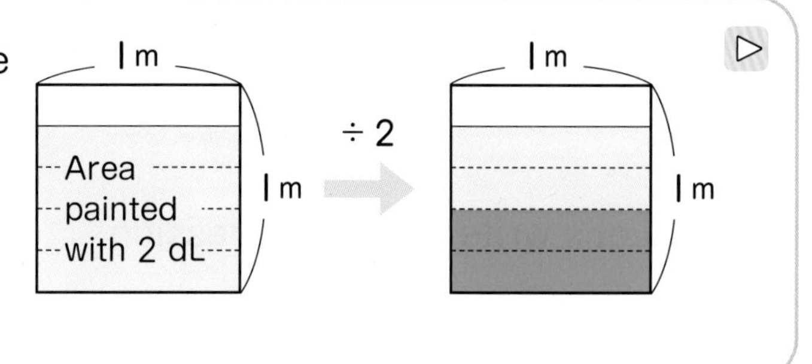

Way to see and think

Sara considers how many units of $\frac{1}{5}$ is $\frac{4}{5}$ has, and takes half of it.

Akari's idea

Since there are 4 sets of ▭ ,

$$\frac{4}{5} \div 2 = \frac{1}{5 \times 2} \times 4$$

$$= \frac{4}{5 \times 2}$$

$$= \frac{\square}{\square}$$

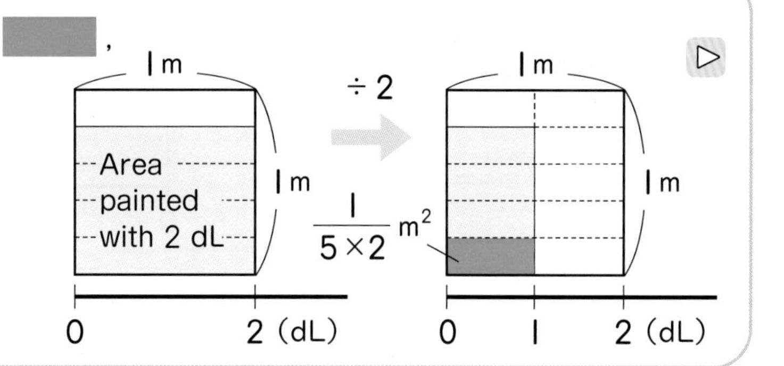

Way to see and think

Akari thinks about how many units of $\frac{1}{5 \times 2}$ there are using a diagram.

Haruto's idea

If I use the rules of division and calculate whole number ÷ whole number,

$$\frac{4}{5} \div 2 = \left(\frac{4}{5} \times 5\right) \div (2 \times 5)$$

$$= 4 \div (2 \times 5)$$

$$= 4 \div (5 \times 2)$$

When represented using fractions,

$$\frac{4}{5} \div 2 = \frac{4}{5 \times 2} = \frac{\square}{\square}$$

Way to see and think

Haruto uses the rules of division, and thinks of it as a calculation of whole numbers.

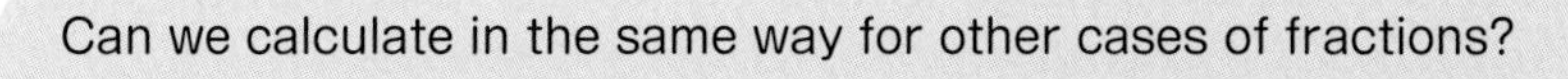

Can we calculate in the same way for other cases of fractions?

2

3 dL of red paint is used to paint a $\frac{4}{5}$ m^2 wall. How many square meters can be painted per 1 dL of this paint?

❶ Let's write a math expression. $\square \div \square$

❷ Let's think about how to calculate.

When we divide the numerator, it will be $\frac{4}{5} \div 3 = \frac{4 \div 3}{5}$, and it is not divisible...

Sara's idea

Since there are four sets of ■,

$\frac{4}{5} \div 3 = \frac{4}{5 \times 3}$

$= \frac{4}{\square}$

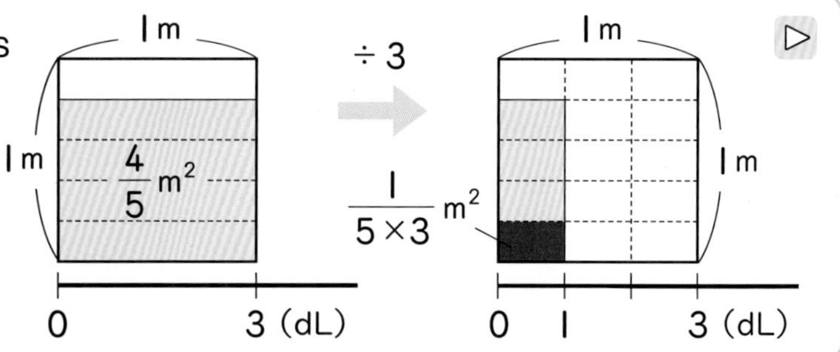

Akari's idea

$\frac{4}{5} \div 3 = \frac{4 \times 3}{5 \times 3} \div 3$

$= \frac{4 \times 3 \div 3}{5 \times 3}$

$= \frac{4}{5 \times 3} = \square$

Way to see and think

As for Akari's idea, since 4 cannot be divided by 3, she first uses the properties of fractions, and then is able to divide.

Summary

When you divide an improper fraction or proper fraction by a whole number, the denominator is multiplied by the whole number and the numerator is left as it is.

$$\frac{b}{a} \div c = \frac{b}{a \times c}$$

❸ Whose idea on the previous page is the same as Sara and Akari's idea in ❷?

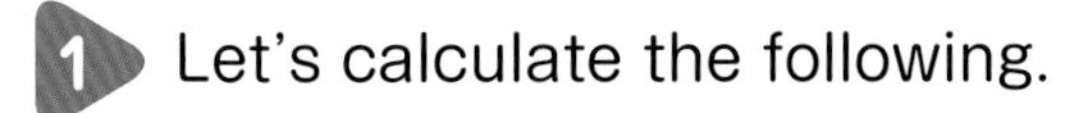

1 Let's calculate the following.

① $\frac{1}{2} \div 4$ ② $\frac{3}{4} \div 2$ ③ $\frac{5}{6} \div 4$ ④ $\frac{7}{8} \div 5$

⑤ $\frac{3}{2} \div 5$ ⑥ $\frac{5}{3} \div 7$ ⑦ $\frac{8}{5} \div 3$ ⑧ $\frac{7}{6} \div 2$

? Should I think in the same way as we reduced the product when we can reduce the quotient?

Division where the quotient can be reduced

3 Let's compare methods Ⓐ and Ⓑ for the calculation of $\frac{10}{7} \div 4$.

Ⓐ $\frac{10}{7} \div 4 = \frac{10}{7 \times 4}$

$= \frac{\overset{5}{\cancel{10}}}{\underset{14}{\cancel{28}}}$

$= \square$

Ⓑ $\frac{10}{7} \div 4 = \frac{\overset{5}{\cancel{10}}}{7 \times \underset{2}{\cancel{4}}}$

$= \square$

Want to know

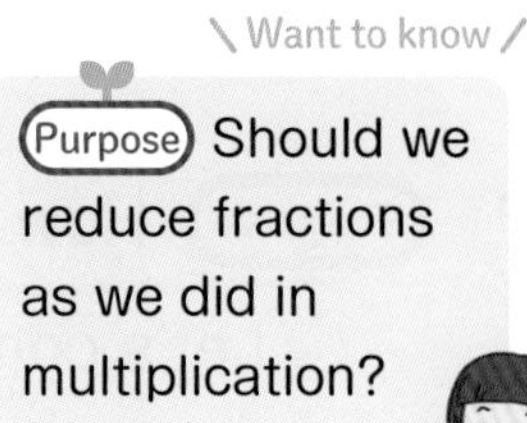

Purpose Should we reduce fractions as we did in multiplication?

Sara

1 There is a field of $\frac{12}{13}$ ha. If this field is equally divided among 6 people, how many hectares will each person receive?

Summary Same as multiplication, it is easier if fractions can be reduced in the middle of the calculation of division.

Haruto

2 Let's calculate the following.

① $\frac{6}{7} \div 3$ ② $\frac{3}{4} \div 12$ ③ $\frac{12}{5} \div 4$ ④ $\frac{8}{3} \div 6$

? What should I do if the dividend is a mixed fraction?

4

There is an iron bar that weighs $2\frac{1}{4}$ kg and has a length of 3 m. How many kilograms is the weight of this iron bar per meter?

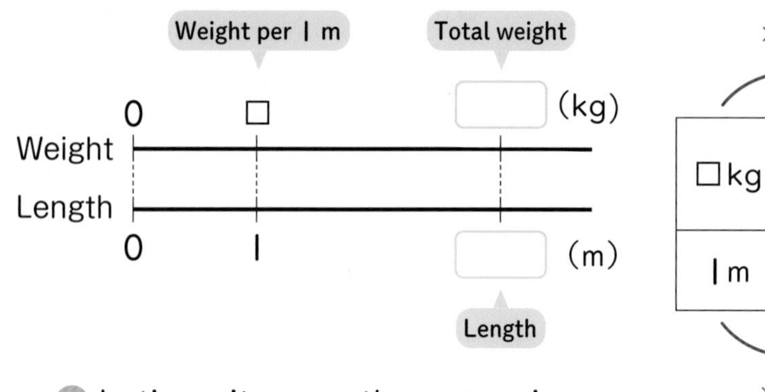

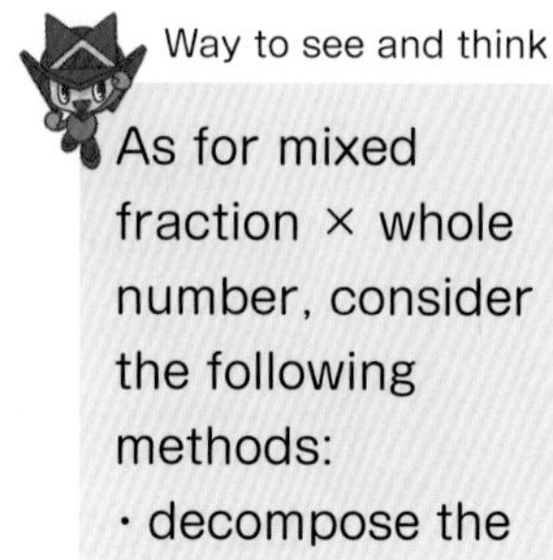

As for mixed fraction × whole number, consider the following methods:
- decompose the mixed fraction into a whole number and fraction.
- change the mixed fraction into an improper fraction.

① Let's write a math expression.

② As for the weight per meter, is it heavier than 1 kg? Is it lighter?

\ Want to think /

Purpose How can we divide a mixed fraction by a whole number?

③ Let's compare the calculation methods of Haruto and Sara.

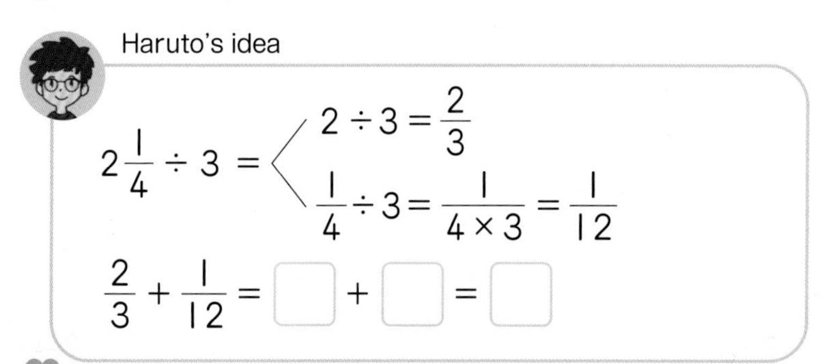

$2\frac{1}{4} \div 3 = \begin{cases} 2 \div 3 = \frac{2}{3} \\ \frac{1}{4} \div 3 = \frac{1}{4 \times 3} = \frac{1}{12} \end{cases}$

$\frac{2}{3} + \frac{1}{12} = \square + \square = \square$

$2\frac{1}{4} \div 3 = \frac{\square}{4} \div 3$

$= \frac{\square}{4 \times 3}$

$= \frac{\square}{4}$

Summary

When you divide a mixed fraction by a whole number, if the mixed fraction is changed into an improper fraction then it can be solved as learned before.

1 Let's calculate the following.

① $1\frac{2}{3} \div 4$ ② $2\frac{5}{8} \div 6$ ③ $2\frac{2}{7} \div 8$ ④ $3\frac{1}{2} \div 7$

CAN What can you do? 

☐ We can understand how to calculate fraction × whole number and fraction ÷ whole number. → p.46, p.52

1 Let's summarize how to calculate fraction × whole number and fraction ÷ whole number.

① $\frac{2}{7} \times 3 = \frac{\square \times \square}{\square} = \square$

② $\frac{5}{7} \div 3 = \frac{\square}{\square \times \square} = \square$

☐ We can calculate fraction × whole number and fraction ÷ whole number. → pp.46 ~ 54

2 Let's calculate the following.

① $\frac{2}{9} \times 4$ ② $\frac{7}{11} \times 5$ ③ $\frac{4}{21} \times 3$ ④ $\frac{7}{6} \times 8$

⑤ $\frac{3}{7} \times 28$ ⑥ $1\frac{5}{12} \times 7$ ⑦ $3\frac{3}{10} \times 30$ ⑧ $2\frac{3}{4} \times 12$

⑨ $\frac{5}{8} \div 3$ ⑩ $\frac{2}{5} \div 7$ ⑪ $\frac{3}{2} \div 2$ ⑫ $\frac{3}{10} \div 6$

⑬ $\frac{4}{5} \div 8$ ⑭ $\frac{10}{7} \div 10$ ⑮ $1\frac{3}{8} \div 3$ ⑯ $2\frac{5}{8} \div 3$

☐ We can create a math expression and find out the answer. → pp.46 ~ 54

3 Let's answer the following questions.

① I drink $\frac{5}{6}$ L of milk every day. What is the amount of milk, in liters, that I drink in 3 days?

② $\frac{7}{6}$ L of milk will be equally divided into 3 bottles. What is the amount of milk, in liters, that is poured in each bottle?

③ $2\frac{4}{7}$ dL of paint was used to paint a 3 m^2 wall. What is the amount of paint, in deciliters, that is used to cover 1 m^2 wall?

Supplementary Problems → p.234 

Which "Way to See and Think Monsters" did you find in "3 Multiplication and Division of Fractions and Whole Numbers"?

I found "Unit" when I was thinking about the calculation of fraction × whole number and fraction ÷ whole number. — Akari

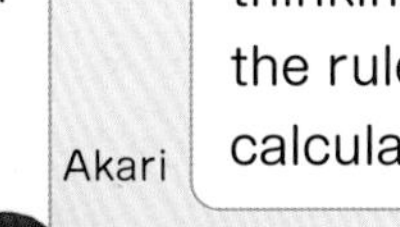

When I was thinking about the rules of calculation... — Sara

Usefulness and Efficiency of Learning

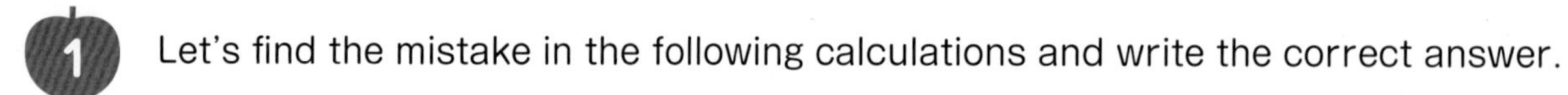

1 Let's find the mistake in the following calculations and write the correct answer.

① $\frac{2}{5} \times 10 = \frac{\overset{1}{\cancel{2}}}{5 \times \underset{5}{\cancel{10}}} = \frac{1}{25}$

② $\frac{8}{7} \div 4 = \frac{7 \times \overset{1}{\cancel{4}}}{\underset{2}{\cancel{8}}} = \frac{7}{2}$

2 Let's answer the following questions.

① There is one card for each of the following numbers [1] ～ [5].

Let's place one card inside □, so that the following math equation holds true.

$$\square\frac{\square}{\square} \times \square = 6\frac{\square}{5}$$

② There is one card for each of the following numbers [1] ～ [4].

Let's place one card inside □, so that the following math equation holds true.

$$\square\frac{\square}{\square} \div \square = \frac{5}{12}$$

3 There is a wire that weighs $1\frac{5}{6}$ g per meter. What is the weight, in grams, for 4 m of this wire?

4 There is a rice field in which $4\frac{2}{3}$ kg of rice can be cultivated for every 4 m^2. Let's answer the following questions.

① What is the weight of rice, in kilograms, that can be cultivated per square meter?

② If this rice field is 300 m^2, what is the weight of rice, in kilograms, that can be cultivated?

With the Way to See and Think Monsters...

Let's Reflect!

Let's reflect on which monster you used while learning " 3 Multiplication and Division of Fractions and Whole Numbers."

Unit

Same Way

We could calculate in the same way as we did in whole numbers when we focus on the number of one unit.

① Let's discuss what you found out looking at the three calculations below.

Ⓐ 4×3　　Ⓑ 0.4×3　　Ⓒ $\frac{4}{5} \times 3$

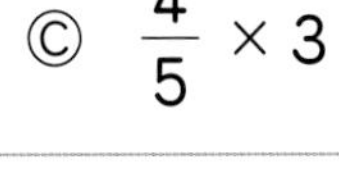

In the whole number x whole number calculation, we can consider that there are (4 x 3) 1s.

Haruto

In the decimal x whole number calculation, we considered that there are (4 x 3) 0.1s.

Sara

We considered that there are (4 x 3) $\frac{1}{5}$s in fraction x whole number.

Yu

Each multiplication can be done by the same calculation, although the numbers for one unit are different.

Akari

Rule

By could apply the rules of operations we have studied so far to find out the answer.

② How did you calculate $\frac{4}{5} \div 3$?

$$\frac{4}{5} \div 3 = \frac{4}{15}$$

↓ × 5　　↓ × 5

$$\left(\frac{4}{5} \times 5\right) \div (3 \times 5) = \frac{4}{15}$$

We used the rule that multiplying a dividend and a divisor by the same number does not change the quotient.

Akari

Solve the ?

I could find out the answers to calculations such as multiplying fractions by whole numbers and dividing fractions by whole numbers by using the calculations I have learned so far.

Haruto

Want to Connect

Can we multiply a fraction with a fraction or divide a fraction by a fraction too?

Sara

Let's solve math problems by using diagrams and tables.

If you don't know the "Total number"

There is an iron bar that weighs $\frac{2}{3}$ kg for 1 m. What is the weight of this iron bar for 4 m?

Let's draw a diagram and think about it with a math equation using words...

Akari

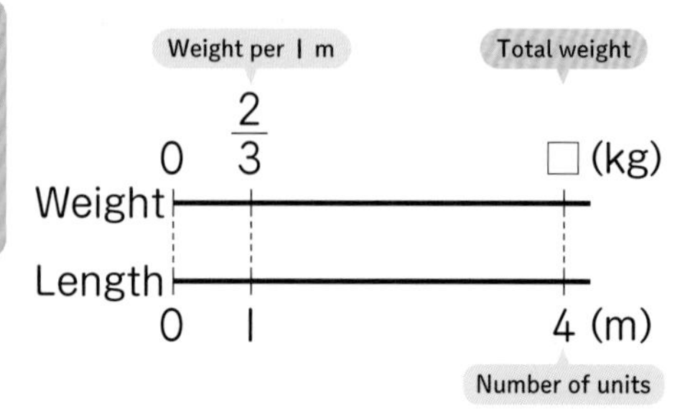

Weight per 1 m × Number of units = Total weight

So,

$\frac{2}{3} \times 4 = \frac{8}{3} = 2\frac{2}{3}$　Answer: $2\frac{2}{3}$ kg

Let's draw a table and think about it using the idea of proportionality...

Haruto

× 4 →

$\frac{2}{3}$ kg	□ kg
1 m	4 m

× 4 →

$\frac{2}{3} \times 4 = \frac{8}{3} = 2\frac{2}{3}$

Answer: $2\frac{2}{3}$ kg

How to draw a diagram

(1) Draw a straight line representing weight and a straight line representing length.

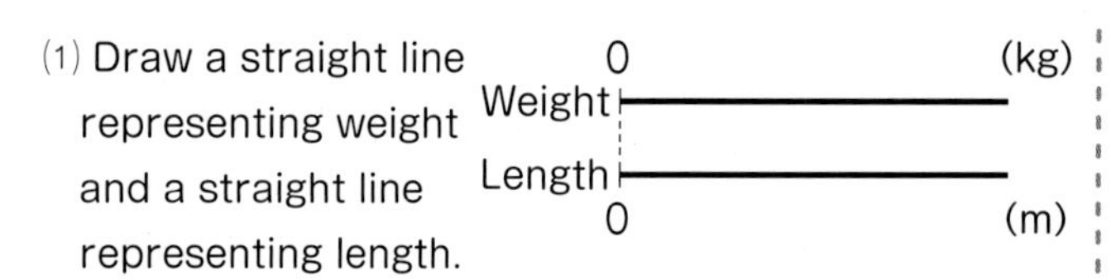

(2) Draw a scale for 1 m, and draw a scale for $\frac{2}{3}$ kg in the corresponding place.

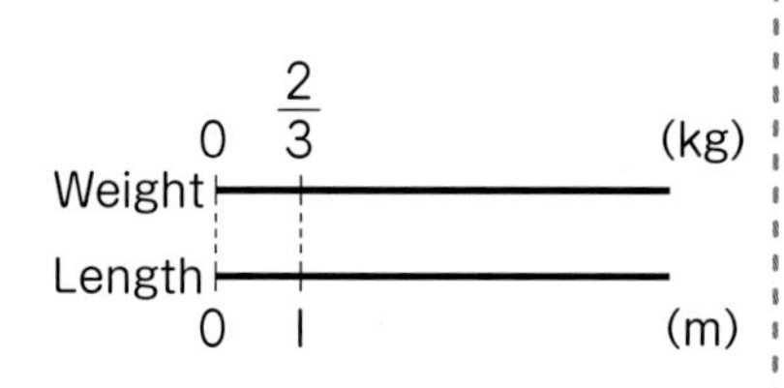

(3) Draw a scale for 4 m and put □ kg in the corresponding place.

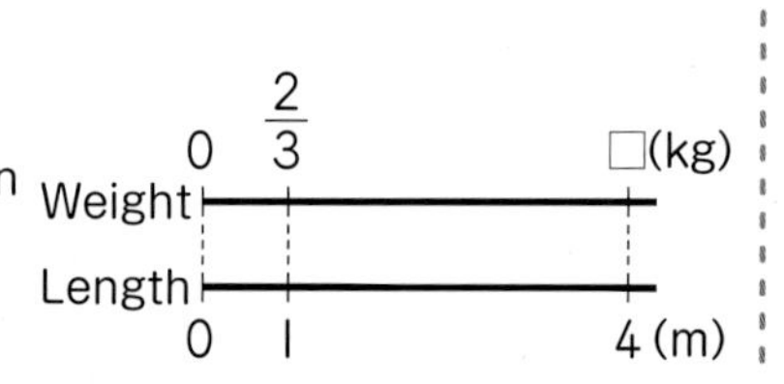

How to make a 4-square relationship table

(1) Draw a table of 4 cells.

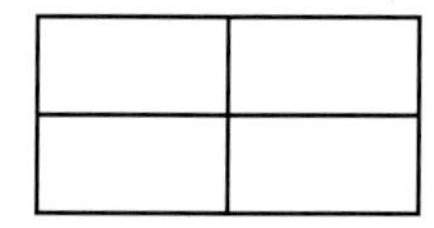

(2) Since the weight per 1 m is $\frac{2}{3}$ kg, write "1 m" and "$\frac{2}{3}$ kg" in the left column cells.

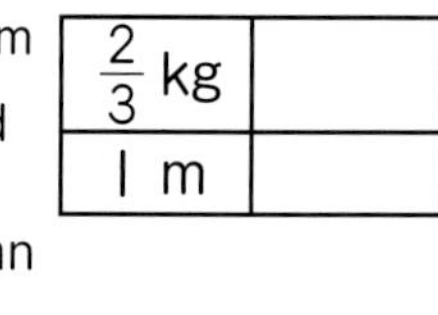

$\frac{2}{3}$ kg	
1 m	

(3) Since we do not know the weight for 4 m, assume it as □ kg and write "4 m" and "□ kg" in the right column cells.

$\frac{2}{3}$ kg	□kg
1 m	4 m

The 4-square relationship table can be written either aligned with the diagram or in the order in which they appear in the problem as shown on the right.

1 m	4 m
$\frac{2}{3}$ kg	□kg

The units in the horizontal cells will be the same.

Sara

In mathematics, it is sometimes easier to understand a problem situation if it is represented by a diagram or table. Let's review the various diagrams we have seen so far.

If you don't know the "Number for each unit"

There is an iron bar that weighs $\frac{9}{5}$ kg for 3 m. What is the weight of this iron bar per meter?

If we use a diagram…

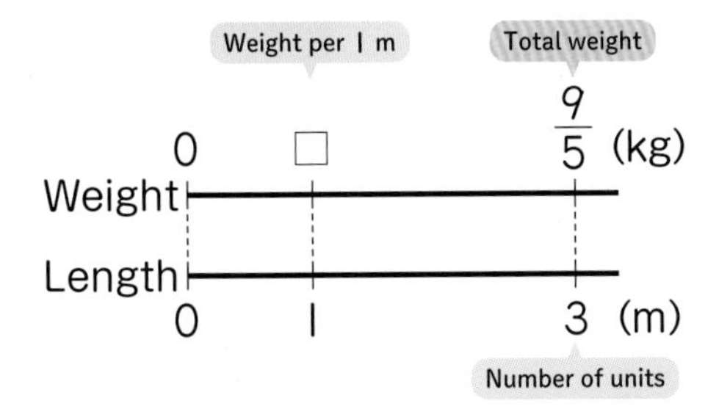

Total weight ÷ Number of units = Weight per 1 m

So,

$\frac{9}{5} \div 3 = \frac{3}{5}$

Answer: $\frac{3}{5}$ kg

If we use a table…

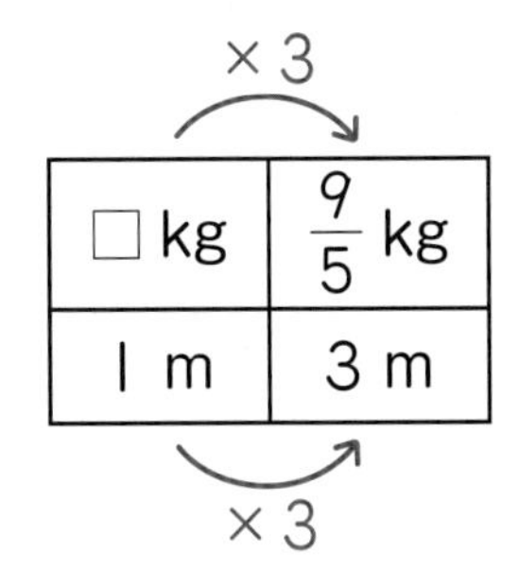

□ kg	$\frac{9}{5}$ kg
1 m	3 m

$\square \times 3 = \frac{9}{5}$

$\frac{9}{5} \div 3 = \frac{3}{5}$

Answer: $\frac{3}{5}$ kg

If you don't know the "Number of units"

There is an iron bar that weighs 2 kg per meter. When the weight of this iron bar is $\frac{4}{7}$ kg, what is the length of this iron bar in meter?

You can think in figures or in tables, whichever is easier to understand.

Yu

If we use a diagram…

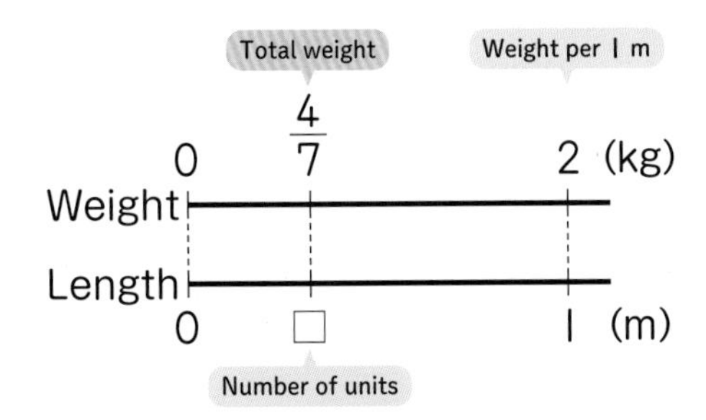

Total weight ÷ Weight per 1 m = Number of units

So,

$\frac{4}{7} \div 2 = \frac{2}{7}$

Answer: $\frac{2}{7}$ m

If we use a table…

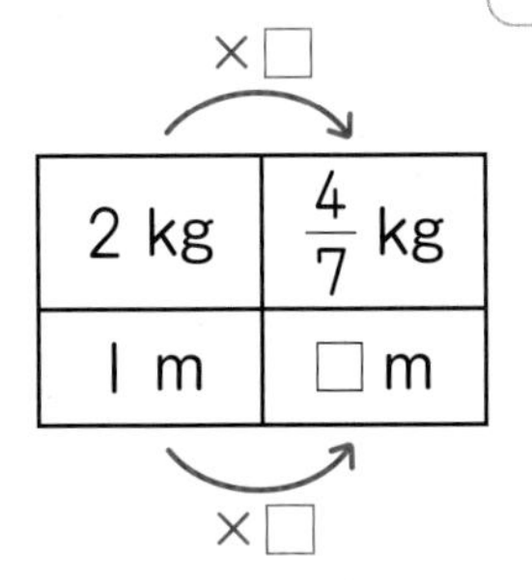

2 kg	$\frac{4}{7}$ kg
1 m	□ m

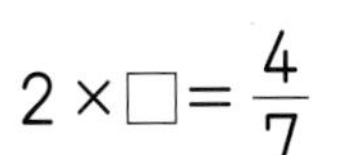

$2 \times \square = \frac{4}{7}$

$\frac{4}{7} \div 2 = \frac{2}{7}$

Answer: $\frac{2}{7}$ m

Find the ?

What is answer to fraction × fraction?

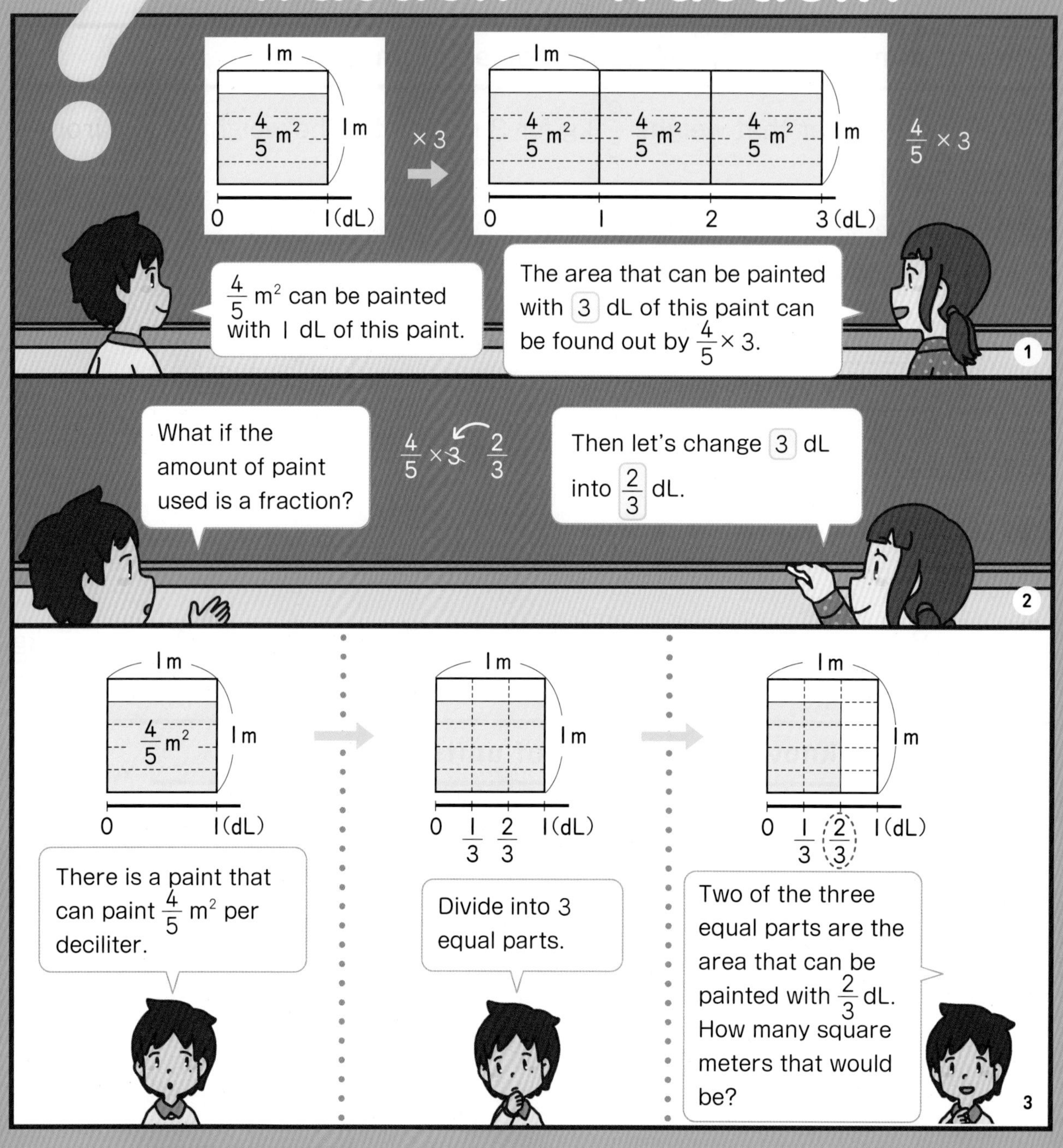

1m
$\frac{4}{5}$m²
1m
0
1(dL)
×3
$\frac{4}{5}$m²
$\frac{4}{5}$m²
$\frac{4}{5}$m²
0
1
2
3(dL)
$\frac{4}{5}$ × 3
$\frac{4}{5}$ m² can be painted with 1 dL of this paint.
The area that can be painted with 3 dL of this paint can be found out by $\frac{4}{5}$ × 3.
1
What if the amount of paint used is a fraction?
$\frac{4}{5}$ × 3 $\frac{2}{3}$
Then let's change 3 dL into $\frac{2}{3}$ dL.
2
$\frac{1}{3}$
$\frac{2}{3}$
There is a paint that can paint $\frac{4}{5}$ m² per deciliter.
Divide into 3 equal parts.
Two of the three equal parts are the area that can be painted with $\frac{2}{3}$ dL. How many square meters that would be?
3

How can we find out the area that can be painted with $\frac{2}{3}$ dL of paint?

4 Fraction × Fraction

Let's think about the meaning of the multiplication of fractions and how to calculate.

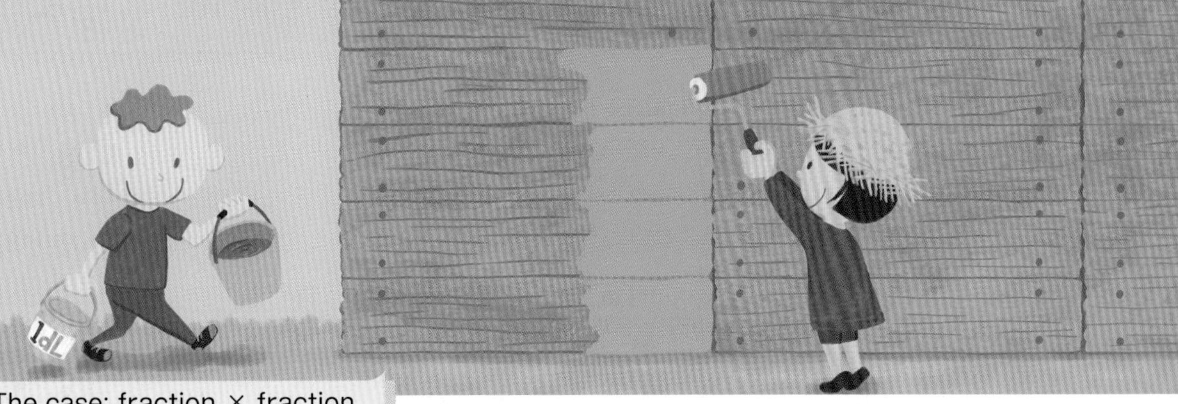

1 The case: fraction × fraction

Operation of proper fractions (1)

1 We want to paint the fence green. A 1 dL of paint can cover $\frac{4}{5}$ m^2. How many m^2 can $\frac{2}{3}$ dL of this paint cover?

① Let's write a math expression.

Painted area / Area painted with 1 dL

	0	□	$\frac{4}{5}$ (m^2)
Painted area			
Amount of paint	0	$\frac{2}{3}$	1 (dL)

Amount of paint

× □

$\frac{4}{5}$ m^2	□m^2
1 dL	$\frac{2}{3}$ dL

× □

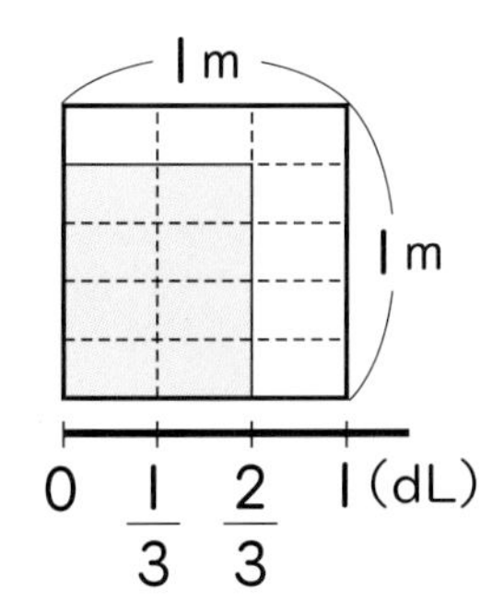

$\frac{2}{3}$ dL is two sets of $\frac{1}{3}$, so we can first think about the area that can be painted with $\frac{1}{3}$ dL.

Sara

 × 

Area painted with 1 dL / Amount of paint

\ Want to think /

? Purpose How can we calculate fractions × fractions?

② How many m^2 can $\frac{1}{3}$ dL of this paint cover? Let's write a math expression.
Let's think about how to calculate based on the diagram below.

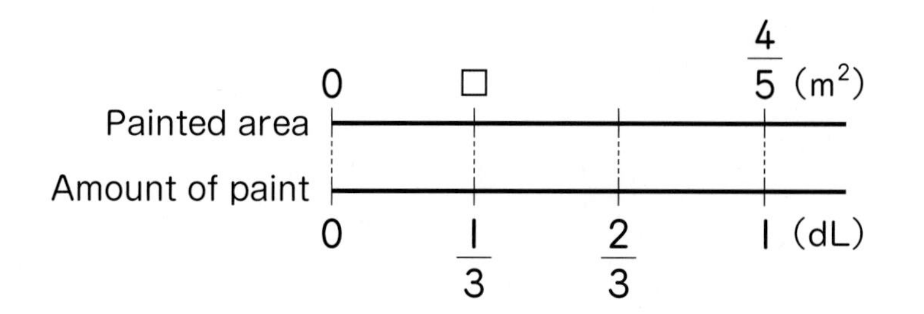

$\frac{4}{5}$ m^2	□ m^2
1 dL	$\frac{1}{3}$ dL

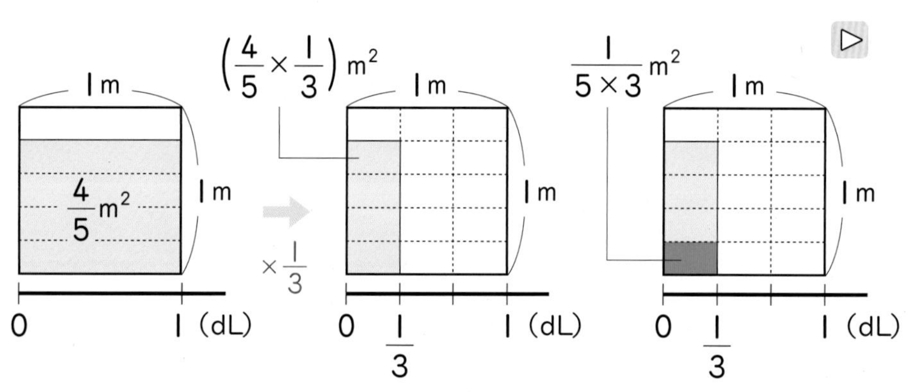

Way to see and think
The multiplication of $\frac{1}{3}$ is the same as the division by 3. So $\frac{4}{5} \times \frac{1}{3}$ is the same as $\frac{4}{5} \div 3$.

There are four $\frac{1}{5 \times 3}$ m^2, so $\frac{4}{5} \times \frac{1}{3} = \frac{\square}{\square \times \square}$

$= \square$

Way to see and think
You can calculate by "How many units" of $\frac{1}{15}$ m^2.

? Since we could do it in the case of $\frac{1}{3}$, so can we also do it for $\frac{2}{3}$?

2

Based on what we have learned so far, let's think about the area that can be painted by $\frac{2}{3}$ dL of this paint.

Akari
Can we think about it based on the area that could be painted with $\frac{1}{3}$ dL?

① Let's think about how to calculate.

② Let's explain the ideas of Akari and Sara and find the same resulting math equation.

Akari's idea

As for the painted area with $\frac{1}{3}$ dL, it was previously done as

$\frac{4}{5} \times \frac{1}{3} = \frac{4}{5 \times 3}$.

$\frac{2}{3}$ dL is two units of that.

$\frac{4}{5 \times 3} \times 2 = \frac{\square \times \square}{5 \times 3}$

$= \square$

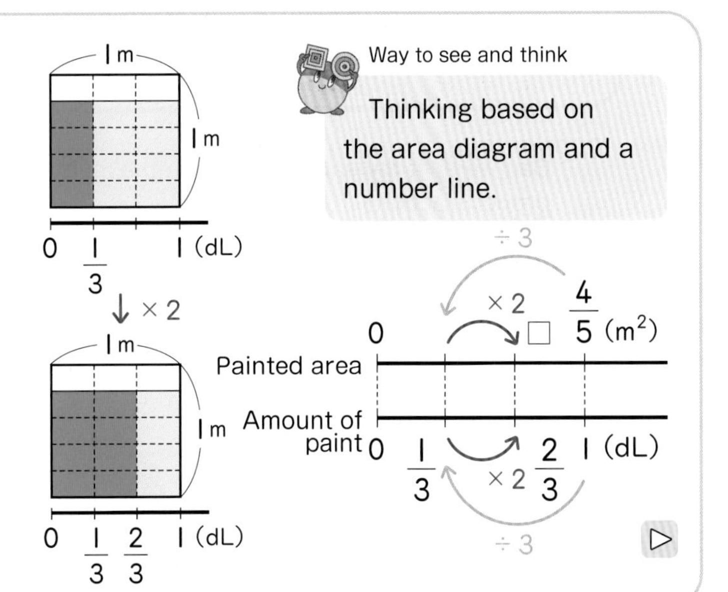

Sara's idea

Using the operation of fraction × whole number and fraction ÷ whole number,

$\frac{4}{5} \times \frac{2}{3} = \frac{4}{5} \times \left(\frac{2}{3} \times 3\right) \div 3$

$= \frac{4}{5} \times 2 \div 3$

$= \frac{4 \times 2}{5} \div 3 = \frac{4 \times 2}{5 \times 3} = \square$

Way to see and think

Use the multiplication rules to change $\frac{2}{3}$ into a whole number.

$\frac{4}{5} \times \frac{2}{3}$ ↓ × 3 $\frac{4}{5} \times \frac{2}{3} \times 3$ (÷ 3)

1 Let's find out the area that can be painted with $\frac{3}{5}$ dL of this paint based on Akari and Sara's ideas.

【Akari's idea】

$\frac{4}{5} \times \frac{3}{5}$

$= \frac{4}{5} \times \square \times 3$

$= \square$

【Sara's idea】

$\frac{4}{5} \times \frac{3}{5}$

$= \frac{4}{5} \times \left(\frac{3}{5} \times \square\right) \div \square$

$= \square$

Summary

When you multiply a proper fraction by a proper fraction, calculate by multiplying the two denominators and the two numerators.

$$\frac{b}{a} \times \frac{d}{c} = \frac{b \times d}{a \times c}$$

Can we do it the same even when the multiplicand and the multiplier is not a proper fraction?

3

How many square meters can be painted with $\frac{4}{3}$ dL of paint that covers $\frac{4}{5}$ m^2 per deciliter?

① Let's write a math expression.

\ Want to think /

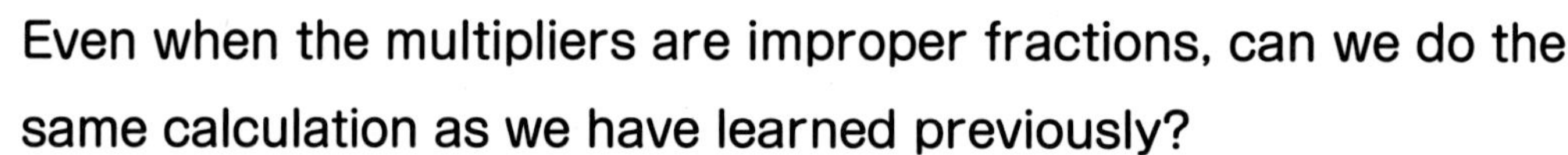

Even when the multipliers are improper fractions, can we do the same calculation as we have learned previously?

② Look at the diagram and let's think about the answer.

③ Let's find out the answer and confirm whether it is the same answer as in ②.

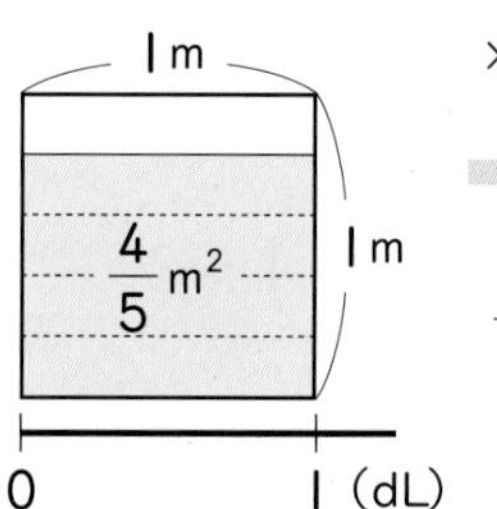

$\times \frac{4}{3}$

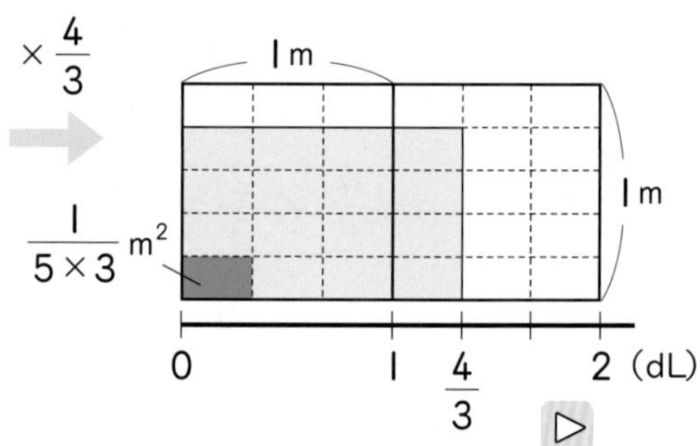

1 1 kg of potatoes contains about $\frac{3}{4}$ L of water. What is the amount of water in liter that $\frac{5}{7}$ kg of potatoes contain?

Even when multipliers are improper fractions, also calculate by multiplying the two denominators and the two numerators.

2 Let's calculate the following.

① $\frac{3}{4} \times \frac{1}{2}$　② $\frac{1}{3} \times \frac{5}{6}$　③ $\frac{3}{5} \times \frac{7}{8}$　④ $\frac{5}{9} \times \frac{2}{7}$

⑤ $\frac{1}{4} \times \frac{5}{3}$　⑥ $\frac{5}{7} \times \frac{3}{2}$　⑦ $\frac{8}{7} \times \frac{9}{5}$　⑧ $\frac{10}{9} \times \frac{4}{3}$

4 Let's think about how to calculate the following.

\ Want to try /

Purpose Let's try various operations.
Sara

❶ $\frac{4}{15}\times\frac{5}{6}=\frac{\overset{2}{\cancel{4}}\times\overset{\square}{\cancel{5}}}{\underset{3}{\cancel{15}}\times\underset{\square}{\cancel{6}}}$

$=\square$

❷ $3\frac{1}{7}\times2\frac{1}{10}=\frac{22}{7}\times\frac{21}{10}$

$=\frac{\overset{11}{\cancel{22}}\times\overset{3}{\cancel{21}}}{\underset{1}{\cancel{7}}\times\underset{5}{\cancel{10}}}$

$=\square$

1 Let's think about how to calculate the following.

① $\frac{5}{8}\times\frac{3}{10}$ ② $3\frac{1}{2}\times1\frac{5}{9}$

As for the multiplication of fractions, change the mixed fractions into improper fractions. Calculation becomes easy if fractions are reduced in the middle of the calculation.

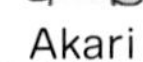
Akari

2 There is 1 L of sand that weighs $1\frac{3}{5}$ kg.
What is the weight in kilograms when there are $3\frac{3}{4}$ L of this sand?

3 Let's think about how to calculate the following.

① $2\times\frac{3}{5}=\frac{2}{\square}\times\frac{3}{5}$

$=\square$

② $\frac{4}{7}\times2=\frac{4}{7}\times\frac{2}{\square}$

$=\square$

As for the multiplication of a whole number and a fraction, if the whole number is changed into the fraction form, it becomes an operation of fraction × fraction.

Haruto

4 Let's calculate the following.

① $\frac{1}{6}\times\frac{6}{7}$ ② $\frac{8}{9}\times\frac{3}{4}$ ③ $\frac{6}{5}\times\frac{5}{12}$

④ $2\frac{5}{8}\times2\frac{2}{9}$ ⑤ $9\frac{1}{3}\times\frac{3}{8}$ ⑥ $\frac{6}{7}\times4\frac{2}{3}$

⑦ $4\times\frac{1}{5}$ ⑧ $4\times\frac{3}{5}$ ⑨ $4\times\frac{6}{5}$

In ⑦, the answer will be smaller than 4. In ⑨, the answer will be larger than 4.

Yu

? In multiplication of fractions, what is the relationship between the size of the multiplier and the product?

5 There is a wire that weighs 10 g per meter. Let's answer the following questions.

❶ How many grams is the weight of a wire with a length of $1\frac{1}{4}$ m or $\frac{2}{5}$ m?

It changes depending on the length.

As for multiplication of decimal numbers, when the multiplier is smaller than 1, the product became smaller than 1.

\ Want to explore /

? Purpose **What kind of relationship exists between the size of the multiplier and the product?**

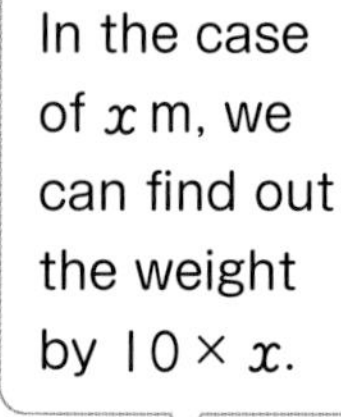

In the case of x m, we can find out the weight by $10 \times x$.

Weight: 0, ☐, 10, ☐ (g)
Length: 0, $\frac{2}{5}$, 1, $1\frac{1}{4}$ (m)

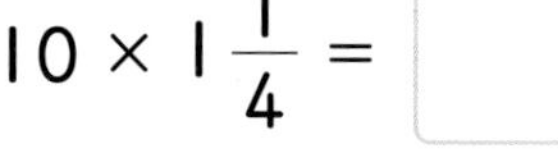

$10 \times 1\frac{1}{4} = ☐$

$10 \times 1 = 10$

$10 \times \frac{2}{5} = ☐$

❷ As for the math expressions $10 \times 1\frac{1}{4}$ and $10 \times \frac{2}{5}$, which has a product that becomes smaller than 10?

! Summary

When the multiplier is a fraction larger than 1, the product becomes larger than the multiplicand. When the multiplier is a fraction smaller than 1, the product becomes smaller than the multiplicand. When the multiplier is 1, the product becomes the same as the multiplicand.

Way to see and think
The same can be said for the multiplication of decimal numbers.

1 When the length of the wire in 5 is $\frac{7}{5}$ m, is it heavier than 10 g? Is it lighter?

2 Let's choose the calculations that have a product smaller than $\frac{3}{4}$ from the following Ⓐ~Ⓓ.

Ⓐ $\frac{3}{4} \times \frac{2}{3}$ Ⓑ $\frac{3}{4} \times \frac{3}{2}$ Ⓒ $\frac{3}{4} \times 1\frac{1}{5}$ Ⓓ $\frac{3}{4} \times \frac{3}{4}$

? Can we do the same in the case of the calculation of 3 fractions?

1 $\frac{3}{4} \times \frac{1}{5} \times \frac{5}{6}$ was calculated. Let's compare the following calculation methods Ⓐ and Ⓑ.

Ⓐ $\frac{3}{4} \times \frac{1}{5} \times \frac{5}{6} = \frac{3 \times 1}{4 \times 5} \times \frac{5}{6}$

$= \frac{3}{20} \times \frac{5}{6}$

$= \frac{\overset{1}{\cancel{3}} \times \overset{1}{\cancel{5}}}{\underset{4}{\cancel{20}} \times \underset{2}{\cancel{6}}}$

$= \frac{1}{8}$

Ⓑ $\frac{3}{4} \times \frac{1}{5} \times \frac{5}{6} = \frac{\overset{1}{\cancel{3}} \times 1 \times \overset{1}{\cancel{5}}}{4 \times \underset{1}{\cancel{5}} \times \underset{2}{\cancel{6}}}$

$= \frac{1}{8}$

\ Want to explain /

Purpose Can you explain how it is calculated?

Yu

1 Let's calculate the following.

① $\frac{3}{4} \times \frac{2}{3} \times \frac{3}{7}$ ② $\frac{5}{6} \times \frac{1}{5} \times \frac{2}{3}$

③ $\frac{3}{8} \times 5 \times \frac{4}{5}$

Summary As for the multiplication of 3 or more fractions, calculate by multiplying the denominators together and the numerators together.

Sara

? Can we use the formula to find out the area and volume of a figure even when the length of the sides is represented by a fraction?

2 Let's find out the area of the rectangle shown on the right.

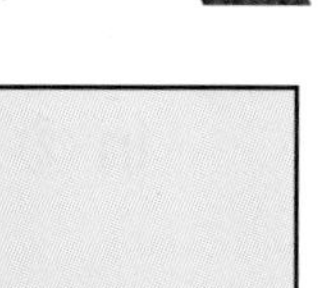

$\frac{5}{7}$ m

$\frac{2}{3}$ m

Yu

We could use the formula in the case of whole numbers and decimal numbers.

The length of the sides is represented by a fraction...

Akari

\ Want to know /

? Purpose **Can we use the formula even when the length of the side is represented by a fraction?**

① Haruto found out the area as shown below. Let's fill in the ☐ with numbers.

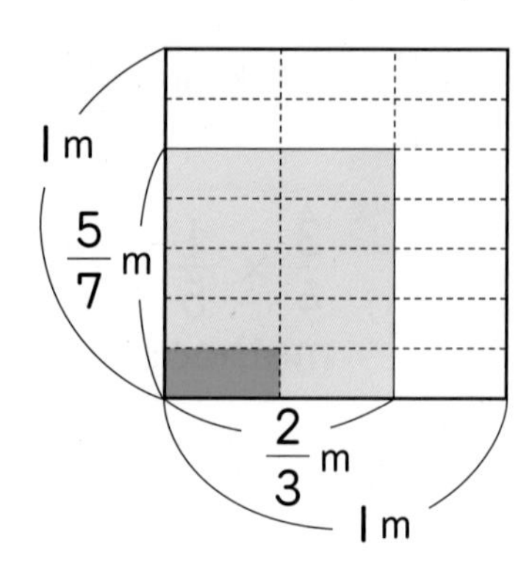

Haruto's idea

The area in ▬ is $\frac{1}{7 \times 3}$ of the square and represents ☐ m^2. As for the area of the colored rectangle, it has (5×2) sets, which is ☐ m^2.

Are the answers for ① and ② the same?

② Let's try to find out the area of the rectangle using the formula.

1 Let's find the volume of the cuboid shown in the diagram on the right.

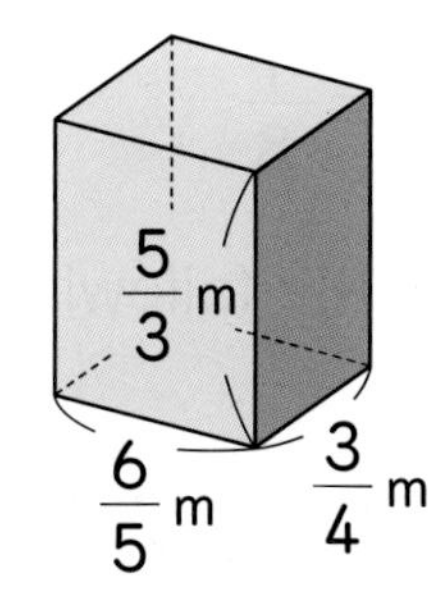

Way to see and think

The volume of a cuboid can be found by length × width × height.

Summary

The area and volume can be found by the formula that applies even when the length of the side is represented by a fraction.

2 Let's find the area of the parallelogram in ① and the volume of the cube in ②.

①

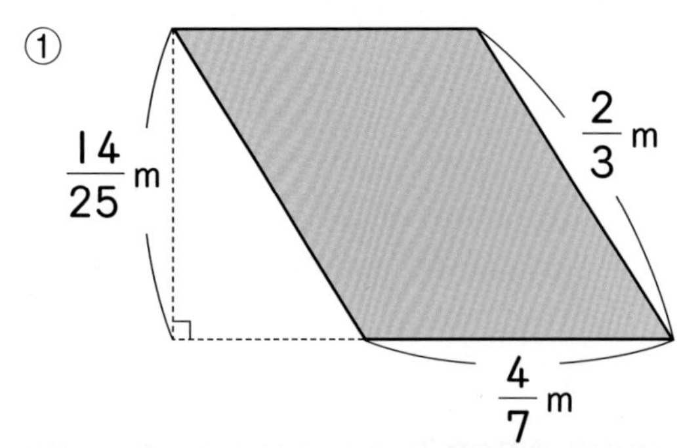

②

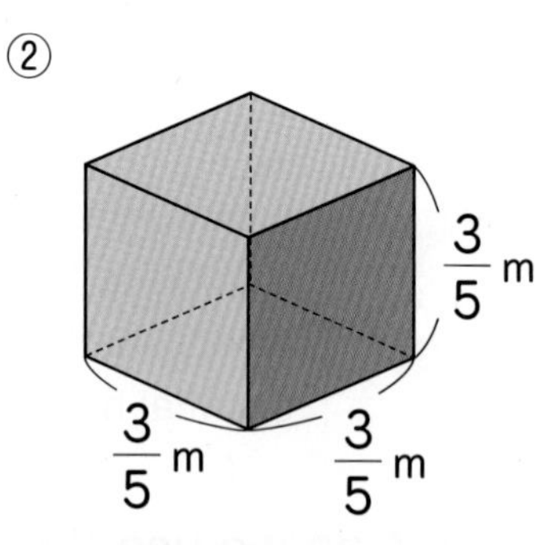

? The rules of operation were valid for whole numbers and decimals, but are they valid for fractions?

3 Rules of operation

\ Want to explore /

Purpose Are the rules of operation valid for fractions?

Akari

1

The following rules of operations are valid for whole numbers and decimal numbers. Let's explore whether these rules are also valid for fractions.

Ⓐ $a \times b = b \times a$ Ⓑ $(a \times b) \times c = a \times (b \times c)$

Ⓒ $(a + b) \times c = a \times c + b \times c$ Ⓓ $(a - b) \times c = a \times c - b \times c$

① Let's explain that the rule of commutativity in rule Ⓐ is valid using the area of the rectangle shown in the figure on the right.

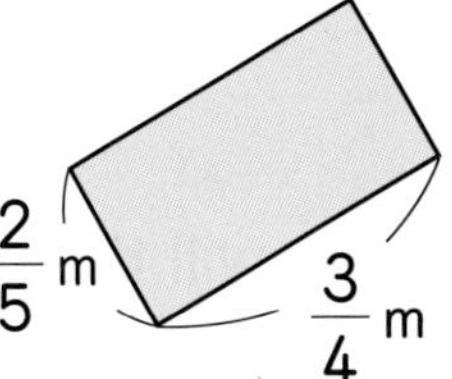

$\frac{2}{5} \times \frac{3}{4} = \square$ $\frac{3}{4} \times \frac{2}{5} = \square$

Way to see and think

Which side is considered as the length?

② Let's explain that the rule of association in rule Ⓑ is valid using the volume of the cuboid shown in the figure on the right.

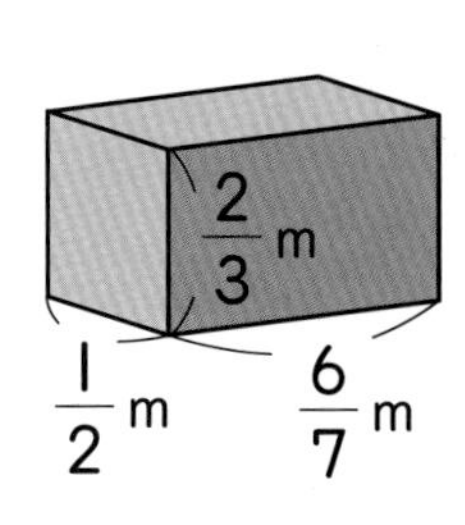

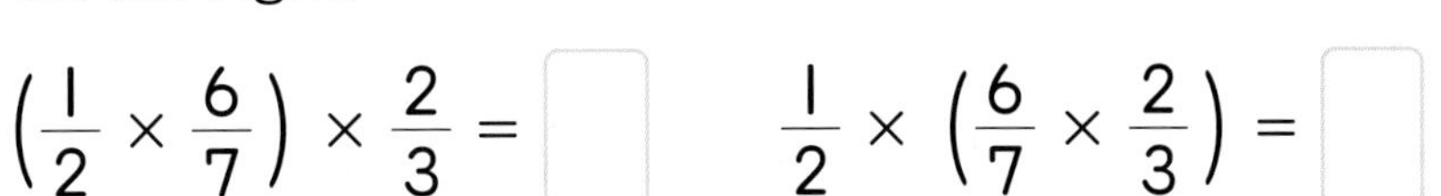

$\left(\frac{1}{2} \times \frac{6}{7}\right) \times \frac{2}{3} = \square$ $\frac{1}{2} \times \left(\frac{6}{7} \times \frac{2}{3}\right) = \square$

③ Using $a = \frac{2}{3}$, $b = \frac{1}{2}$, $c = \frac{6}{7}$, let's confirm that the rule of distribution in rule Ⓒ and Ⓓ can be used.

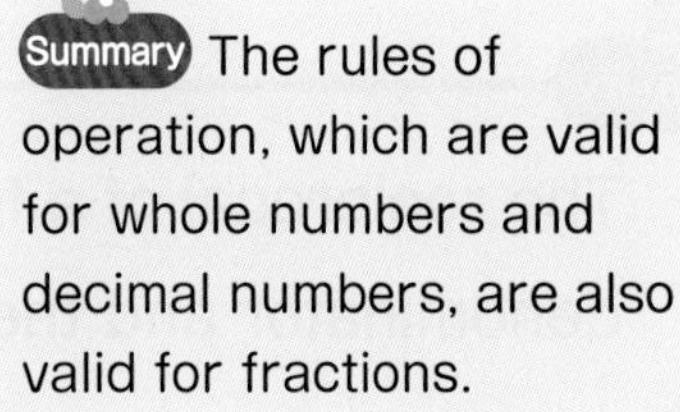

Summary The rules of operation, which are valid for whole numbers and decimal numbers, are also valid for fractions.

Sara

1 Sara calculated $\frac{3}{5} \times \frac{2}{3} + \frac{2}{5} \times \frac{2}{3}$ as follows.
Let's explain Sara's idea.

Sara's idea

$\frac{3}{5} \times \frac{2}{3} + \frac{2}{5} \times \frac{2}{3} = (\square + \square) \times \frac{2}{3} = \square \times \frac{2}{3} = \frac{2}{3}$

4 Reciprocal

Reciprocal ↓

There are 2 cards for each number from $\boxed{1}$ to $\boxed{9}$.
Let's answer the following questions.

❶ Let's use the cards to complete the following calculation.

$$\frac{\square}{\square} \times \frac{\square}{\square} = 1$$

In the case of $\frac{\boxed{3} \times \square}{\boxed{5} \times \square}$, reduce so that both the numerator and the denominator will be 1.

\ Want to think /

When will the product of two fractions be 1?

❷ Let's discuss what you noticed by looking at the multiplicand and the multiplier of multiplications where the product is 1.

When the product of two numbers is 1, one number is called the **reciprocal** of the other number.

The reciprocal of $\frac{2}{3}$ is $\frac{3}{2}$. The reciprocal of $\frac{3}{2}$ is $\frac{2}{3}$.

The reciprocal of a fraction is a fraction where the denominator and the numerator are exchanged.

$$\frac{b}{a} \to \frac{a}{b}$$

1 Let's find out the reciprocals of 0.4 and 6.

2 Let's find out the reciprocals of the following numbers.

① $\frac{4}{5}$ ② $\frac{10}{3}$ ③ $\frac{1}{8}$

④ $1\frac{5}{6}$ ⑤ 0.6 ⑥ 11

Way to see and think

When thinking about the reciprocals of whole numbers and decimal numbers, you can represent them in the form of fractions.

CAN What can you do?

☐ We can calculate multiplication that has fractions as multipliers. → pp.61 ~ 65

1 Let's calculate the following.

① $\frac{1}{5} \times \frac{3}{4}$　② $\frac{2}{5} \times \frac{6}{7}$　③ $\frac{5}{6} \times \frac{2}{3}$

④ $\frac{9}{14} \times \frac{7}{18}$　⑤ $2\frac{5}{6} \times \frac{2}{17}$　⑥ $1\frac{2}{3} \times 1\frac{1}{5}$

⑦ $\frac{15}{8} \times \frac{6}{5}$　⑧ $7 \times \frac{4}{5}$　⑨ $6 \times \frac{9}{8}$

☐ We understand the relationship between the multiplier and the product. → p.66

2 Which calculation has a product smaller than 5?

Ⓐ $5 \times 1\frac{1}{12}$　Ⓑ $5 \times \frac{5}{6}$　Ⓒ $5 \times \frac{4}{3}$　Ⓓ $5 \times \frac{9}{10}$

☐ We understand the formulas studied previously and the rules of operations that are valid for fractions. → pp.67 ~ 69

3 Let's find out the area of the following figures.

① Trapezoid

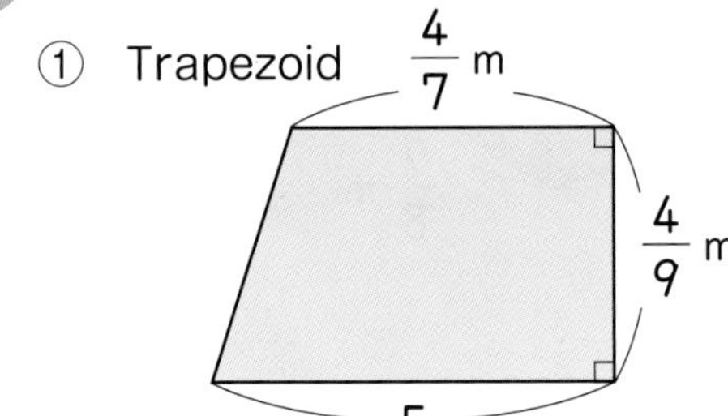

② Rectangle

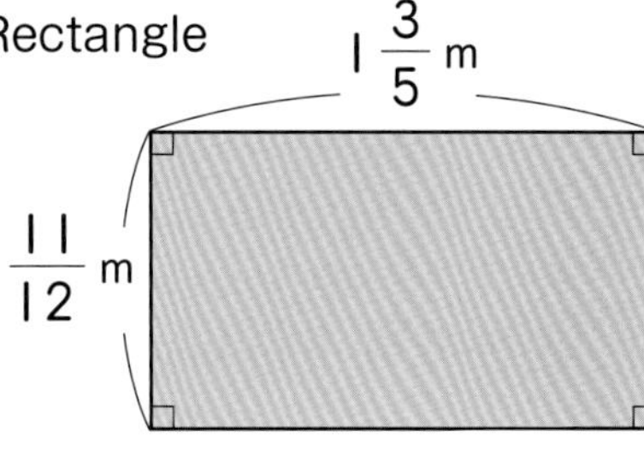

☐ We understand the meaning of reciprocals and how to find them. → p.70

4 Let's find out the reciprocals of the following numbers.

① $\frac{1}{3}$　② $\frac{7}{2}$　③ $\frac{5}{6}$　④ $1\frac{1}{2}$　⑤ 9　⑥ 0.7

Supplementary Problems → p.235

Which "Way to See and Think Monsters" did you find in "4 Fraction × Fraction"?

I found "Unit" when I was thinking about how to calculate.

Haruto

I found other monsters when I was calculating.

Yu

Utilize Usefulness and Efficiency of Learning

1 There is a rice field that produces $\frac{4}{7}$ kg of rice per square meter.
What is the weight of rice in kilograms will $\frac{5}{8}$ m^2 of this field produce?

2 Let's find out the area of the following figures.

①

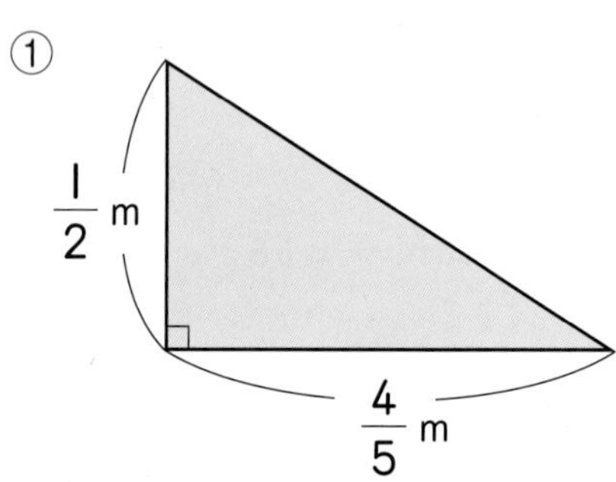

②

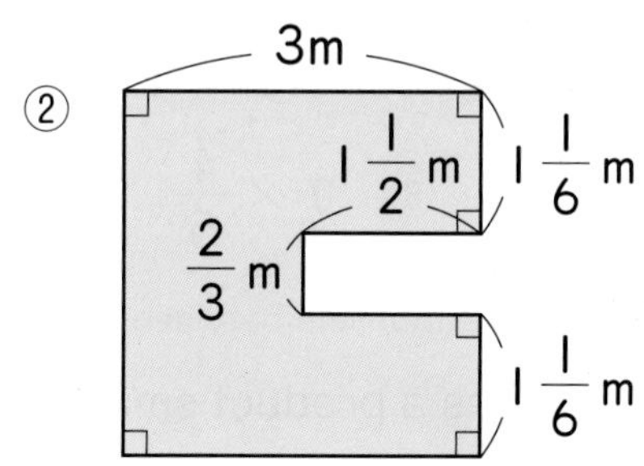

3 Let's find out the volume of the cuboid on the right.

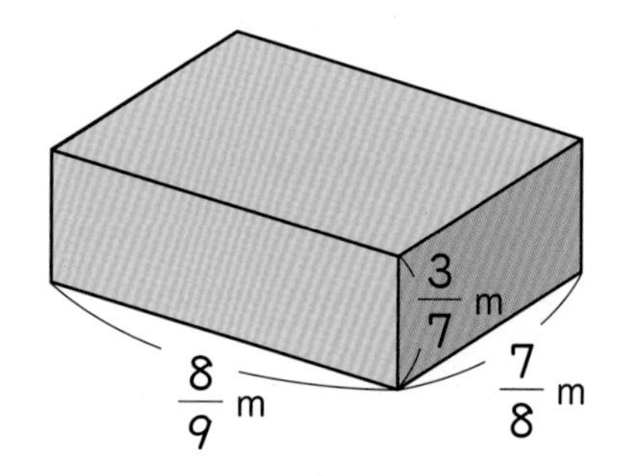

4 There are 2 cards for each number from [2] to [7].

[2] [2] [3] [3] [4] [4] [5] [5] [6] [6] [7] [7]

Let's create math expressions by placing cards inside each □.

① Let's create math expressions with answer 1.

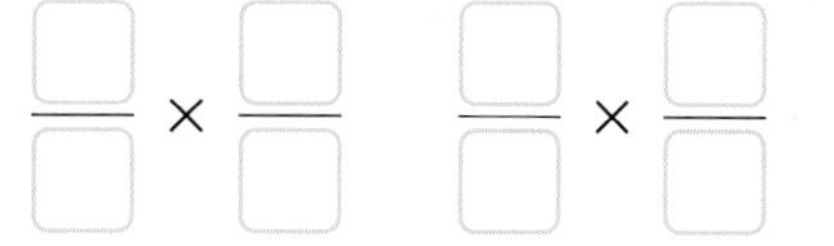

② Let's create math expressions with answer 2.

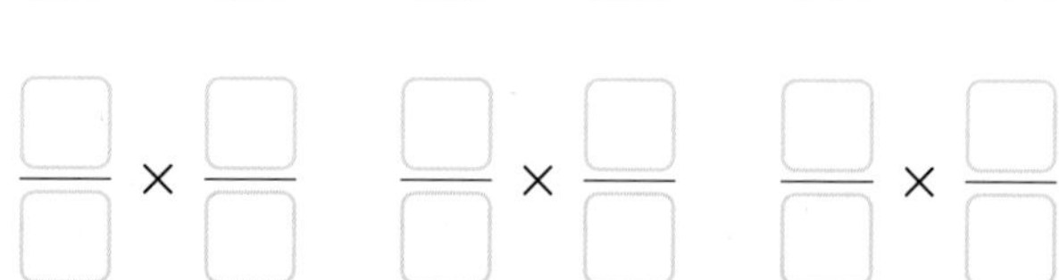

With the Way to See and Think Monsters...

Let's Reflect!

Let's reflect on which monster you used while learning "4 Fraction × Fraction."

Other Way

We could find out the area by representing in a diagram and a number line.

As for the calculation for $\frac{4}{5}\times\frac{2}{3}$, first, we divided $\frac{4}{5}$ into three equal parts, and then doubled it.

Akari

【Finding out the area that can be painted with $\frac{2}{3}$ dL of paint that covers $\frac{4}{5}$ m^2 per deciliter】

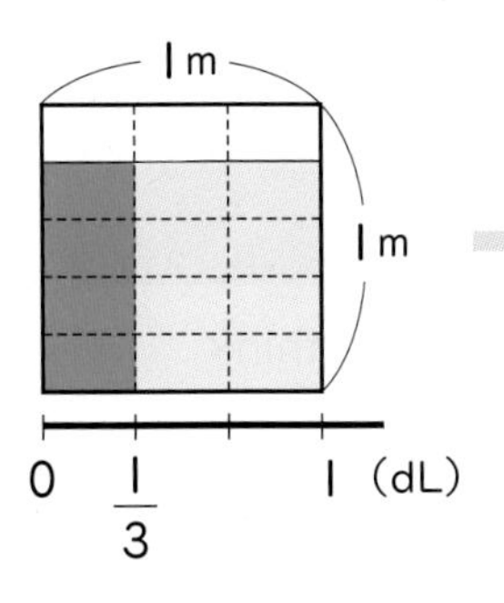

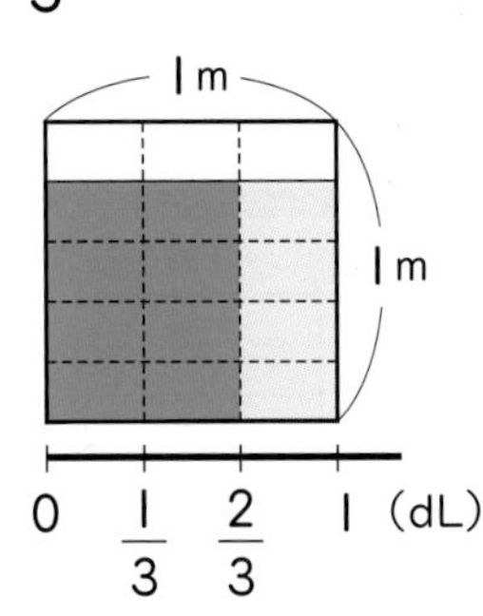

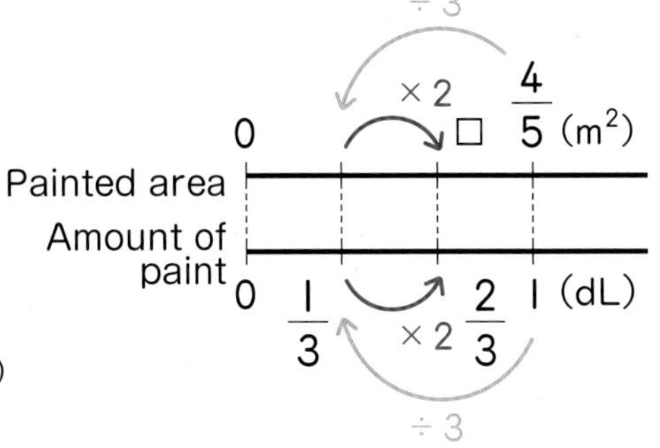

Rule

Using the rules of calculation, we could rearrange the calculations into the ones we had learned so far.

$$\frac{4}{5}\times\frac{2}{3}=\frac{4}{5}\times\left(\frac{2}{3}\times 3\right)\div 3$$

$$=\frac{4}{5}\times 2\div 3$$

$$=\frac{4\times 2}{5}\div 3=\frac{4\times 2}{5\times 3}=\square$$

I rearranged it to fraction ÷ whole number by using the rules of multiplication.

Sara

Solve the ?

We could solve fraction × fraction based on the calculation we have learned so far.

Yu

Want to Connect

Can we calculate fraction ÷ fraction in the same way?

Sara

Find the

How much can we paint with 1 dL?

When $\frac{2}{5}$ m^2 can be painted with 2 dL of paint, how many square meters can be painted per deciliter of this paint?
1 m
1 m
$\frac{2}{5}$ m^2
0
2 (dL)
÷ 2
1 m
1 m
0
1
2 (dL)
$\frac{2}{5} \div 2$
It can be found by $\frac{2}{5} \div 2$
What if the amount of paint is a fraction?
$\frac{2}{5} \div 2$ $\frac{3}{4}$
Let's change 2 dL into $\frac{3}{4}$ dL
1 m
1 m
$\frac{2}{5}$ m^2
0
$\frac{3}{4}$ (dL)
Since $\frac{2}{5}$ m^2 can be painted with $\frac{3}{4}$ dL of paint...
1 dL is larger than $\frac{3}{4}$ dL, so...
1 m
1 m
0
$\frac{3}{4}$
(dL)
I think it's more towards the right.
Is 1 dL around here?
1 m
1 m
0
$\frac{3}{4}$
1 (dL)
1 dL will be around here.
It's easier to find out in the case of $\frac{1}{4}$.
1 m
1 m
0
$\frac{1}{4}$
$\frac{3}{4}$ (dL)

How can we find out the area that can be painted with 1 dL?

5 Fraction ÷ Fraction

Let's think about the meaning of the division of fractions and how to calculate.

1 The case: fraction ÷ fraction

1 $\frac{2}{5}$ dL of yellow paint was used to paint a $\frac{3}{4}$ m^2 wall. How many square meters can be painted per deciliter of this paint?

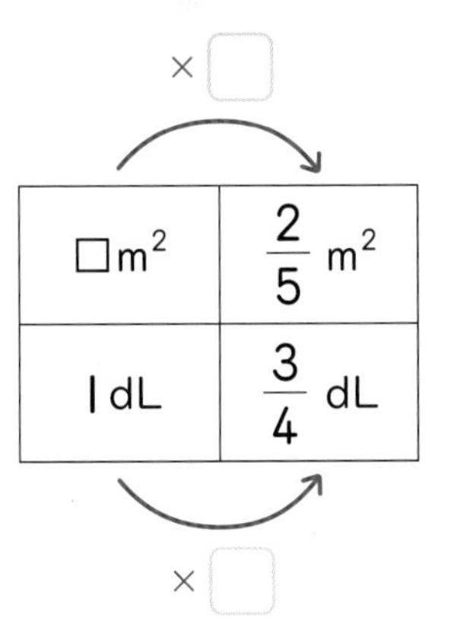

① Let's write a math expression.

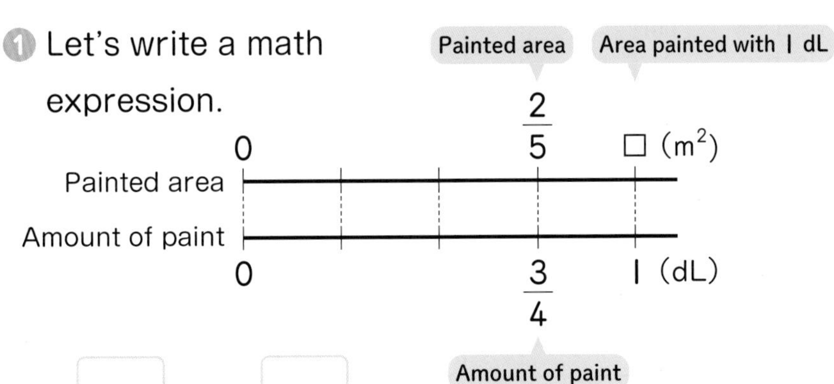

□m^2	$\frac{2}{5}$ m^2
1 dL	$\frac{3}{4}$ dL

□ ÷ □

Painted area　Amount of paint

Way to see and think

Since you know "Total measurement" and "Number of units," you can use a division expression.

Way to see and think

If you can paint x m^2 per deciliter, it can be represented by $x \times \frac{3}{4} = \frac{2}{5}$.

\ Want to think /

? **Purpose** What should we do to divide a fraction by a fraction?

② Let's think about how to calculate.

Akari

First, we need to think about how many m^2 can be painted with $\frac{1}{4}$ dL, then...

If we think about the relationship between $\frac{3}{4}$ dL and 1 dL...

Sara

If we think by using a diagram, it will be like this.

Haruto

Divide $1\,m^2$ into 5 equal parts.

The area that can be painted with $\frac{3}{4}$ dL is $\frac{2}{5}\,m^2$

Area that can be painted with 1 dL

1 m

1 m

0 $\frac{3}{4}$ (dL)

0 $\frac{1}{4}$ $\frac{2}{4}$ $\frac{3}{4}$ 1 (dL)

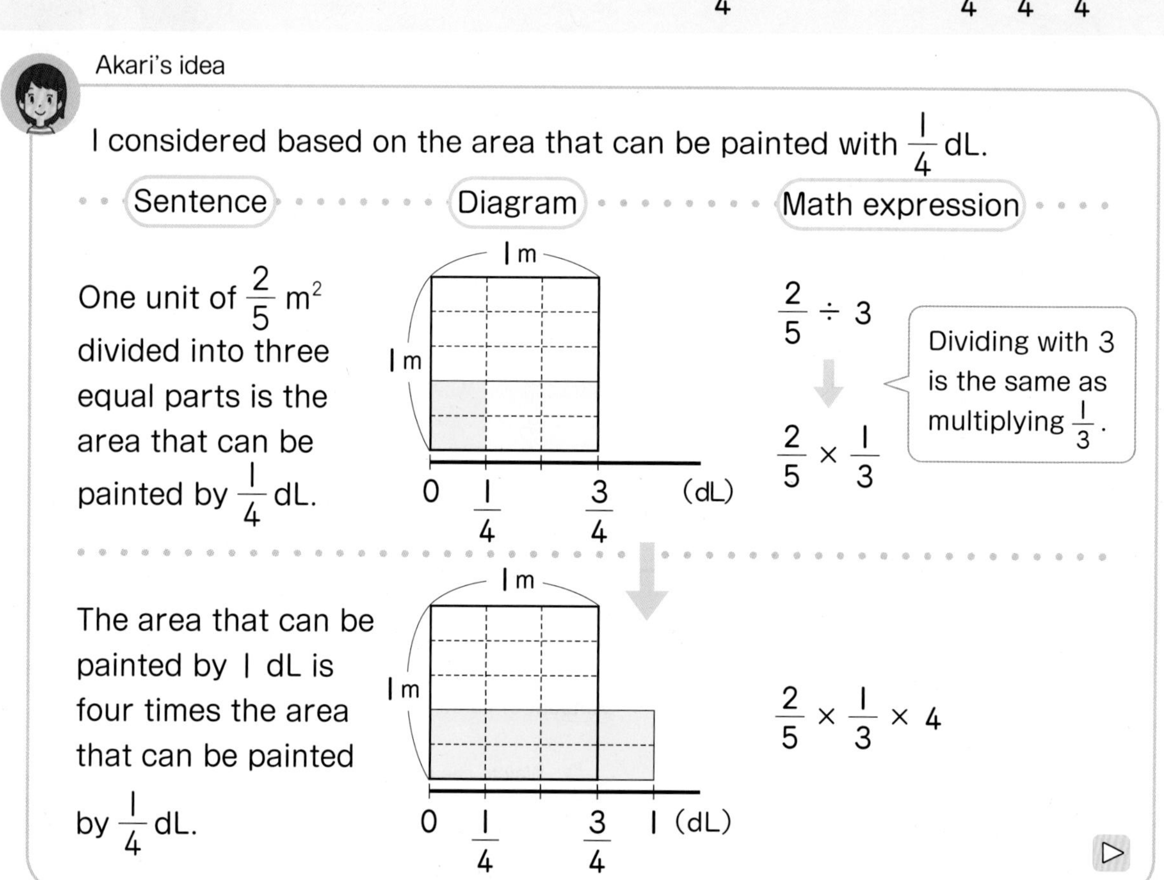

Akari's idea
I considered based on the area that can be painted with $\frac{1}{4}$ dL.
Sentence
Diagram
Math expression
One unit of $\frac{2}{5}\,m^2$ divided into three equal parts is the area that can be painted by $\frac{1}{4}$ dL.
1 m
1 m
0 $\frac{1}{4}$ $\frac{3}{4}$ (dL)
$\frac{2}{5} \div 3$
$\frac{2}{5} \times \frac{1}{3}$
Dividing with 3 is the same as multiplying $\frac{1}{3}$.
The area that can be painted by 1 dL is four times the area that can be painted by $\frac{1}{4}$ dL.
1 m
1 m
0 $\frac{1}{4}$ $\frac{3}{4}$ 1 (dL)
$\frac{2}{5} \times \frac{1}{3} \times 4$

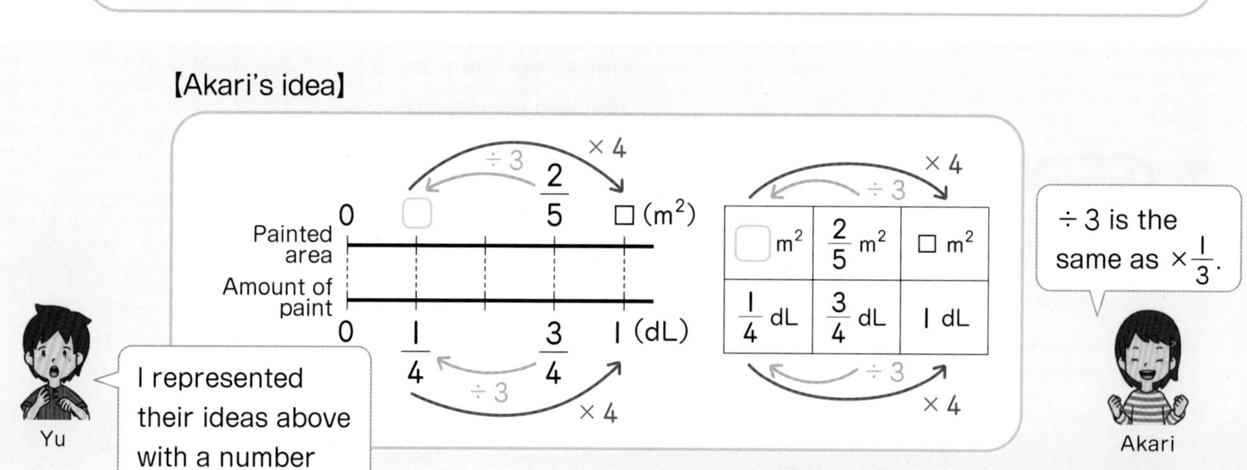

m²	$\frac{2}{5}$ m²	□ m²
$\frac{1}{4}$ dL	$\frac{3}{4}$ dL	1 dL

÷ 3 is the same as $\times\frac{1}{3}$.

Akari

I represented their ideas above with a number line and a table.

Yu

Sara's idea

To make the amount of paint 1 dL, multiply $\frac{4}{3}$.

So we can multiply $\frac{4}{3}$ to the painted area.

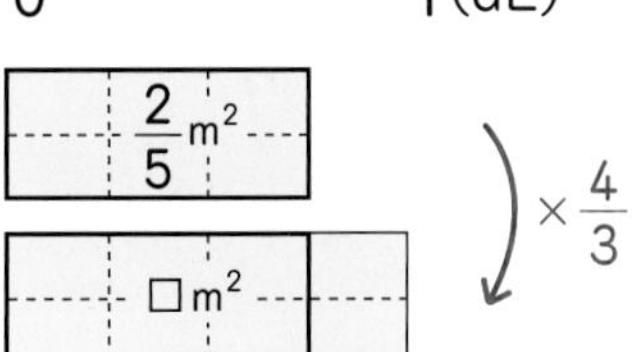

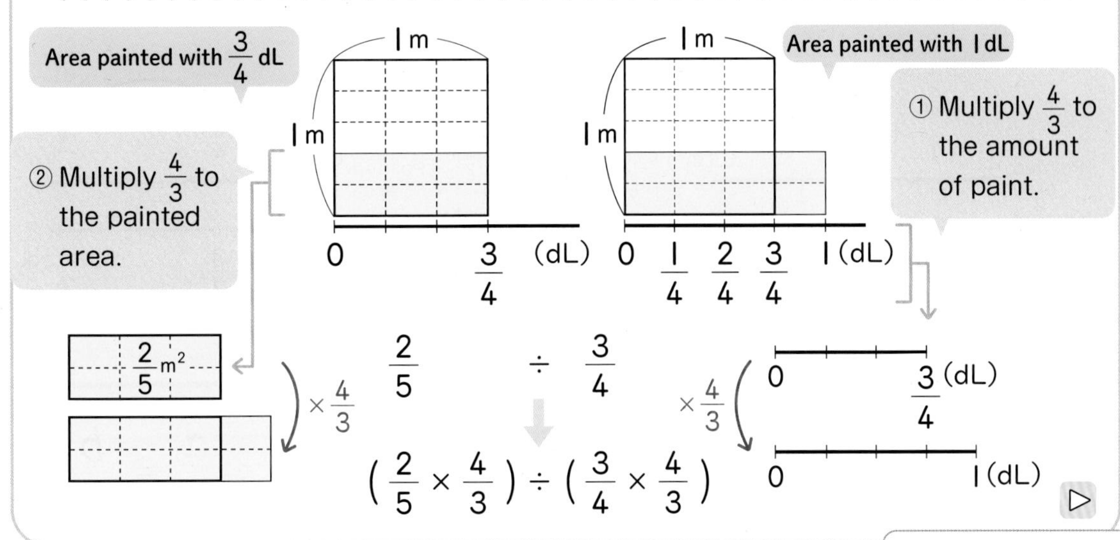

Akari: In Sara's idea, the quotient remains the same because both the dividend and the divisor are multiplied by $\frac{4}{3}$. This is using the division rule.

By calculating the divisor, we are finding out the area per 1 dL.

$\left(\frac{2}{5}\times\frac{4}{3}\right)\div 1$

Yu

Since the divisor is 1, the calculation is the same as multiplying $\frac{4}{3}$ to $\frac{2}{5}$.

Haruto

【 Sara's idea 】

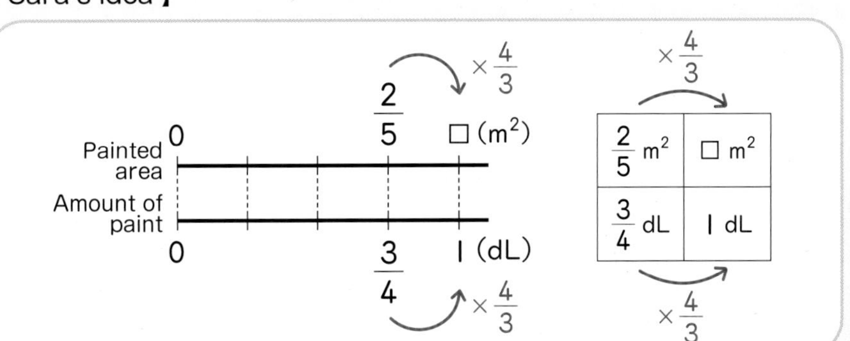

$\frac{2}{5}$ m²	□ m²
$\frac{3}{4}$ dL	1 dL

It's easier to see the different way of thinking by comparing with a number line and a table.

Sara

1 As for the problem in page 75, let's find out the area that can be painted by 1 dL of paint when the amount of paint used is $\frac{3}{7}$ dL, using the ideas of Akari and Sara on the previous page.

【Akari's idea】

$$\frac{2}{5} \div \frac{3}{7}$$
$$= \frac{2}{5} \times \square \times \square$$
$$= \square$$

【Sara's idea】

$$\frac{2}{5} \div \frac{3}{7}$$
$$= \left(\frac{2}{5} \times \square\right) \div \left(\frac{3}{7} \times \square\right)$$
$$= \square$$

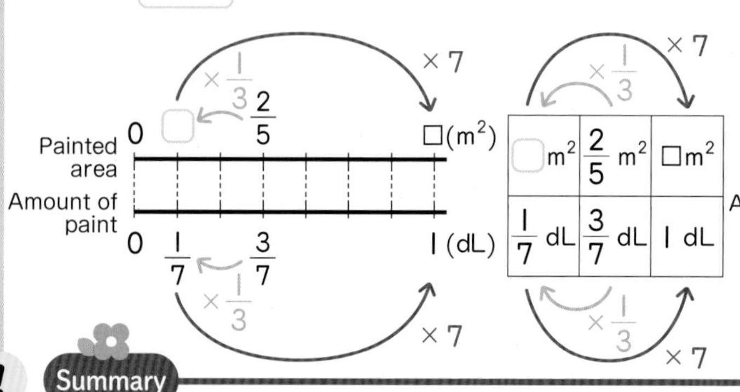

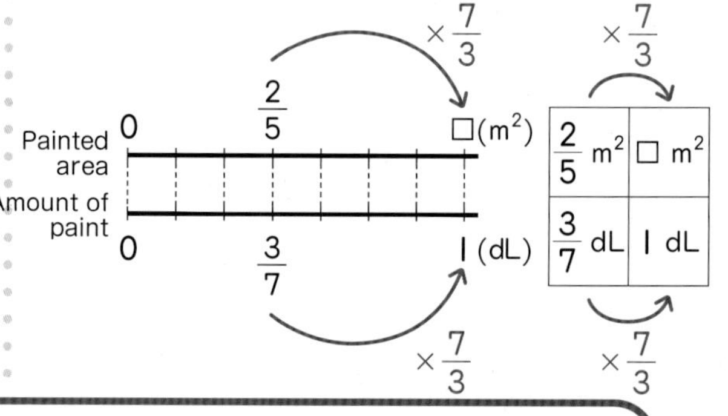

Summary

When you divide a proper fraction by a proper fraction, calculate by multiplying the dividend by the reciprocal of the divisor.

$$\frac{b}{a} \div \frac{d}{c} = \frac{b}{a} \times \frac{c}{d}$$

2 Let's explain how to calculate the following.

Want to try

Purpose Let's explore various divisions.

Yu

If the fraction can be reduced, then let's reduce.

① $\frac{8}{3} \div \frac{12}{5} = \frac{8}{3} \times \frac{\square}{\square}$
$= \square$

② $\frac{15}{16} \div \frac{5}{4} = \frac{15}{16} \times \frac{\square}{\square}$
$= \square$

③ $4 \div \frac{2}{5} = \frac{\square}{1} \div \frac{2}{5}$
$= \frac{\square}{1} \times \frac{5}{2}$
$= \square$

④ $\frac{3}{7} \div 6 = \frac{3}{7} \div \frac{\square}{1}$
$= \frac{3}{7} \times \frac{1}{\square}$
$= \square$

① $\frac{1}{4} \div \frac{1}{3}$ ② $\frac{2}{7} \div \frac{3}{4}$ ③ $\frac{2}{3} \div \frac{7}{8}$ ④ $\frac{3}{5} \div \frac{5}{6}$

⑤ $\frac{16}{7} \div \frac{4}{9}$ ⑥ $\frac{4}{3} \div \frac{2}{3}$ ⑦ $8 \div \frac{2}{3}$ ⑧ $\frac{8}{9} \div 4$

3

$1\frac{1}{3}$ dL of green paint was used to paint a $\frac{2}{5}$ m^2 wall. How many square meters can be painted per deciliter of this paint?

\ Want to think /

Purpose Can we think in the same way as multiplication?

Haruto

❶ Let's write a math expression.

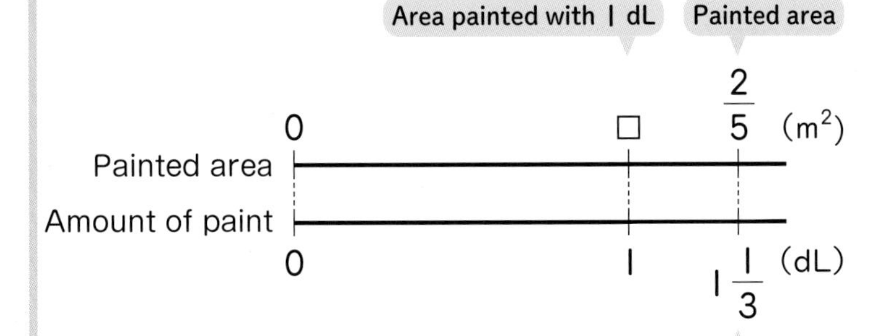

× □

□m^2	$\frac{2}{5}$ m^2
1 dL	$1\frac{1}{3}$ dL

× □

❷ Let's think about how to calculate.

$$\frac{2}{5} \div 1\frac{1}{3} = \frac{2}{5} \div \frac{4}{3}$$
$$= \frac{2}{5} \times \frac{3}{4}$$
$$= \square$$

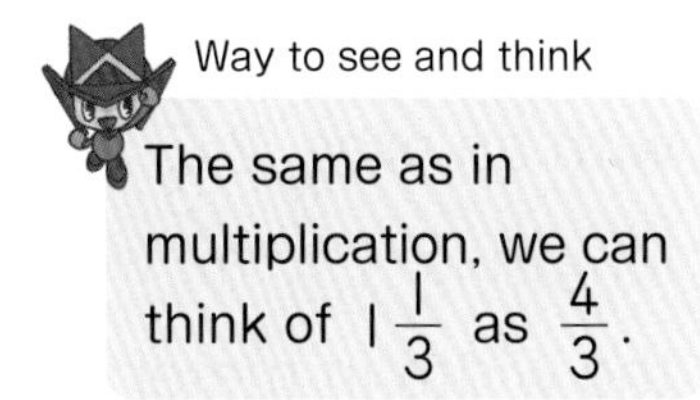

Way to see and think
The same as in multiplication, we can think of 1 1/3 as 4/3.

Summary In division of fractions, we can rearrange the mixed fractions to improper fractions.

1 Let's calculate the following.

① $\frac{5}{8} \div 1\frac{4}{5}$ ② $\frac{4}{7} \div 1\frac{3}{5}$ ③ $\frac{3}{4} \div 2\frac{2}{5}$ ④ $\frac{1}{2} \div 1\frac{1}{6}$

4 Let's try to think about the various questions shown below.

\ Want to try /

Purpose Can you find out the answer for various questions by thinking about the appropriate math expression?

Yu

❶ There are $1\frac{4}{5}$ L of milk. If a family drinks $\frac{3}{5}$ L of milk at a time, how many times can they drink the milk?

Amount of milk: 0, $\frac{3}{5}$, $1\frac{4}{5}$ (L)
Number of times: 0, 1, ☐ (times)

$\frac{3}{5}$ L	$1\frac{4}{5}$ L
1 time	☐ times

Way to see and think

Since you know the "Total measurement" and "Measurement per unit quantity," you can use a division expression.

❷ There is a wire that weighs $4\frac{1}{2}$ g per meter. If the total weight of the wire is 24 g, What is the length of this wire in meters?

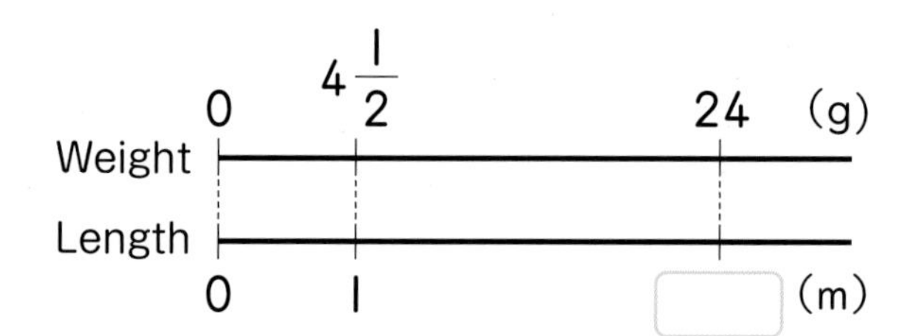

$4\frac{1}{2}$ g	24 g
1 m	☐ m

Way to see and think

Use the formula to find out the area of a rectangle.

❸ There is a rectangular cloth with an area $2\frac{2}{3}$ m^2. If the length is $1\frac{7}{9}$ m, what is the width in meters?

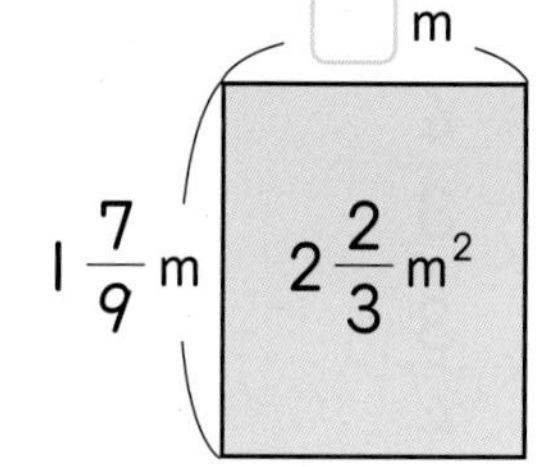

Summary By using the size per unit amount or using area formulas, you can find out the appropriate way of calculation.

Sara

1 Let's calculate the following.

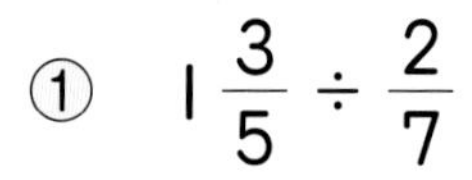

① $1\frac{3}{5} \div \frac{2}{7}$ ② $1\frac{1}{4} \div \frac{5}{8}$ ③ $4\frac{2}{3} \div 1\frac{1}{5}$ ④ $2\frac{1}{3} \div 1\frac{5}{9}$

? In division of fractions, what is the relationship between the size of the divisor and the size of the quotient?

5

There is a thin wire with a length of $1\frac{4}{5}$ m that weighs 24 g and a thick wire with a length of $\frac{3}{5}$ m that weighs 24 g. What is the weight in grams of each wire per meter?

① Let's find out the weight per meter of each wire.

【Thin wire】

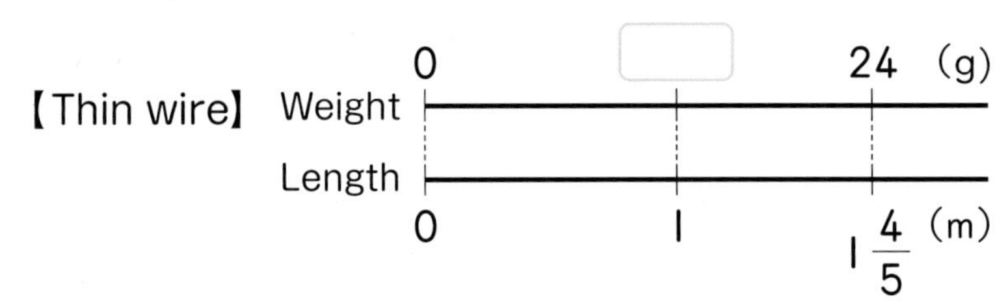

☐ g	24 g
1 m	$1\frac{4}{5}$ m

【Thick wire】

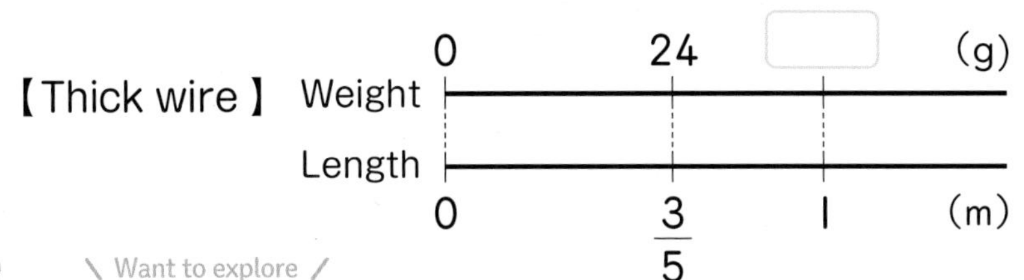

☐ g	24 g
1 m	$\frac{3}{5}$ m

\ Want to explore /

? **Purpose** **What kind of relationship exists between the size of the divisor and the quotient?**

② When does the quotient become larger than 24? When does it become smaller than 24?

! **Summary**

When the divisor is a fraction larger than 1, the quotient becomes smaller than the dividend. When the divisor is a fraction smaller than 1, the quotient becomes larger than the dividend. When the divisor is 1, the quotient becomes the same as the dividend.

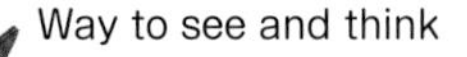

Way to see and think

We can say the same for the division of decimal numbers.

1 The weight of a $\frac{7}{3}$ m wire is 24 g. Is the weight per meter of this wire heavier than 24 g? Is it lighter than 24 g?

2 Which of the following calculations has the quotient larger than 7? Let's explain the reasons.

Ⓐ $7 \div \frac{3}{4}$ Ⓑ $7 \div 1\frac{2}{3}$ Ⓒ $7 \div \frac{2}{3}$ Ⓓ $7 \div 7\frac{7}{8}$

1 Deciding a math expression

\ Want to think /

1 Read the following problems and let's write a math expression. Then let's find the answer.

Purpose Can we think about what to find out and then out it in a math expression?

Akari

① There is a metal bar with a length of $\frac{4}{3}$ m that weighs $\frac{9}{5}$ kg. What is the weight in kilograms for 1 m of this metal bar?

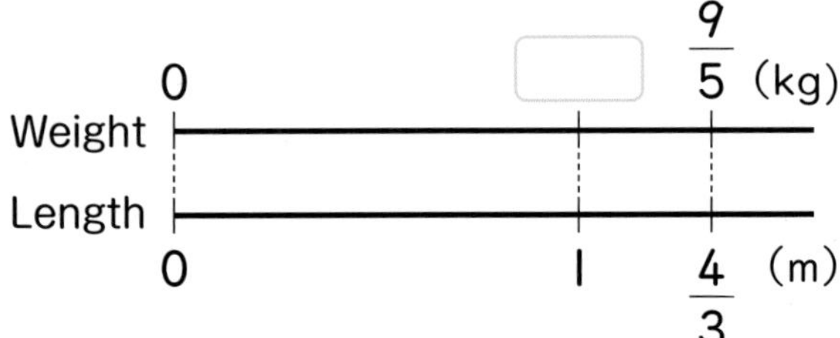

kg	$\frac{9}{5}$ kg
1 m	$\frac{4}{3}$ m

② There is a metal bar with a length of 1 m that weighs $\frac{5}{3}$ kg. What is the weight in kilograms for $\frac{5}{2}$ m of this metal bar?

0 $\frac{5}{3}$ (kg)
Weight
Length
0 1 $\frac{5}{2}$ (m)

$\frac{5}{3}$ kg	kg
1 m	$\frac{5}{2}$ m

1 Let's think about the problem that was created by Asahi.

> $\frac{6}{7}$ L of water is used to water 1 m^2 of a field. If $\frac{2}{3}$ m^2 of a field is watered, ☐ L of water are needed. Let's find out the number that applies for ☐.

Summary The math expression differs depending on what we want to find out.

Haruto

① Let's answer the problem created by Asahi.

② Let's change the number inside each ☐ and create a multiplication or division problem.

Way to see and think

Replace the sentence into a table or diagram and decide which values you want for the problem.

CAN What can you do?

☐ We understand how to calculate divisions of fractions. → p.78, p.79

1 Let's fill in each ☐ with a number.

① $\frac{7}{14} \div \frac{3}{5} = \frac{7}{14} \times \square$ ② $3 \div \frac{4}{7} = 3 \times \square$

☐ We can calculate divisions that have fractions as divisors. → pp.75 ~ 80

2 Let's calculate the following.

① $\frac{2}{5} \div \frac{3}{7}$ ② $\frac{1}{5} \div \frac{9}{10}$ ③ $\frac{4}{9} \div \frac{2}{3}$

④ $\frac{3}{4} \div \frac{15}{16}$ ⑤ $9 \div \frac{5}{6}$ ⑥ $4 \div \frac{8}{9}$

⑦ $2\frac{2}{9} \div \frac{2}{7}$ ⑧ $5\frac{1}{4} \div \frac{3}{8}$ ⑨ $\frac{1}{6} \div 1\frac{1}{18}$

⑩ $3\frac{1}{3} \div 1\frac{5}{7}$ ⑪ $4\frac{1}{6} \div 2\frac{1}{2}$ ⑫ $1\frac{1}{14} \div 1\frac{2}{7}$

☐ We understand the relationship between the quotient and the divisor. → p.81

3 Which calculation has a quotient larger than 5?

Ⓐ $5 \div \frac{2}{3}$ Ⓑ $5 \div 1\frac{1}{2}$ Ⓒ $5 \div \frac{5}{4}$ Ⓓ $5 \div \frac{7}{9}$

☐ We can create a division expression and find out the answer. → p.82

4 Let's answer the following questions.

① A tape that is $1\frac{4}{5}$ m long is cut into pieces with a length $\frac{3}{10}$ m. How many pieces with a length of $\frac{3}{10}$ m can be cut?

② Ryunosuke read 16 pages of a book today. This represents $\frac{2}{29}$ of the book he is reading. How many pages are there in total for this book?

Supplementary Problems → p.236

Which "Way to See and Think Monsters" did you find in "5 Fraction ÷ Fraction"?

I found "Unit" when I was thinking about how to calculate.

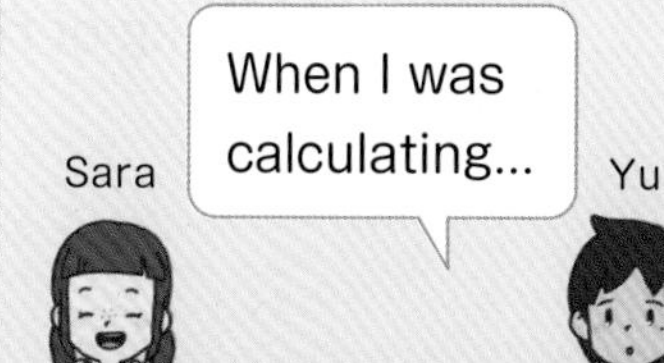

Usefulness and Efficiency of Learning

1 There is a parallelogram with an area of 12 cm^2. If the length of the base is $4\frac{4}{5}$ cm, What is the length of the height in centimeters?

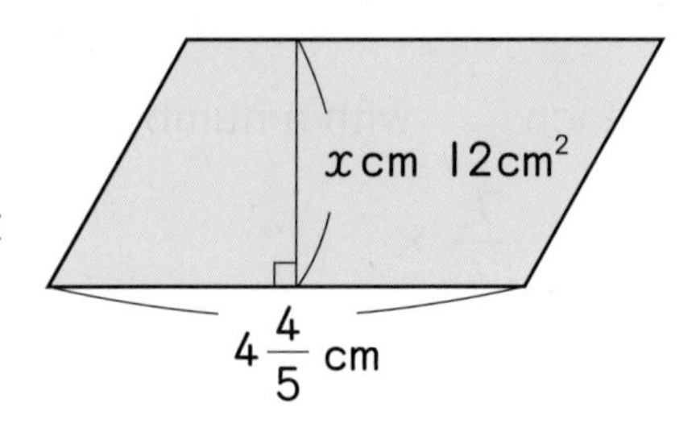

2 Let's answer the following questions.

① There is $\frac{2}{3}$ L of paint that weighs $\frac{3}{4}$ kg. What is the weight in kilograms for 1 L of this paint?

② There is a $2\frac{1}{4}$ m checkered cloth. This length is $\frac{3}{8}$ times of a polka dot cloth. What is the length of the polka dot cloth in meters?

3 Let's choose the ones that becomes a division expression from Ⓐ to Ⓔ.

Ⓐ A $\frac{4}{5}$ m aluminium bar weighed $\frac{2}{3}$ kg. What is the weight per meter of this aluminium bar in kilograms?

Ⓑ If a $12\frac{1}{2}$ m rope is cut into pieces with a length of $1\frac{1}{4}$ m, how many pieces can it be cut into?

Ⓒ There is an oil that weighs $\frac{6}{7}$ kg per liter. What is the weight in kilograms for $\frac{1}{3}$ L of this oil?

Ⓓ A $\frac{2}{3}$ m^2 wall can be painted with $\frac{4}{5}$ L of paint. What is the area that can be painted with 1 L of this paint in square meters?

Ⓔ I bought $\frac{4}{5}$ kg of rice that had a price of 540 yen per kilogram. How much was the cost?

With the Way to See and Think Monsters...

Let's Reflect!

Let's reflect on which monster you used while learning "5 Fraction ÷ Fraction."

Other Way

We tried to find out the area by representing the problem in a diagram and a number line.

【How to find the paintable area per 1 dL when $\frac{3}{4}$ dL of paint is used to paint $\frac{2}{5}$ m^2】

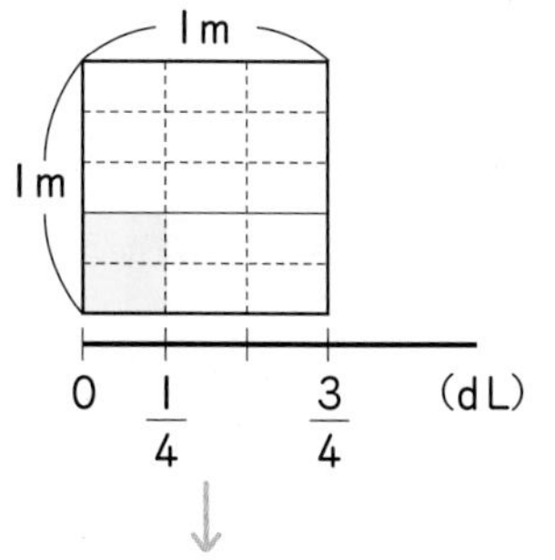

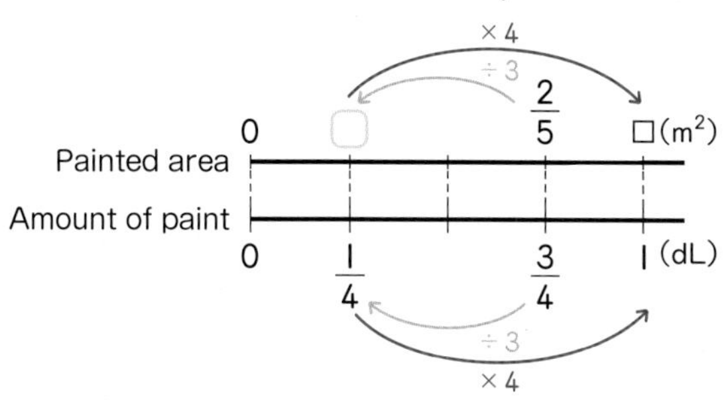

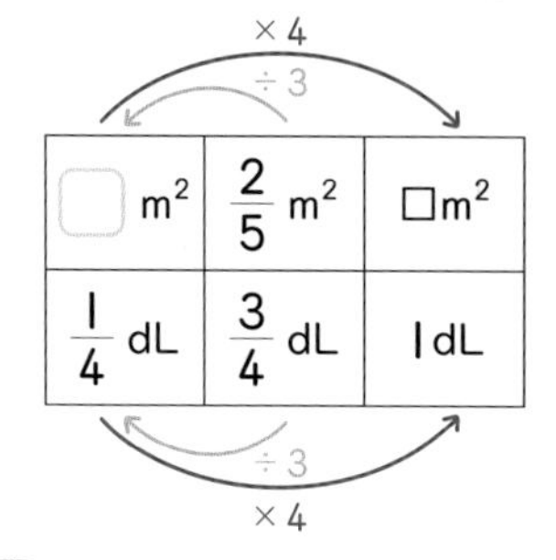

m²	$\frac{2}{5}$ m²	□m²
$\frac{1}{4}$ dL	$\frac{3}{4}$ dL	1 dL

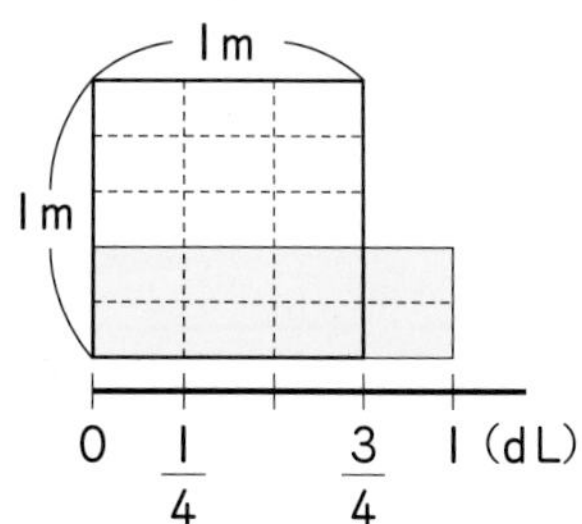

I first thought about the case of $\frac{1}{4}$ and then made it four times of that.

Akari

Sara

Since we multiplied $\frac{4}{3}$ to make $\frac{3}{4}$ into 1, we multiplied the area by $\frac{4}{3}$ as well.

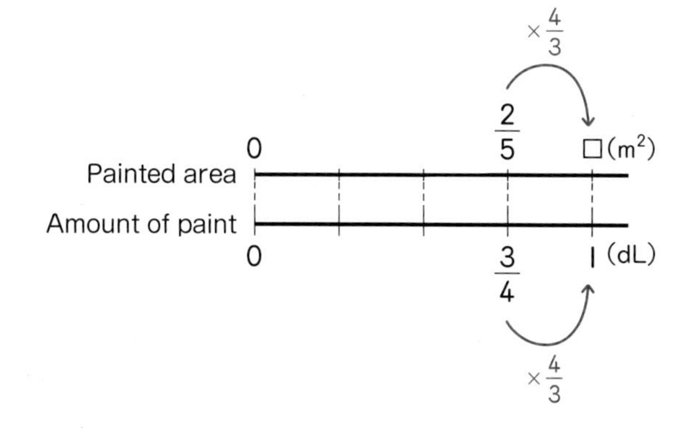

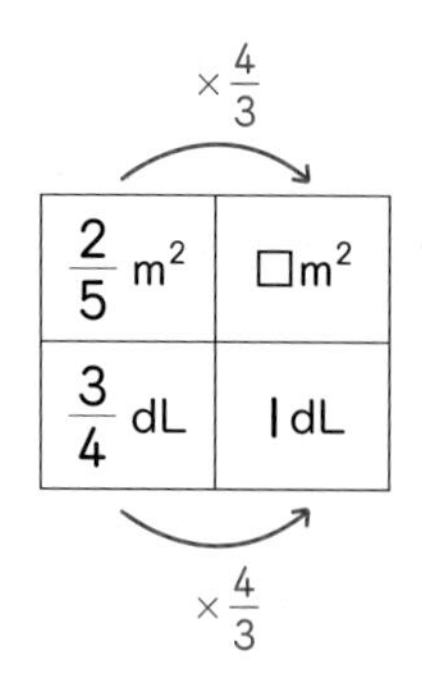

$\frac{2}{5}$ m²	□m²
$\frac{3}{4}$ dL	1 dL

Solve the ?

We could solve fraction ÷ fraction based on the calculation we have learned so far.

Yu

Want to Connect

Can we calculate in the same way even when the fractions and decimal numbers are mixed?

Haruto

Find the ?

Is your physical strength decreasing?

\ Want to compare /

How can we compare whether the physical strength has declined or not?

Data Arrangement

Let's explore the distribution of data and its representative values.

1 Representative values

1 The following tables summarize the results for the new physical fitness test of the current 6th graders and the 6th graders 15 years ago. Let's explore whether we can say the following: "The physical strength of the current 6th graders has declined from the 6th graders 15 years ago."

Current 6th graders

Number	Score	Number	Score
1	52	11	38
2	43	12	49
3	44	13	61
4	63	14	62
5	60	15	63
6	49	16	50
7	50	17	63
8	63	18	48
9	50	19	65
10	51	20	70

6th graders 15 years ago

Number	Score	Number	Score	Number	Score
1	58	11	48	21	54
2	50	12	54	22	62
3	54	13	54	23	52
4	53	14	55		
5	50	15	57		
6	51	16	58		
7	65	17	46		
8	60	18	59		
9	52	19	60		
10	61	20	52		

If we compare the highest score and the lowest score...

Yu

The number of total students differs, so...

Akari

If we have different total numbers, we can compare the mean.

Sara

① What is the maximum score for each group? What is the minimum score for each group? Can we say which group has higher physical strength by looking at these two facts?

❷ Let's find out the mean for the records of each group.

The mean of the records of the current 6th grade students

☐ ÷ ☐ = ☐

Total score | Number of children | Mean of the current 6th grade students

The mean of the records of the 6th grade students 15 years ago

☐ ÷ ☐ = ☐

Total score | Number of children | Mean of 6th grade students 15 years ago

The mean we have learned so far is called the **mean value**.

Mean value = Total of data values ÷ Number of data

❸ Based on what you found out in ❷, which group's record is better?

Summary

By representing the data in mean value, we can compare the data.

❹ Let's discuss whether we can say that "The physical strength of the current 6th graders has declined from the 6 graders 15 years ago" based on what we have found out so far.

Can we say that the group which includes the child who got the highest score has a better record?
Since there are various records, I don't think we can just simply compare by the maximum score or the mean value.
Can we say that the higher the mean value is, the better the record is?
Can we represent the distribution of the scores?

? Are there ways to compare other than the maximum score or the mean value?

2

Let's examine and compare how the records of the new physical fitness test of the 6th graders now and 15 years ago, respectively.

Records of the new physical fitness test of the current 6th graders

37 38 39 40 41 42 43 44 45 46 47 48 49 50 51 52 53 54 55 56 57 58 59 60 61 62 63 64 65 66 67 68 69 70 71 72 (points)

① The diagram above is made based on the table on page 87. What does the number written below the line represent? What does ● represent?

The graph shown above is called a **dot plot**. Vertically, you can see the amount of data. Horizontally, you can understand how the data is distributed.

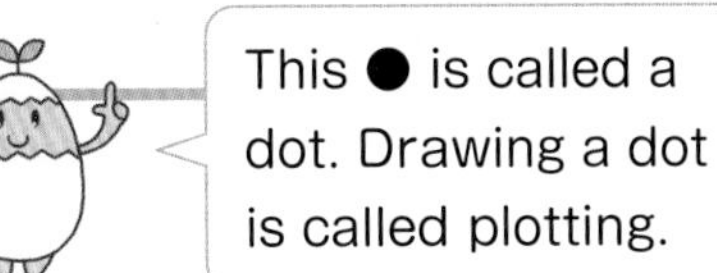

This ● is called a dot. Drawing a dot is called plotting.

Way to see and think

We can see the amount of data and how the data is distributed easily by representing in a dot plot.

\ Want to think /

? **What can we know from a graph representing data using a dot plot?**

② Let's represent the records of the new physical fitness test of the 6th graders 15 years ago based on the table on page 87.

Records of the new physical fitness test of the 6th graders 15 years ago

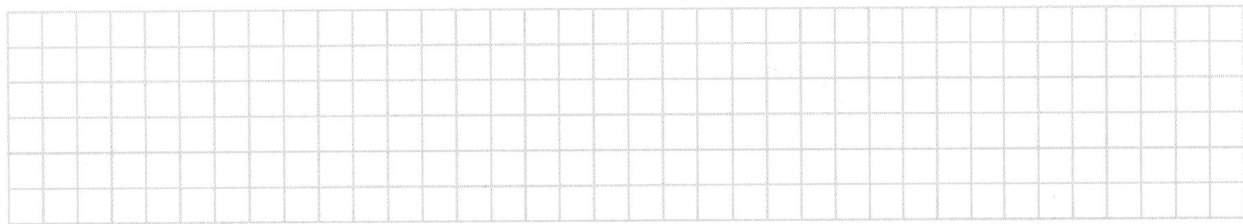

37 38 39 40 41 42 43 44 45 46 47 48 49 50 51 52 53 54 55 56 57 58 59 60 61 62 63 64 65 66 67 68 69 70 71 72 (points)

1 Let's look at the dot plots in the previous page and answer the following questions.

① In each dot plot, insert a ↑ representing the mean value found out on page 88.

② Among the current 6th graders, which score has the highest number of children?

Among the data, the most frequent value is called the **mode value**. ▷

③ Let's answer the mode value in the 6th graders 15 years ago.

④ As for the 6 graders 15 years ago, when the scores are aligned in size order, what is the score in the middle?

There are 23 children in the 6th graders 15 years ago, so the 12th child's score is right in the middle.
Akari

When the data is aligned in size order, the value that is located in the middle is called the **median value**. The median value can be found as follows:

when the number of data is odd…the value in the exact middle.

when the number of data is even…the mean value of the two values in the middle. ▷

⑤ Let's find out the median value for the current 6th graders and that of 15 years ago.

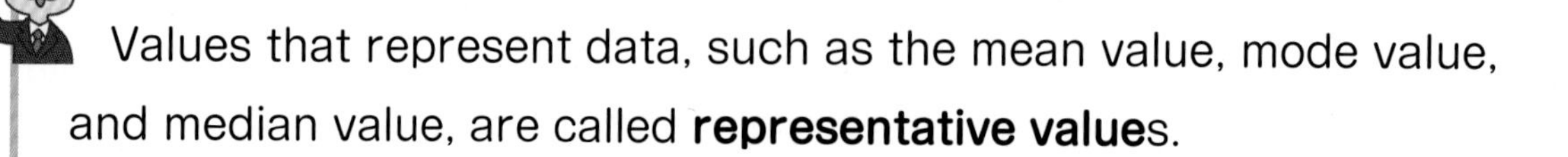

Values that represent data, such as the mean value, mode value, and median value, are called **representative values**.

Summary

Sometimes it is easier to compare data by representing it with dot plot or finding out the representative values.

2 Answer the following questions about the records of the new physical fitness test for the current 6th graders and the 6th graders 15 years ago by organizing the various representative values and the characteristics of the distribution that can be seen from the dot plots.

① Let's fill in the blanks of the following table based on what you have investigated fo far.

	Current 6th graders	6th graders 15 years ago
Maximum score		
Minimum score		
Mean value		
Mode value		
Median value		
Characteristics of distribution seen from the dot plots		

② Based on table ①, what do you think about whether or not it can be said that "Strength of the current 6th graders has declined from the 6th graders 15 years ago"? Choose one from the following Ⓐ~Ⓒ and explain the reason.

Ⓐ Declined.　Ⓑ Have not declined.　Ⓒ Can't say either way.

By looking at the maximum score and the mode value, I don't think it has declined...
By looking at the mean value and the median value, it seems that it has declined...
We have been focusing on the total scores, but maybe we need to look into the details...
How about focusing on one event and comparing it?

? Can we investigate in detail about the fitness test?

I want to find out about the physical strength focusing on different events.
How about investigating throwing a softball?

2 Frequency distribution table and histogram

1 **The following table summarizes the records at throwing a softball of the current 6th graders and the 6th graders 15 years ago. Let's answer the following questions.**

Records at throwing a softball of the current 6th graders (m)

38	15	14	11	17	48	30	19	21
28	24	19	32	32	34	37	7	39
22	18							

Records at throwing a softball of the 6th graders 15 years ago (m)

9	12	38	28	46	18	18	20	21
23	14	24	25	26	15	34	33	29
23	43	17	22	37				

The score was the quantity thrown one by one.

The distance is a continuous quantity.

When throwing a softball, 33 m 15 cm and 33 m 98 cm are both considered 33 m.

Want to know

? **Purpose** **How can we summarize the distribution of continuous quantities?**

① The records at throwing a softball of the current 6th graders are summarized in the table on the right so that you can see the total distribution. Let's complete the blank spaces of the table.

② From what number to what number has the largest number of children?

③ How many children threw the softball greater than or equal to 20 m and less than 25 m?

Records at throwing a softball of the current 6th graders

Distance (m)	Number of children
greater than or equal to 5 ～ less than 10	1
10 ～ 15	2
15 ～ 20	
20 ～ 25	
25 ～ 30	
30 ～ 35	4
35 ～ 40	3
40 ～ 45	
45 ～ 50	1
Total	20

A section (delimiter) such as "greater than or equal to 30 m and less than 35 m" is called a **class**, and the size of the section (delimiter) is called a **class interval**. The number of data counted for each class is called **frequency**. A table that shows the distribution by class or frequency is called a **frequency distribution table**.

4. Which class has a number of 4 children?
5. Let's complete the frequency distribution table for the records at throwing a softball of the 6th graders 15 years ago on the right.

Records at throwing a softball of the 6th graders 15 years ago

Distance (m)	Number of children
greater than or equal to 5 ~ less than 10	
10 ~ 15	
15 ~ 20	
20 ~ 25	
25 ~ 30	
30 ~ 35	
35 ~ 40	
40 ~ 45	
45 ~ 50	
Total	

Summary

If we want to summarize distributed data that has continuous quantities such as a soft ball throwing record, you can decide the class interval and use a frequency distribution table.

6. When comparing the following records for the current 6th graders and the 6th graders 15 years ago, which group has a larger number?
   - Ⓐ number of children with a distance greater than or equal to 25 m
   - Ⓑ number of children with a distance less than 15 m
   - Ⓒ number of children with a distance greater than or equal to 20 m and less than 30 m
7. Let's discuss which group we can say has a better record based on the data investigated so far.

It's difficult to decide only with the frequency distribution table.

Can we represent it with a dot plot?

Should we find the representative values?

But since the distances are continuous quantities...

? Can't we make the frequency distribution table easier to understand?

2

In order to examine the distribution of the records at softball throwing of the current 6th graders on page 92, the following graph was drawn based on the frequency distribution table.
Let's think about the following questions.

Records at throwing a softball of the current 6th graders

Distance (m)	Number of children
greater than or equal to 5 ～ less than 10	1
10 ～ 15	2
15 ～ 20	5
20 ～ 25	3
25 ～ 30	1
30 ～ 35	4
35 ～ 40	3
40 ～ 45	0
45 ～ 50	1
Total	20

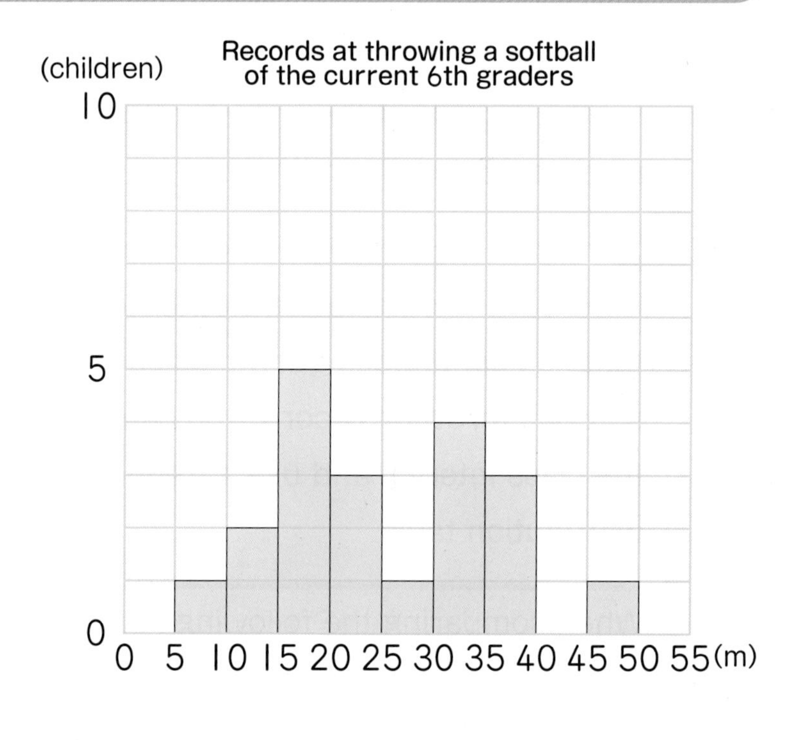

1. How many children threw the softball greater than or equal to 30 m and less than 35 m?
2. From what number to what number is a class with a number of two children?

A graph like the one above is called a **histogram**. If you look at the histogram, it is easy to understand the distribution. In a histogram, the horizontal axis represents the class interval and the vertical axis represents how many children there are in that class.

What is the difference between bar graphs?

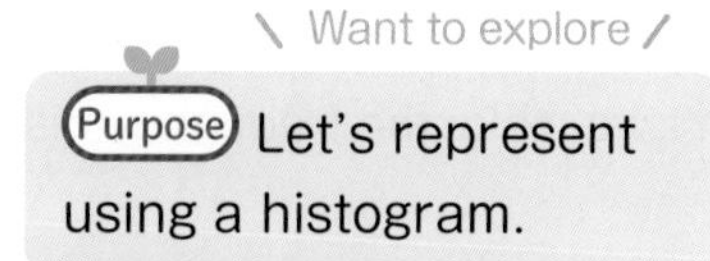

Want to explore
Purpose Let's represent using a histogram.

3 Let's draw a histogram for the 6th graders 15 years ago.

Records at throwing a softball of the 6th graders 15 years ago

Distance (m)	Number of children
greater than or equal to 5 ～ less than 10	1
10 ～ 15	2
15 ～ 20	4
20 ～ 25	6
25 ～ 30	4
30 ～ 35	2
35 ～ 40	2
40 ～ 45	1
45 ～ 50	1
Total	23

Records at throwing a softball of the 6th graders 15 years ago

(children)
10
5
0
0 5 10 15 20 25 30 35 40 45 50 55 (m)

4 Let's compare both histograms and discuss how they are distributed.

5 Which class has the largest number of children in each group? What is the percentage of the children in each class, compared to that of the total? Round off to the tens place.

6 Which class does the median value of the two groups belong to, respectively?

7 Let's discuss whether or not it can be said that "the physical strength of the current 6th graders has declined from the 6th graders 15 years ago" based on what we have investigated so far.

What can we find out from the frequency distribution table and the histogram?

Yu

I want find out the representative values using the results of softball throwing.

Sara

What would the result be in other events?

Haruto

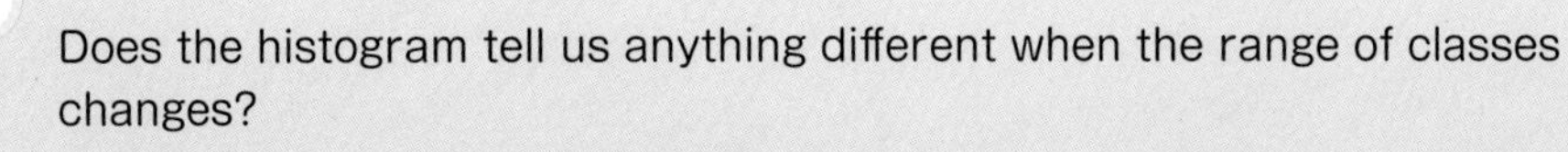

?
Does the histogram tell us anything different when the range of classes changes?

If the class interval changes?

Let's consider what happens when you change the class interval in the frequency distribution table for the throwing records of the 6th graders 15 years ago.

Records at throwing a softball of 6th graders 15 years ago (m)

9	12	38	28	46	18	18	20	21
23	14	24	25	26	15	34	33	29
23	43	17	22	37				

Records at throwing a softball of 6th graders 15 years ago

Distance (m)	Number of children
greater than or equal to 5 ~ less than 10	1
10 ~ 15	2
15 ~ 20	4
20 ~ 25	6
25 ~ 30	4
30 ~ 35	2
35 ~ 40	2
40 ~ 45	1
45 ~ 50	1
Total	23

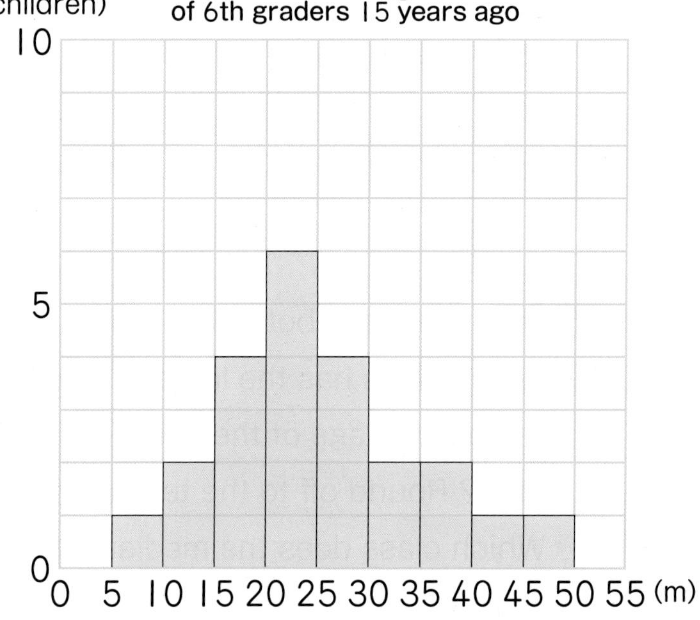

① Create a frequency distribution table with different class intervals as shown in the next page. Let's fill in the blank spaces and draw a histogram.

② Let's look at the histogram on the next page and discuss what you notice.

The class intervals are 3 m and 10 m.

Akari

Yu

If you change the class interval, the impression is completely different.

If you don't consider the class interval well, you will not know how the data is distributed.

Sara

Records at throwing a softball of 6th graders 15 years ago

Distance (m)	Number of children
greater than or equal to 9 ~ less than 12	1
12 ~ 15	2
15 ~ 18	
18 ~ 21	
21 ~ 24	
24 ~ 27	
27 ~ 30	
30 ~ 33	0
33 ~ 36	2
36 ~ 39	2
39 ~ 42	0
42 ~ 45	1
45 ~ 48	
Total	23

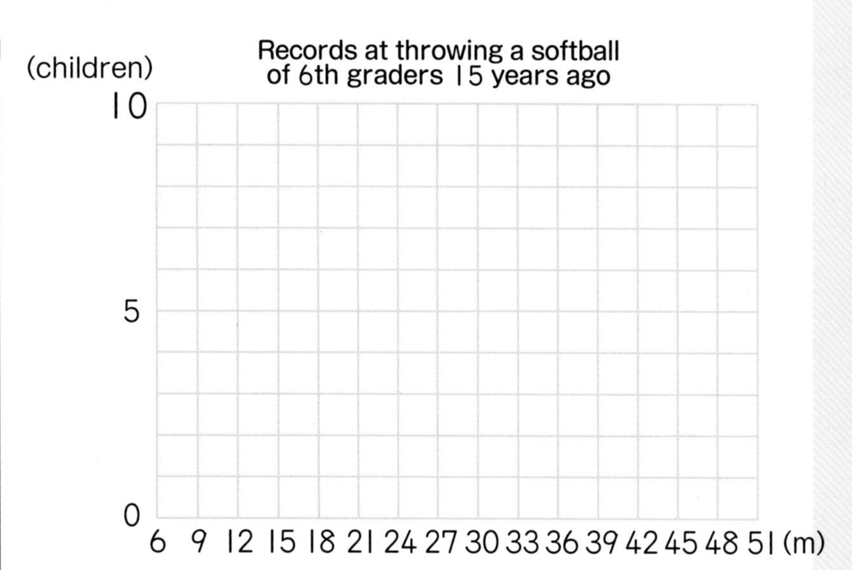

Records at throwing a softball of 6th graders 15 years ago

Distance (m)	Number of children
greater than or equal to 0 ~ less than 10	
10 ~ 20	6
20 ~ 30	
30 ~ 40	
40 ~ 50	2
Total	23

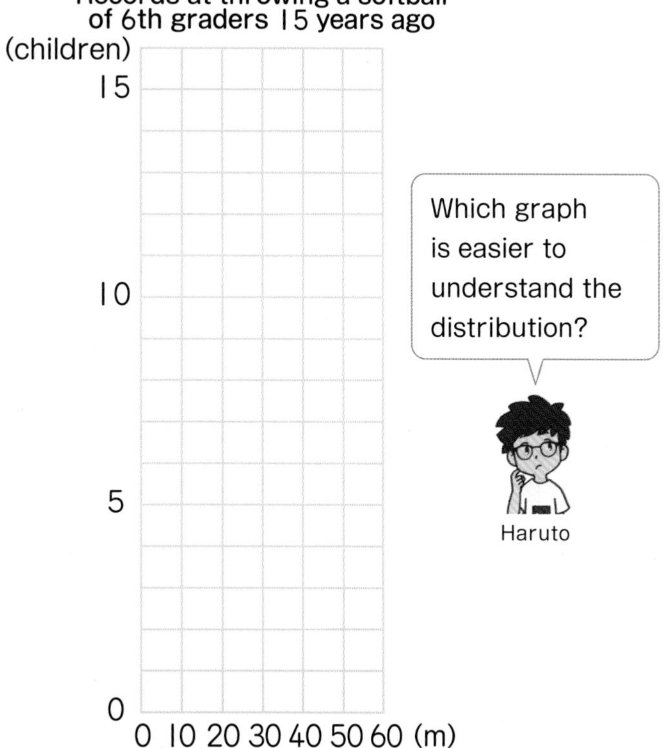

CAN What can you do?

☐ We understand the representative value, dot plot, frequency distribution table, and histogram.

→ pp.87 ~ 95

1 The table shows the records for raising the body in the new physical fitness test of Yumi's class.

Records for raising the body

Number	Times	Number	Times	Number	Times	Number	Times
1	22	6	13	11	22	16	24
2	11	7	24	12	27	17	26
3	10	8	15	13	20	18	26
4	15	9	28	14	22	19	23
5	21	10	17	15	29	20	22

① Let's draw a dot plot and find the mode value, median value, and mean value.

Records for raising the body

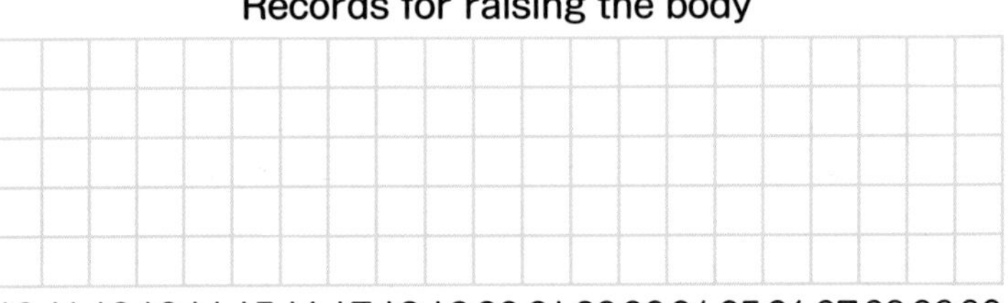

10 11 12 13 14 15 16 17 18 19 20 21 22 23 24 25 26 27 28 29 30 (times)

② Let's complete the frequency distribution table. Let's draw a histogram.

Records for raising the body

Class (times)	Number of children
greater than or equal to 5 ~ less than 10	
10 ~ 15	
15 ~ 20	
20 ~ 25	
25 ~ 30	
Total	

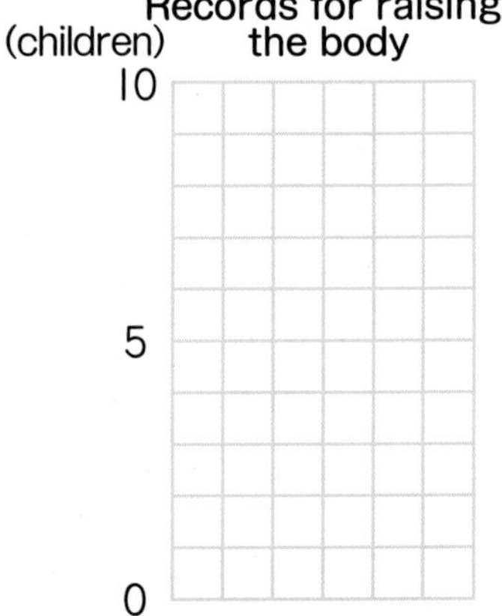

③ Let's look at the frequency distribution table in ② and answer.

ⓐ How many children raised their body greater than or equal to 15 times and less than 20 times?

ⓑ From the best record, which class does the 7th place child belong to?

Supplementary Problems → p.237

Which "Way to See and Think Monsters" did you find in "6 Data Arrangement"?

When I was comparing the data, I found "Other Way."

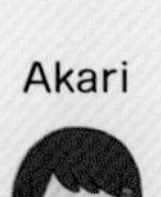

Akari

When I was explaining...

Haruto

Usefulness and Efficiency of Learning

The following table shows the results of examining the school's commuting time for Class 1 and Class 2.

Commuting time for Class 1

Number	Time (min)	Number	Time (min)
1	20	11	28
2	22	12	20
3	19	13	20
4	21	14	22
5	15	15	26
6	13	16	11
7	25	17	24
8	18	18	23
9	20	19	27
10	16	20	10

Commuting time for Class 2

Number	Time (min)	Number	Time (min)
1	28	11	16
2	15	12	25
3	30	13	23
4	23	14	22
5	13	15	13
6	15	16	29
7	21	17	18
8	11	18	23
9	10	19	23
10	25		

① Let's find out the mode value, median value, and mean value for each group. As for the mean value, round off to the hundredths place.

② The diagramth on the right are represented as the histograms based on the tables above. Let's answer the following questions about each group.

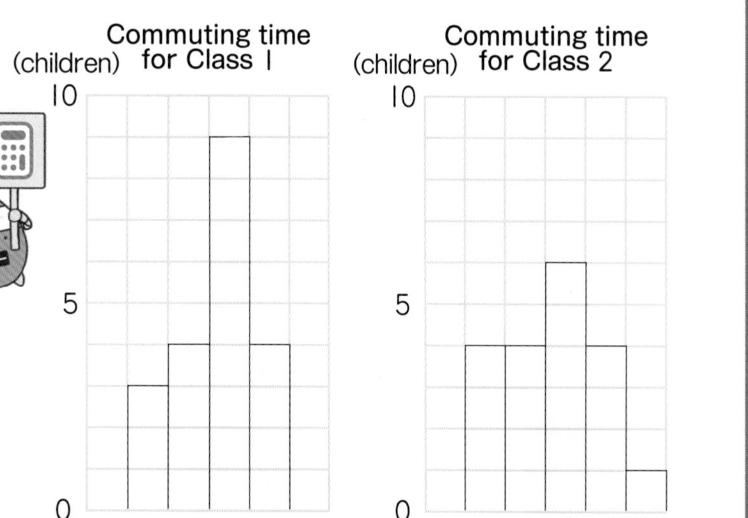

ⓐ How many children does the class greater than or equal to 20 min and less than 25 min have in each group? What is the percentage of the children in this class, compared to the total? If the answer is not divisible, round off to the tenths place.

ⓑ Which class does the 8th place child from the shortest commuting time belong to?

③ Let's compare the answers from ① and histograms from ②, and write down what you noticed.

With the Way to See and Think Monsters...

Let's Reflect!

Let's reflect on which monster you used while learning " 6 Data Arrangement."

Other Way

By representing the examined data as representative values, dot plots, and histograms, we were able to find the characteristics of the data.

① We looked at the new physical fitness test scores of the current 6th graders and the 6th graders 15 years ago and summarized them in the table on the right and graph. Let's discuss what we can find out.

	Current 6th graders	6th graders 15 years ago
Maximum score	70	65
Minimum score	38	46
Mean value	54.7	55
Mode value	63	54
Median value	51.5	54

Records of the new physical fitness test of the current 6th graders

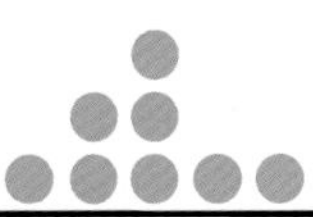

37 38 39 40 41 42 43 44 45 46 47 48 49 50 51 52 53 54 55 56 57 58 59 60 61 62 63 64 65 66 67 68 69 70 71 72 (points)

Comparing with the mean value, I think that the physical strength hasn't changed that much.

I considered that the distribution is larger now, and that it could not be said either that the physical strength had declined or had not declined.

Let's deepen. → p. 244

Solve the ?

By examining the data and representing them with representative values and graphs, I was able to express the characteristics of what I wanted to know.

Sara

Want to Connect

Can we investigate various things using the tables and graphs we have learned so far?

Math Patrol

A number of similar terms such as "mean value," "mode value," and "median value" were used. It is important to check the meanings of the terms repeatedly to make sure you understand the correct meaning.

① The shoe sizes of 35 children were examined. We looked at how many were in each size category. The mode value among these 35 children was 25.5 cm.

Choose one from the following ⓐ ～ⓔ that you can say is always true.

ⓐ Of the 35 shoes, the largest size is 25.5 cm.

ⓑ Of the 35 shoes, the smallest size is 25.5 cm.

ⓒ Dividing the total size of 35 children's shoes by 35, we get 25.5 cm.

ⓓ If we put the 35 shoes in order from the smallest to the largest, the 18th shoe from the smallest to largest is 25.5 cm.

ⓔ The size most commonly worn among the 35 children is 25.5 cm.

In decreasing order, the middle value is....

Sara

Yu

Mode value means that...

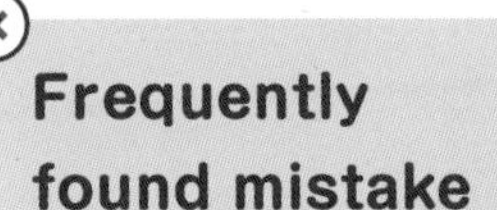

Frequently found mistake
The representative values are indistinguishable.

Be careful!
"The mode value is 25.5 cm" means that "The most frequent value is 25.5 cm".

② The following records are the results of 9 people performing the repetitive horizontal jump for 20 seconds, sorted from the least number of repetitions. Find the median value of the repetitive horizontal jump.

37　41　43　45　47　50　50　50　51

Frequently found mistake
Find the mean value and use it as the answer.

Be careful!
The median value is the value in the middle. When the number of data is odd, it is the value exactly in the middle, and when the number of data is even, it is the average of the two values in the middle.

Let's review the meanings of "mean value," "mode value," and "median value" on pages 88 and 90.

Problem Let's try to explore the data gathered at the library.

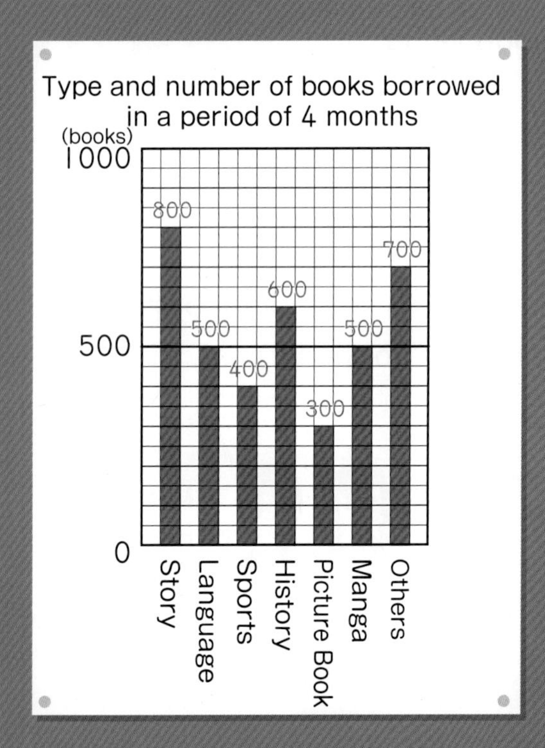

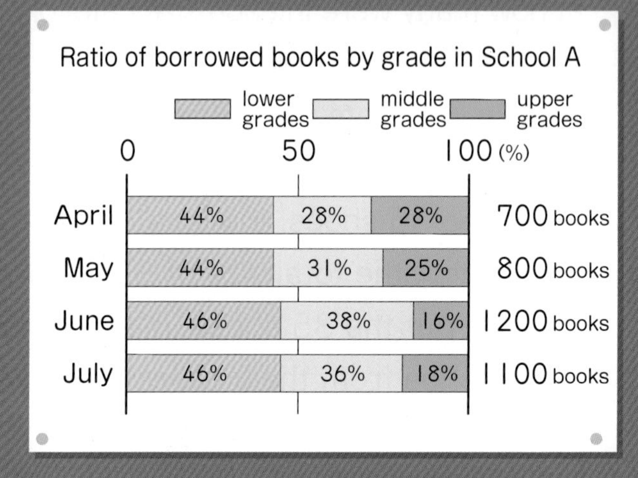

(Bar Graph)

- Story was the most borrowed books
- The second borrowed was History
- The third were Language and Manga.

With a bar graph, the amount can be understood at a glance.

(Strip Graph)

- The percentage for lower grades has hardly changed.
- The percentage for middle grades is increasing.
- Although the percentage for upper grades has decreased, the number of books has hardly changed.

April $700 \times 0.28 = 196$

July $1100 \times 0.18 = 198$

With a strip graph, the percentages can be understood immediately.

Looking at the bar graph, we can understand what kind of book was borrowed the most.

We can immediately understand percentages in a strip graph or circle graph. A strip graph is useful for comparing percentages.

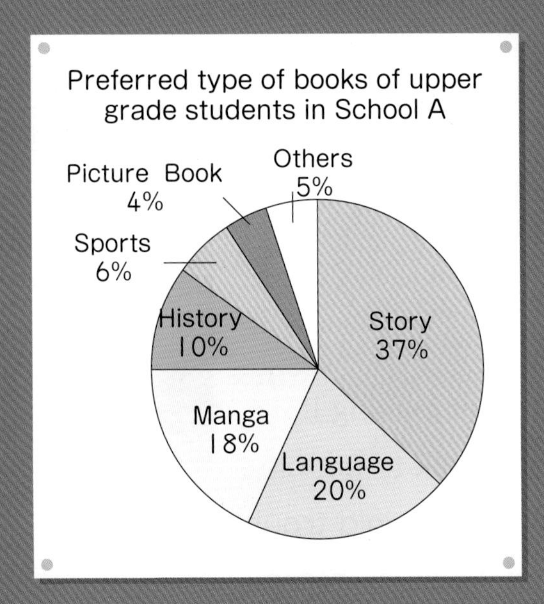

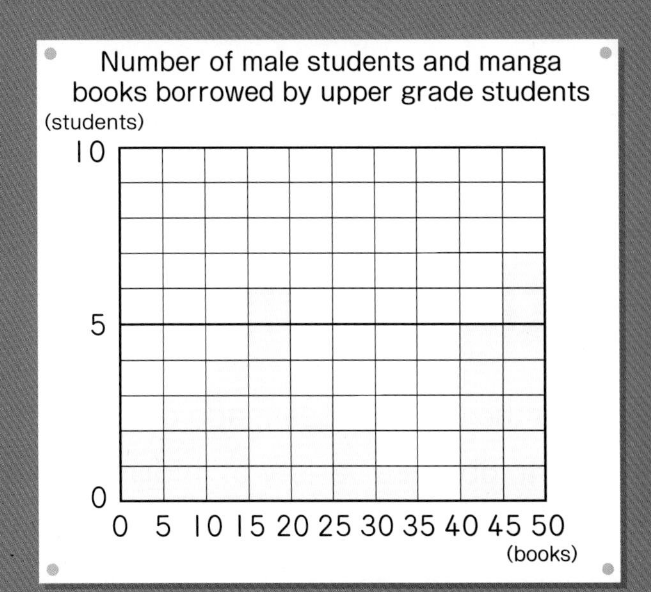

(Circle Graph)

- The number of students who wanted more books of Story is the highest.
- Language and Manga are popular among upper grade students.

With a circle graph, the percentages can also be understood immediately, and can be easily compared.

(Histogram)

- Most boys borrowed between 45 ~ 50 books.
- There were 12 boys who borrowed more than 40 manga books.
- The second highest number of boys borrowing books was between 15 ~ 20 books.

With a histogram, the data distribution can be understood.

Summary

Our understanding changes depending on the graph we see.

↓

Our understanding changes depending on the graph.

The histogram lets you understand how each child borrowed books.

Want to Connect

I want to try to explore the type of books borrowed by upper grade girls using a histogram.

Utilizing Math for SDGs

Let's get into digital citizenship!

9 INDUSTRY, INNOVATION AND INFRASTRUCTURE

16 PEACE, JUSTICE AND STRONG INSTITUTIONS

While the Internet has made our lives better by making it easy to obtain a variety of information and to talk to people far away, it can also cause problems. In order to avoid trouble, it is necessary to understand and utilize various types of information, and to work with it in a safe and responsible manner.
This is called "digital citizenship," which is considered to have the following nine elements. Let's learn these so that we can use the Internet properly and make the best use of it.

9 elements of "digital citizenship"

① Everyone should be able to use digital.

② To acquire the correct knowledge to buy and sell safely on the Internet.

③ Share information on the Internet correctly and securely.

④ Use information on the Internet in a courteous and responsible manner.

⑤ Use and respond to changing technology with digital devices.

⑥ Use digital devices in a healthy way, both physically and psychologically.

⑦ Do not steal or damage other people's property found on the Internet.

⑧ Have freedom and responsibility on the Internet.

⑨ Be careful how you handle your information and protect yourself.

(%)

	Elementary school students	Junior high school students	High school students
I have received rude or harassing messages, e-mails, or posts.	2.2	4.2	6.0
I have sent or written rude or harassing messages, e-mails, or posts.	0.5	0.9	1.5
I have posted information about myself on social networking sites where others can see it.	1.2	6.3	14.0
I have spent too much money on games and apps.	2.4	3.6	6.3
I have received spam messages or e-mails.	3.9	14.9	33.4
I have exchanged messages, e-mails, etc. with people I met on the Internet.	5.2	13.4	24.2
I have been so absorbed in the Internet that I have had trouble concentrating on my studies and have had trouble sleeping.	10.9	17.9	27.4

Source: "2020 Survey of Internet Use by Youth" (Cabinet Office)

① The table above summarizes a survey of elementary, middle, and high school students in 2020 about their experiences on the Internet. Let's consider what digital citizenship can prevent.

② Let's discuss how to avoid trouble when using the Internet.

Think back on what you felt through this activity, and put a circle.

Let's reflect on yourself!

	Strongly agree	Agree	Don't agree
① We learned that we have "digital citizenship" to protect ourselves.			
② I realized that it is important to use the Internet properly.			
③ I could read the table.			

	Strongly agree
④ I am proud of myself because I did my best.	

Let's praise yourself with some positive words for trying hard to learn!

What is the running order?

Yu, Sara, and Haruto have joined the same relay team.
First of all, we need to decide the running order.
1
It seems there are various running orders.
What kind of running order do we have?
2
Mmm...
Yu Sara Haruto
Sara Haruto Yu
Haruto Yu Sara
Sara Yu Haruto
There is another one.
Did we cover all the running orders?
3

Purpose

\ Want to explore /

How can we count without missing or overlapping?

7 Ways of Ordering and Combining

Let's arrange without missing or overlapping.

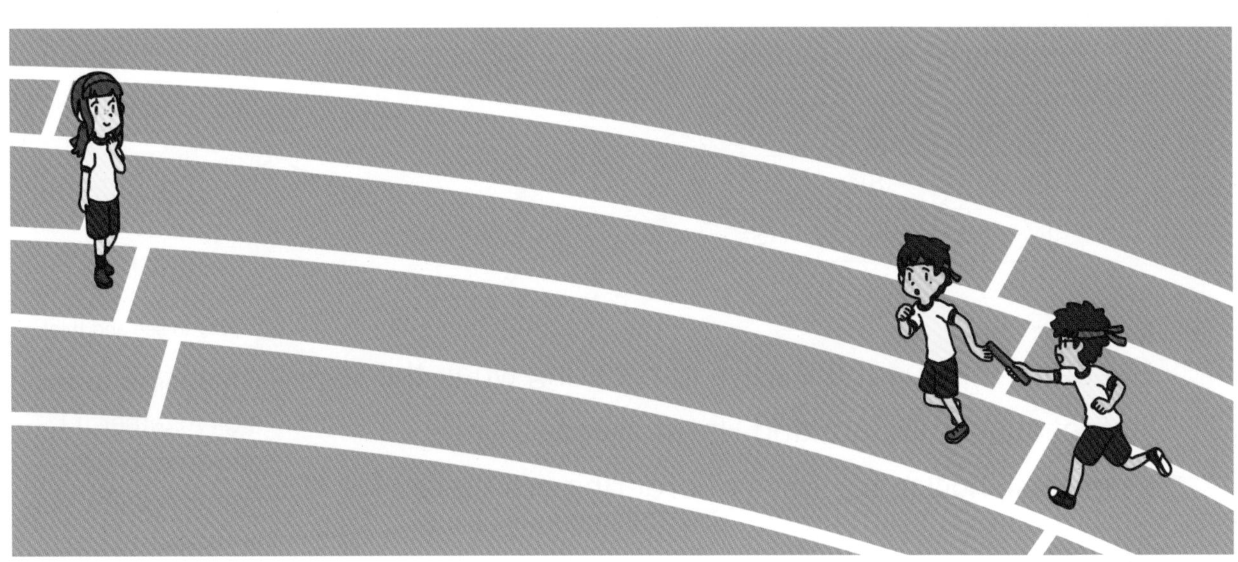

1 Ways of ordering

Yu, Sara, and Haruto are relay athletes. Let's explore how many ways there are as for how to decide the running order.

First, "Yu, Sara, Haruto."
Next, "Sara, Haruto, Yu."
Also, "Haruto, Yu, Sara," "Haruto, Sara, Yu." "Yu, Sara, Haruto."
Mmm, are there still more?

Akari

Is there a way missing or overlapping?

Yu

① Consider the case where Yu runs first. How will the running order of Sara and Haruto be decided? Let's think about the ways of ordering.

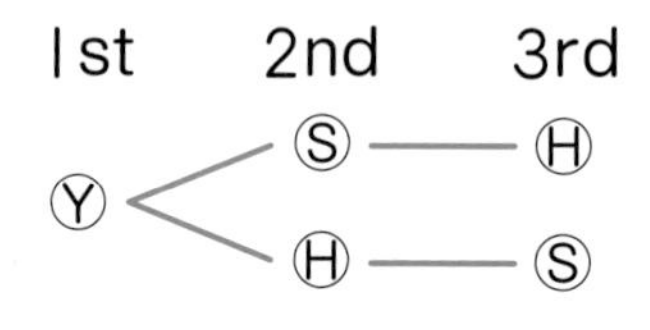

Way to see and think
It's easy to represent in a diagram if you use symbols.

Yu… Ⓨ Sara…Ⓢ Haruto… Ⓗ

Haruto's idea

I considered using a table.

1st	2nd	3rd
Ⓨ	Ⓢ	Ⓗ
Ⓨ	Ⓗ	Ⓢ

Sara's idea

I considered using a diagram.

1st 2nd 3rd

Ⓨ < Ⓢ — Ⓗ / Ⓗ — Ⓢ

② Let's think about the ways of ordering when the first runner is Sara and Haruto respectively.

③ In total, how many ways of ordering are there?

Way to see and think

It's easier to put it all together if I decide who runs first and then organize regularly.

Summary

You can count how many ways there are without missing or overlapping by using tables and diagrams.

1 Akari was joined the three members in 1, and there are four members in the relay team. How many ways of ordering are there in total?

2 At the amusement park, you ride one time on the Go Karts, Ferris Wheel, and Merry Go Round. As for the riding order, how many ways are there in total?

Yu

If we think by using tables and diagrams like we did in 1 …

1st	2nd	3rd	4th
Ⓨ	Ⓢ	Ⓗ	Ⓐ
Ⓨ	Ⓢ	Ⓐ	Ⓗ

1st 2nd 3rd 4th

Ⓨ < Ⓢ < Ⓗ — Ⓐ / Ⓐ — Ⓗ; Ⓗ; Ⓐ

? Can we find out in the same way in other cases?

2

There is one card for each of the following numbers: 1, 3, 5, 7. From these cards, use 3 cards to create 3-digit whole numbers. How many whole numbers can you make in total?

Hundreds	Tens	Ones

① Let's consider ways of ordering when 1 is in the hundred place.

〈 Using a table 〉

Hundreds	Tens	Ones
1	3	5
1	3	
1		
1		
1		
1		

〈 Using a diagram 〉

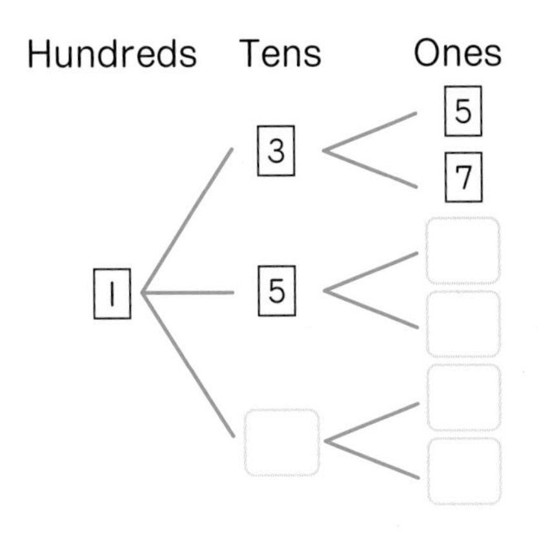

\ Want to explore /

Purpose What should I do if I choose several cards?
Akari

② How many 3-digit whole numbers be made in total? Let's explain the thinking method.

Summary I can use tables and diagrams to help me organize my choices.
Yu

1 Takuto, Naho, Hayato, and Mai are deciding the leader and the subleader for their group. How many ways of deciding are there in total?

2 There is one card for each of the following numbers: 0, 2, 4, 6. From these cards, use 3 cards to create 3-digit whole numbers. How many whole numbers can you make in total?

If you place 0 in the hundreds place, you can't make a 3-digit whole number.

? Can we explore further in other situations?

3

You are shooting soccer penalty kicks. When 3 consecutive penalty kicks are shot, what different cases are there?

Either you goal or you don't. There are two ways each.

Akari

\ want to explore /

Purpose How can we organize the results when each result is two different ways?

Sara

❶ Place ○ in case of goal and × in case of miss. Let's explore using a table and a diagram, the case in which the 1st kick is a goal.

〈 Using a table 〉

①	②	③
○	○	
○	○	
○	×	
○	×	

〈 Using a diagram 〉

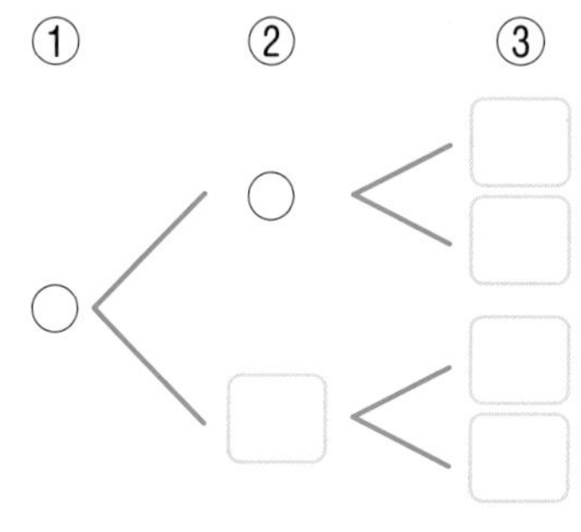

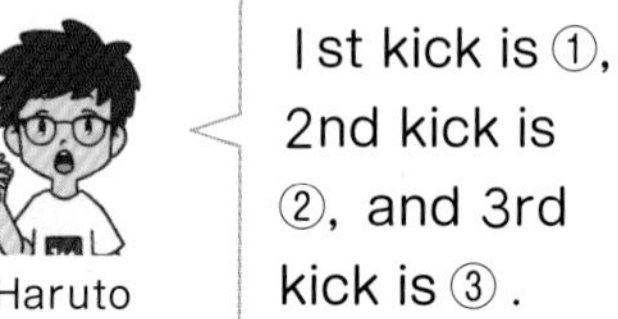

Haruto

1st kick is ①, 2nd kick is ②, and 3rd kick is ③.

Way to see and think

It's easier to summarize if I decide whether a goal was scored or not with the first kick, and then organize the others.

❷ Let's also explore using a table and a diagram the case in which the 1st kick is a miss.

❸ As for the results of the penalty kicks, how many ways are there in total?

1 A 500-yen coin is thrown three consecutive times. As for the front and the back of the coin coming out, how many ways are there in total?

Front

Back

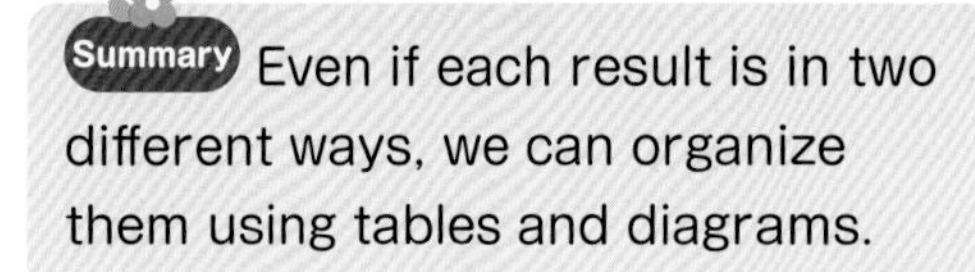

Summary Even if each result is in two different ways, we can organize them using tables and diagrams.

Yu

That's it!

Password

Passwords that utilize numerals are used in various places in daily life.
If you create a password using three numerals from 0 to 9, how many ways are there in total?

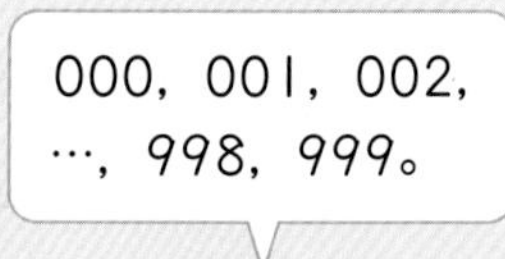

2 Ways of combining

1 Four teams will play basketball games. If each team competes with the other teams only one time, how many games will be played in total?

Team A Team B Team C Team D

First, A vs B, C vs D.
A will play a game with B, C, and D. B will play a game with A, C, and D.
Next, we'll look at teams that have not yet played.
But then A vs B would be played twice....

\ Want to know /

How can we count without missing or overlapping?

1 Let's explain the ideas of the following children.

Team A··· Ⓐ Team B··· Ⓑ
Team C··· Ⓒ Team D··· Ⓓ

Haruto's idea

First, draw the same diagram as when you think the ways of ordering.
Next, in the case of the same combination, I have to [ ] one.
Lastly, I count the remaining number of games and find the [ ] .

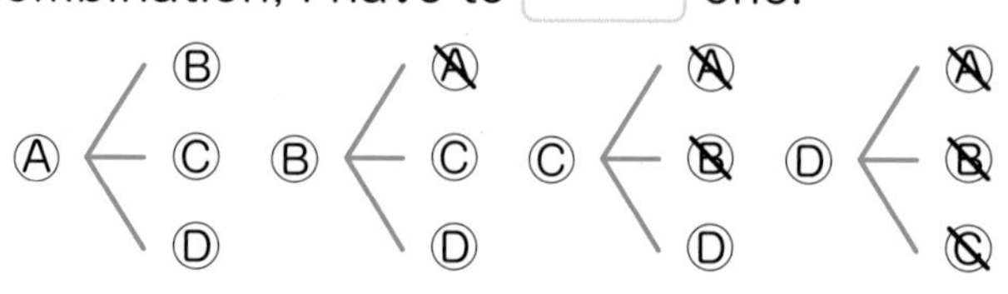

Sara's idea

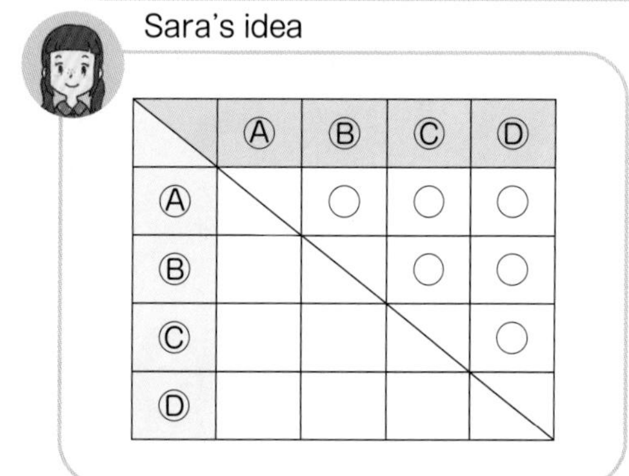

	Ⓐ	Ⓑ	Ⓒ	Ⓓ
Ⓐ		○	○	○
Ⓑ			○	○
Ⓒ				○
Ⓓ				

Yu's idea

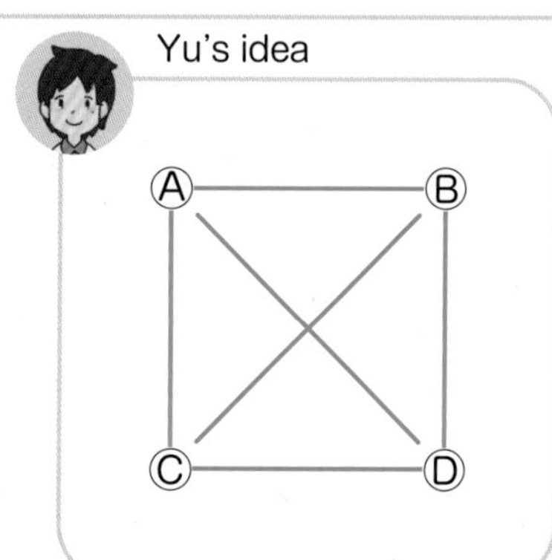

Way to see and think

I can organize them by using tables and diagrams, just as I did when I thought about how to order them.

2 How many ways of combining the games are there in total?

1 Five teams will play baseball games. If each team competes with the other teams only one time, how many games will be played in total?

Summary

The same as when exploring ways of ordering, use a table or a diagram to erase one of the repeated combinations and then count the total cases.

? Even in the case of choosing several, can we think of combinations as we did in ordering?

Words

【 First....Next....Lastly.... 】

These are convenient words to use when explaining in order.

2

2 types of pastries were bought from the following 5 types. How many combinations are there in total?

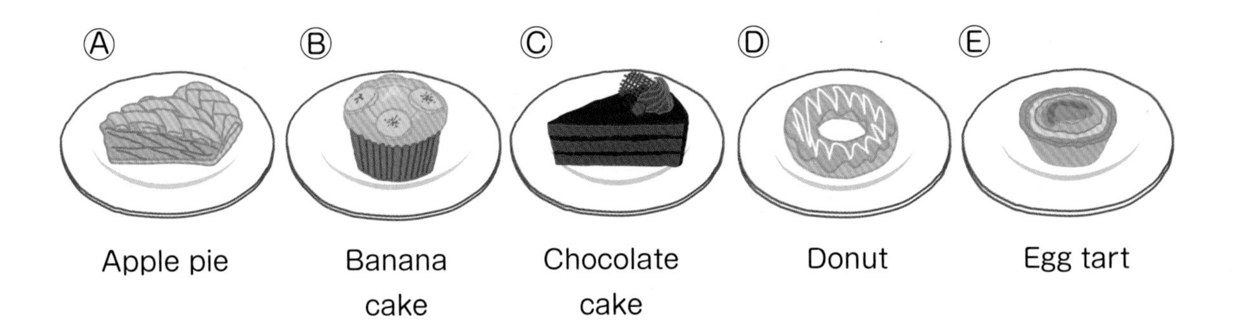

① Let's draw a table or a diagram to explore the existing cases as for how to choose 2 types of pastries.

② How many combinations are there in total?

\ Want to explore /

Purpose How can we examine when we choose some from several?

Yu

1 3 members of the breeding committee are chosen among the 4 children: Kaori, Genta, Hitomi, and Issei. How many ways in total are there to choose the members?

Akari

1st is Kaori, 2nd is Genta, and continue choosing...

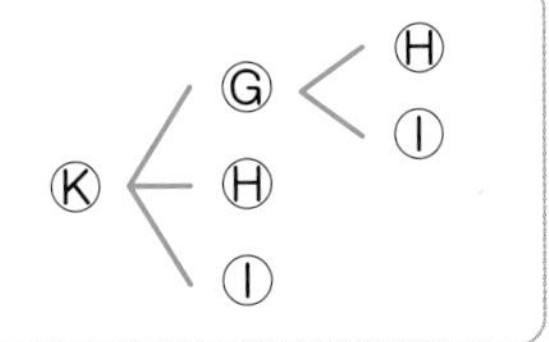

Since choosing three people is the same as choosing the remaining person...

Haruto

Summary Sometimes it is easy to find ways of combination by organizing them with a focus on what is not chosen.

Sara

2 There is one card for each of the following numbers: 1, 2, 3, 4, 5. From these 5 cards, choose 4 cards to find out the sum. As for the sum, how many can you find in total?

What can you do? 

☐ We can count the ways of ordering without missing or overlapping. → pp.107 ～ 108

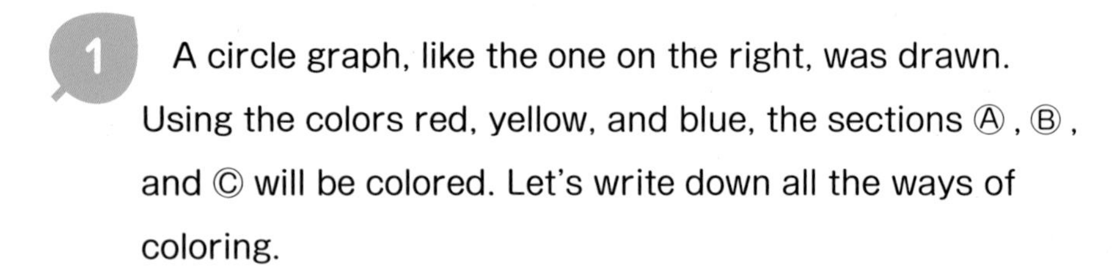

1 A circle graph, like the one on the right, was drawn. Using the colors red, yellow, and blue, the sections Ⓐ, Ⓑ, and Ⓒ will be colored. Let's write down all the ways of coloring.

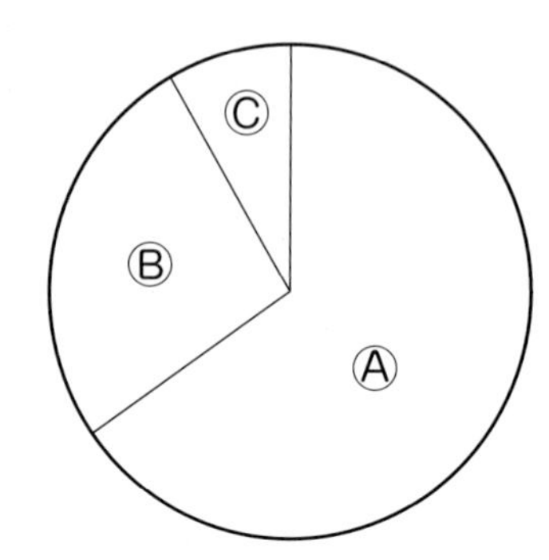

☐ We understand the ways of combination. → p.113

2 There is one coin for each of the following: 1-yen, 10-yen, 50-yen, and 100-yen. From these 4 coins, choose 3 coins and find out the total amount of money. Let's write down all the different amounts of money.

☐ We can create a combination of numbers. → p.109, p.113

3 There are three number cards: [3], [4], [5]. Let's answer the following questions.

① Write down all the 3-digit numbers that can be created with these cards. How many can you make in total?

② As for choosing 2 cards out of 3, how many combinations are there? Let's write all the combinations.

③ When you make a 2-digit number choosing 2 cards out of 3, what is the third largest number you can make?

Supplementary Problems → p. 238

Which "Way to See and Think Monsters" did you find in "7 Ways of Ordering and Combining"?

When I was trying to count without missing or overlapping, I found "Other Way".

Yu

I found other monsters, too!

Sara

Usefulness and Efficiency of Learning

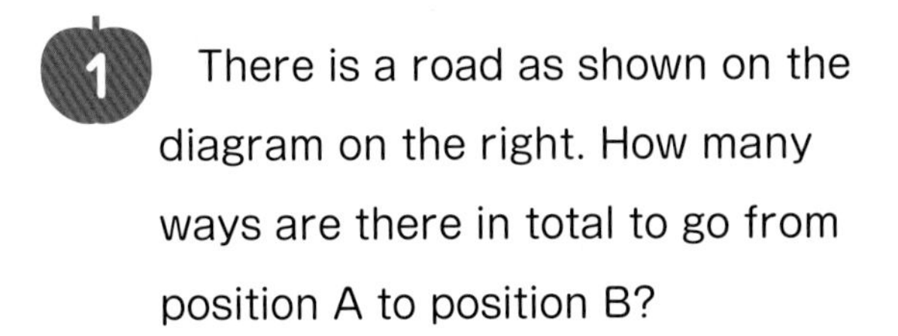

1 There is a road as shown on the diagram on the right. How many ways are there in total to go from position A to position B?

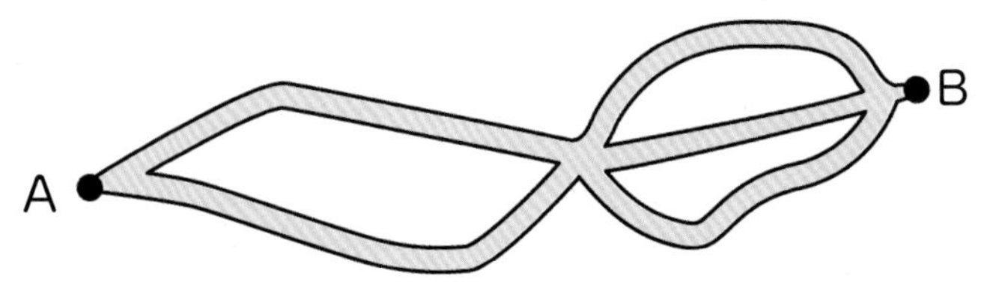

2 There is one card for each of the following numbers: 0, 1, 2, 3. From these 4 cards, create 4-digit whole numbers. Let's answer the following questions.

① How many whole numbers can you create? Write down all of them.

② From the created whole numbers, how many are even numbers? Write down all the even numbers in ascending order.

3 Taishi, Rena, Sakito, and Manaka are sitting on a bench. As for the sitting ways in which Taishi and Manaka sit next to each other, how many are there in total?

If we consider Taishi and Manaka as one pair, then we can think of it as a group of 3 elements.

That's it!

Number of games in a knock-out competition?

A knock-out competition is also referred to as a tournament.

Eight teams will play a soccer knock-out competition. What is the total number of games until the winning team is decided?

I can find out by 8 − 1.

Why?

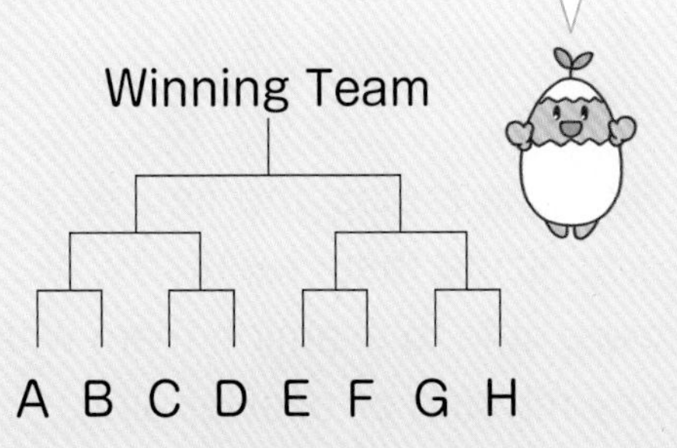

With the Way to See and Think Monsters...

Let's Reflect!

Let's reflect on which monster you used while learning " 7 Ways of Ordering and Combining."

Other Way

By replacing each person or thing with a symbol and representing them in a table or diagram, **I was able to organize them without missing or overlapping.**

【How to find out the running order of the 3 relay athletes, Yu, Sara, and Haruto】

1st	2nd	3rd
Ⓨ	Ⓢ	Ⓗ
Ⓨ	Ⓗ	Ⓢ
Ⓢ	Ⓨ	Ⓗ
Ⓢ	Ⓗ	Ⓨ
Ⓗ	Ⓨ	Ⓢ
Ⓗ	Ⓢ	Ⓨ

1st　2nd　3rd

Ⓨ < Ⓢ — Ⓗ / Ⓗ — Ⓢ

Ⓢ < Ⓨ — Ⓗ / Ⓗ — Ⓨ

Ⓗ < Ⓨ — Ⓢ / Ⓢ — Ⓨ

After deciding on the first runner, I thought about representing it in a table or diagram.

Akari

【How to find out how many basketball games are played in total when 4 teams play one game each with other teams】

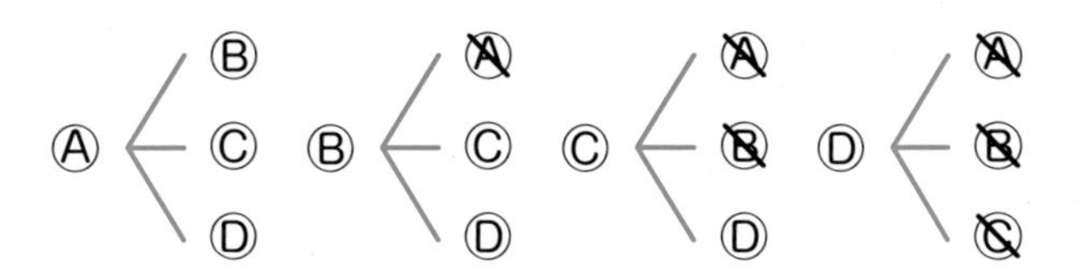

After representing them on the diagram, I thought about erasing one of them.

Yu

Solve the ?

By using tables and figures and organizing them regularly, we were able to examine them without any missing or overlapping.

Haruto

Even when the numbers get large, can we also think by using a table and a diagram?

Akari

Calculations with Decimal Numbers and Fractions

Let's think about how to operate mixed calculations of decimal numbers and fractions.

1 Mixed calculations of decimal numbers and fractions

Addition and subtraction ↓

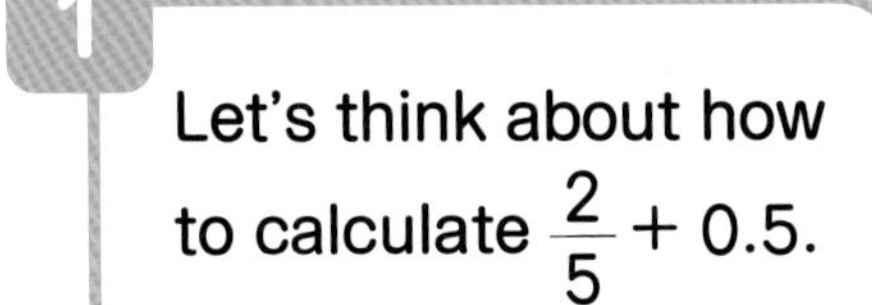

1
Let's think about how to calculate $\frac{2}{5} + 0.5$.

Fractions and decimal numbers are mixed.

I could do fractions only or decimal numbers only...

\ Want to think /

? **Purpose** How can we calculate when decimal numbers and fractions are mixed?

1. Let's calculate by aligning to decimal numbers.

$\frac{2}{5} = 2 \div \square = \square$ $\square + 0.5 = \square$

2. Let's calculate by aligning to fractions.

$0.5 = \frac{5}{\square} = \square$ $\frac{2}{5} + \square = \square$

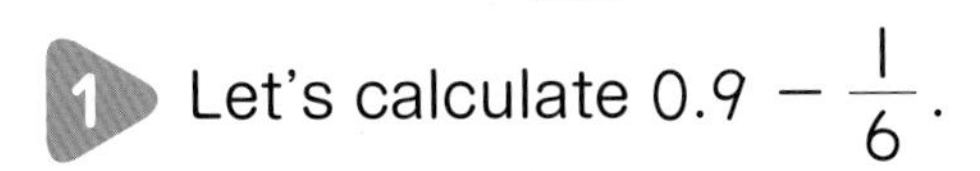

① Let's calculate by aligning to decimal numbers.

$\frac{1}{6} = 1 \div \square = 0.1666\cdots$ $0.9 - 0.167 = \square$

↓

0.167

Since it's not divisible, it's not possible to calculate exactly using decimal numbers.

② Let's calculate by aligning to fractions.

$0.9 = \frac{\square}{\square}$ $\frac{\square}{\square} - \frac{1}{6} = \square$

Summary

As for addition or subtraction that includes decimal numbers and fractions, the calculation is done after aligning to decimal numbers or fractions. When the numbers after the decimal point continue infinitely, the calculation should be performed by aligning to fractions.

2 Let's calculate the following.

① $0.6+\frac{4}{9}$ ② $0.7+\frac{4}{5}$ ③ $\frac{3}{7}+0.4$ ④ $\frac{2}{3}+0.45$

⑤ $\frac{7}{8}-0.3$ ⑥ $1\frac{4}{7}-0.4$ ⑦ $\frac{8}{7}-0.25$ ⑧ $\frac{1}{5}-0.12$

? Can we do multiplication and division by aligning to decimal numbers or fractions?

Multiplication and division →

2

Let's think about how to calculate the following.

❶ $\frac{5}{9}\div\frac{3}{4}\times\frac{7}{10}$ ❷ $7\times\frac{1}{6}\div 1.4$

Want to think

Purpose Can multiplication and division with mixed decimals and fractions be calculated in the same way?

❶ $\frac{5}{9}\div\frac{3}{4}\times\frac{7}{10}=\frac{5}{9}\times\frac{\square}{\square}\times\frac{7}{10}$

$=\frac{5\times\square\times 7}{9\times\square\times 10}$

$=\square$

❷ $7\times\frac{1}{6}\div 1.4=\frac{7}{\square}\times\frac{1}{6}\div\frac{\square}{10}$

$=\frac{7}{\square}\times\frac{1}{6}\times\frac{\square}{\square}$

$=\frac{7\times 1\times\square}{\square\times 6\times\square}$

$=\square$

If we use reciprocal for fraction ÷ fraction...

1 Let's find out the area of the triangle shown on the right.

① Let's write a math expression.

② Let's calculate.

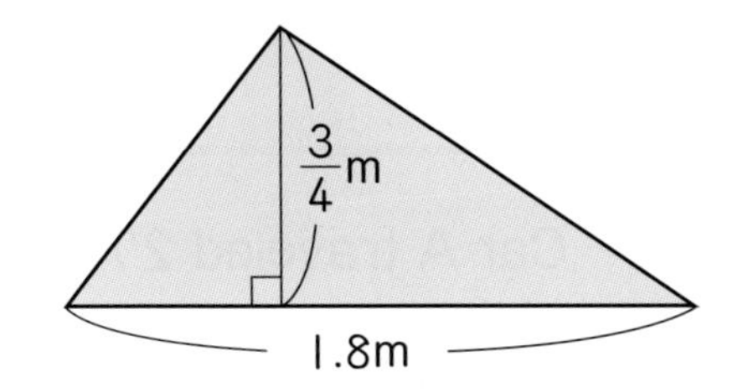

Summary

As for multiplication or division that includes decimal numbers and fractions, the calculation is done after aligning to fractions. As for math expressions that incorporate division and multiplication, if the divisor is changed for the multiplication of its reciprocal, then the math expression will only include multiplications.

2 Let's calculate $0.3 \times 0.48 \div 0.45$.

① Let's calculate by decimal numbers.

② Let's calculate using fractions.

$$0.3 \times 0.48 \div 0.45 = \frac{3}{\square} \times \frac{48}{\square} \div \frac{45}{\square}$$

$$= \frac{3}{\square} \times \frac{48}{\square} \times \frac{\square}{45}$$

$$= \frac{3 \times 48 \times \square}{\square \times \square \times 45}$$

$$= \square$$

Which calculation method is easier?

Way to see and think

When we replace decimals with fractions, we can combine them into one or reduce them at once.

3 Let's calculate the following using fractions.

① $\frac{1}{3} \div 0.4 \times \frac{5}{3}$ ② $0.8 \times \frac{3}{5} \div 0.36$ ③ $\frac{3}{7} \div 0.75 \times \frac{9}{14}$

④ $0.7 \times 0.35 \div 0.25$ ⑤ $0.5 \div 0.21 \times 0.7$

⑥ $0.2 \div 0.5 \div 0.4$ ⑦ $14 \div 6 \times 3$ ⑧ $27 \div 48 \times 32$

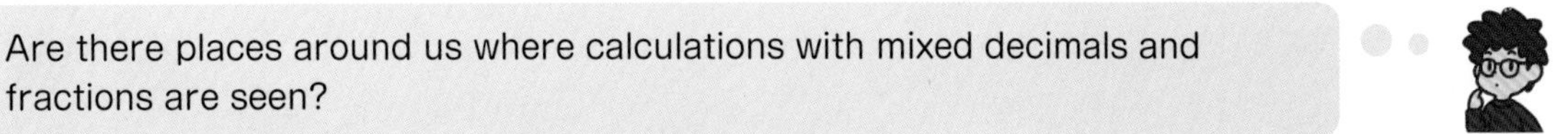

Are there places around us where calculations with mixed decimals and fractions are seen?

2 Various problems

\ Want to try /

Purpose Let's think about the problems around you.

Haruto

1 **Car A traveled 270 km using 15 L of gasoline. Car B traveled 372 km using 24.8 L of gasoline. How many liters of gasoline did each car need to travel 100 km?**

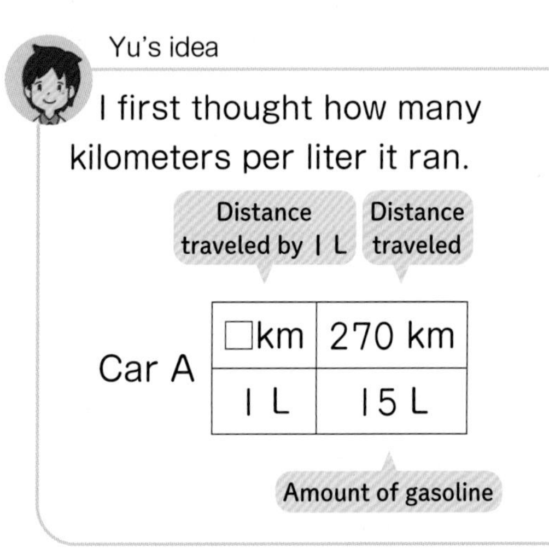

Yu's idea
I first thought how many kilometers per liter it ran.
Distance traveled by 1 L
Distance traveled
Car A
□km
270 km
1 L
15 L
Amount of gasoline

Akari's idea

I thought how many liters of gasoline were used per kilometer.

Car A	□ L	15 L
	1 km	270 km

1 I purchased an 800 yen pencil case that had a discount of 15%. How much was this pencil case?

2 Let's look at the picture on the right and try to think about our body.

① About how many kilograms is the weight of the brain of a person that weighs 36 kg?

② About $\frac{1}{7}$ of the bones are in the head. About how many bones are there in the whole body?

③ About how many kilograms is the water inside the body of a person that weighs 45 kg?

Number of bones in the head is 29.

Weight of the brain is about $\frac{1}{45}$ of the body weight.

The water in the body is about $\frac{2}{3}$ of the body weight.

That's it!

How to represent time: Let's try to represent by decimal numbers and fractions

Can we represent 30 minutes in terms of hours?
In the case of whole numbers, the unit of measurement was 10. When 10 ones are gathered together it became 10, and when 10 hundreds are gathered together it became 1000. Meanwhile, the unit of time changes when 60 gathers, as 60 seconds = 1 minute, 60 minutes = 1 hour.
For example, since 30 minutes is half of 60 minutes, 60 minutes can be expressed as $\frac{1}{2}$ hour in fractions and 0.5 hour in decimals.

① How many hours can 15 minutes be represented?

In fraction, $\frac{15}{60} = \frac{\square}{\square}$, so $\frac{\square}{\square}$ hours.

In decimals, $\square$ hours.

If we divide by 60, we can represent time by fractions and decimals.

② How many minutes is $\frac{1}{5}$ of an hour? What is the number of hours in decimals?

In terms of minutes, $\square \times \frac{1}{\square} = \square$, so $\square$ minutes.

In terms of decimals, $\square$ hours.

③ How many hours are the following Ⓐ to Ⓓ? Represent them as fractions and decimals.

Ⓐ 18 minutes Ⓑ 36 minutes

Ⓒ 90 minutes Ⓓ 102 minutes

The relationship between units of time is shown in the table on the right. Although decimals cannot always be used to represent time, fractions can be used to represent time in a variety of ways.

Time	Minutes	Seconds
24	1440	86400
1	60	3600
$\frac{1}{60}$	1	60
$\frac{1}{3600}$	$\frac{1}{60}$	1

CAN What can you do?  

☐ We can find out the sum, difference, product, and quotient of a mixed calculation of decimal numbers and fractions. → pp.117 ~ 119

1 Let's find out the sum, difference, product, and quotient of the following pairs of decimals and fractions. As for the quotient, divide the number on the left by the number on the right.

① $\frac{1}{3}$, 0.2　　② 3.5, $2\frac{1}{3}$

☐ We can calculate addition and subtraction where decimal numbers and fractions are mixed. → pp.117 ~ 118

2 Let's calculate the following.

① $0.8 + \frac{1}{3}$　② $\frac{2}{5} + 2.6$　③ $\frac{5}{7} - 0.09$　④ $0.32 - \frac{1}{5}$

☐ We understand how to calculate multiplications and divisions where decimal numbers and fractions are mixed. → pp.118 ~ 119

3 Let's calculate the following using fractions.

① $\frac{1}{5} \div 0.6 \times \frac{2}{3}$　② $36 \div 27 \times 16$　③ $0.9 \times \frac{2}{7} \div 0.18$

④ $\frac{5}{12} \div 0.25 \div \frac{3}{10}$　⑤ $0.2 \div 0.16 \div 0.35$　⑥ $0.7 \div 0.35 \div 0.5$

☐ We can solve problems using calculations of fractions. → p.120

4 There is a rhombus, like the one shown on the right, that has an area of 4 cm^2. How many centimeters is the length of the other diagonal?

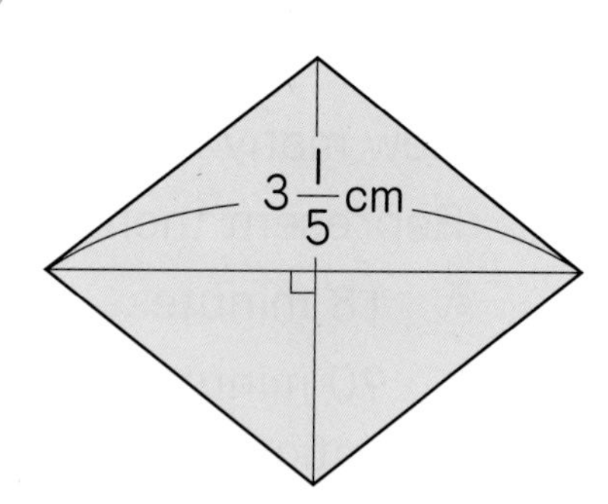

Supplementary Problems → p.239

Which "Way to See and Think Monsters" did you find in " 8 Calculations with Decimal Numbers and Fractions"?

When I was thinking about the calculation where decimal numbers and fractions are mixed, I found "Other Way."

Haruto

I found other monsters, too!

Akari

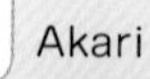

Math Patrol

The range of numbers that can be calculated has expanded from whole numbers to decimals and fractions. Calculations in which whole numbers, decimals, and fractions are mixed in a single expression seem to be getting more complicated. However, if you think about it one by one, you will find that you can do the calculation in the way you have learned so far.

① Let's calculate $\frac{2}{9} \times 3$.

Let's reduce the product if possible.

Since it's fraction × whole number, so...

Sara

Frequently found mistake

- Multiply 3 to the denominator.
- Multiply 3 to the numerator, but forget to reduce.

Be careful!

In case of fractions × whole numbers, multiply the whole number of the multiplier directly to the numerator. Even if you reduce in the middle of a calculation, be sure to check it again in the end.

$$\frac{2}{9} \times 3 = \frac{2 \times \overset{1}{\cancel{3}}}{\underset{3}{\cancel{9}}} = \frac{2}{3}$$

② Let's calculate $\frac{2}{5} \times 0.6$.

Frequently found mistake

- Multiply 0.6 to the denominator.
- Rearrange to decimal number × decimal number, but put the decimal point in the wrong place.

Be careful!

$\frac{2}{5}$ can be represented with decimal numbers as, $0.4 \times 0.6 = 0.24$.

0.6 can be represented with fraction as, $\frac{2}{5} \times \frac{6}{10} = \frac{6}{25}$.

Either method is acceptable, so use the method that is convenient for you.

Multiply to the numerator, whether the multiplier is a whole number or a decimal.
When calculating with decimals, be careful of the position of the decimal point.

Reflect

Connect

ADVANCED
Junior High School

Problem Let's represent math equations created with 3, 5, and + − × ÷ on the number line.

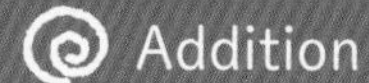

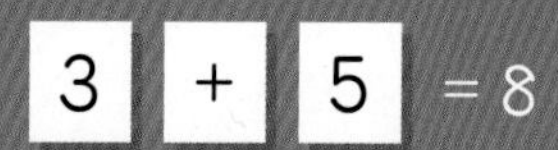

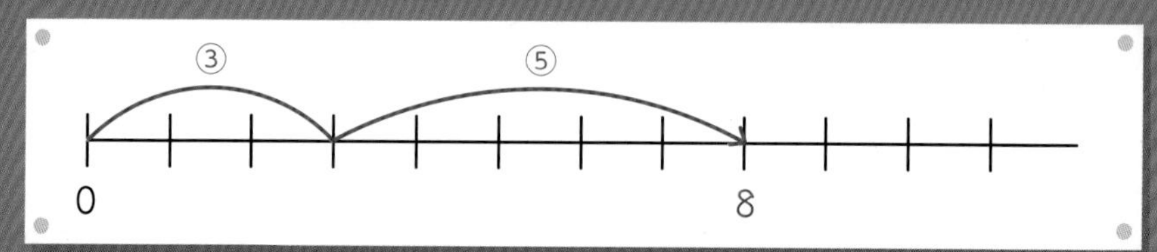

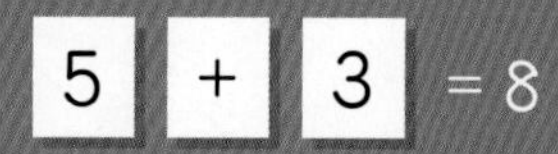

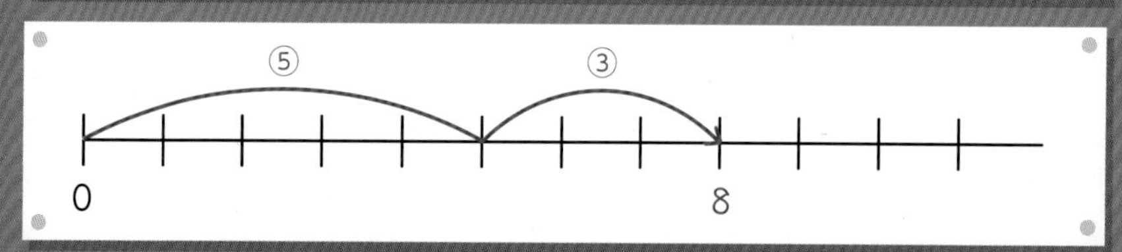

Subtraction

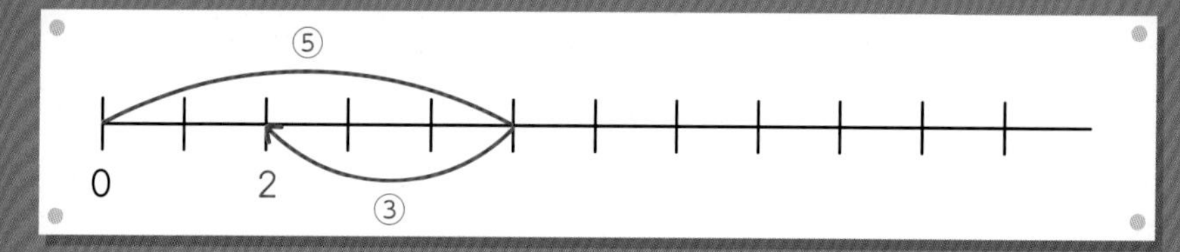

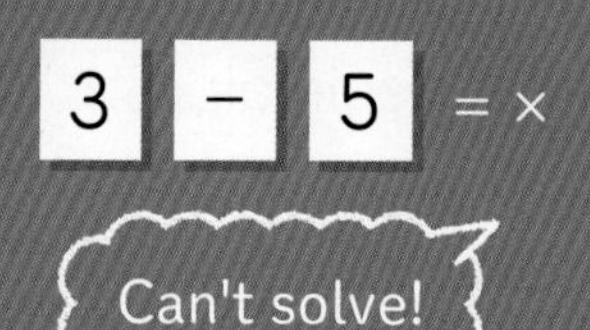

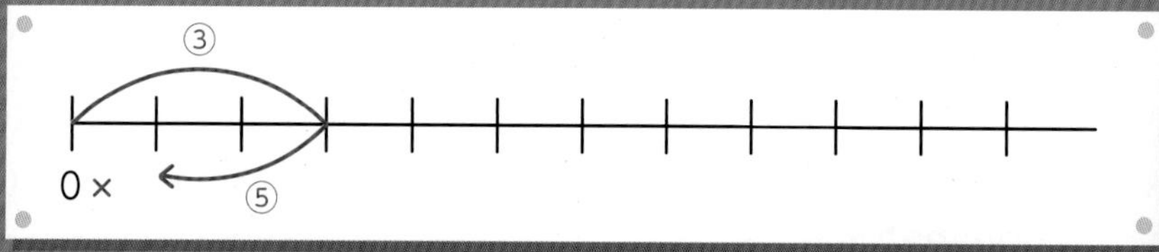

Both math equations are valid except for subtraction.

For both addition and multiplication, the math equations have the same answer.

There is no answer for one of the subtractions.

= 15

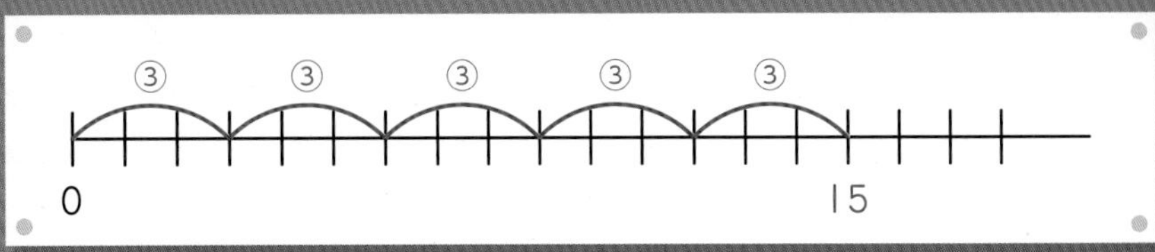

= 15

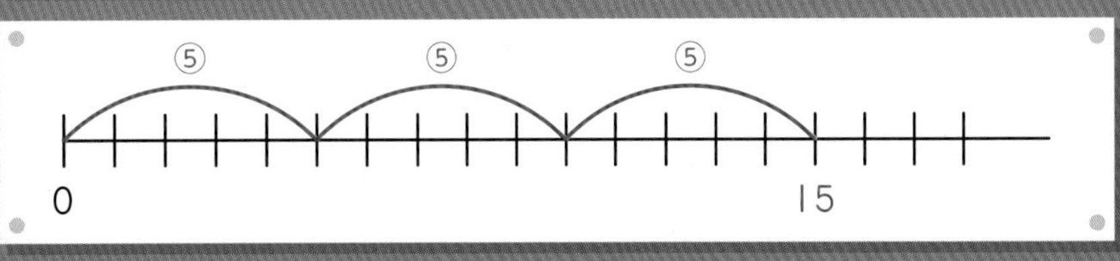

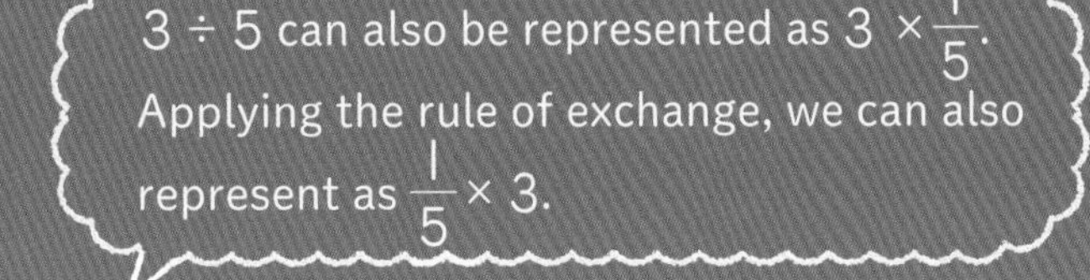

3 ÷ 5 can also be represented as $3 \times \frac{1}{5}$.
Applying the rule of exchange, we can also represent as $\frac{1}{5} \times 3$.

Division

$= 0.6 = \frac{6}{10} = \frac{3}{5}$

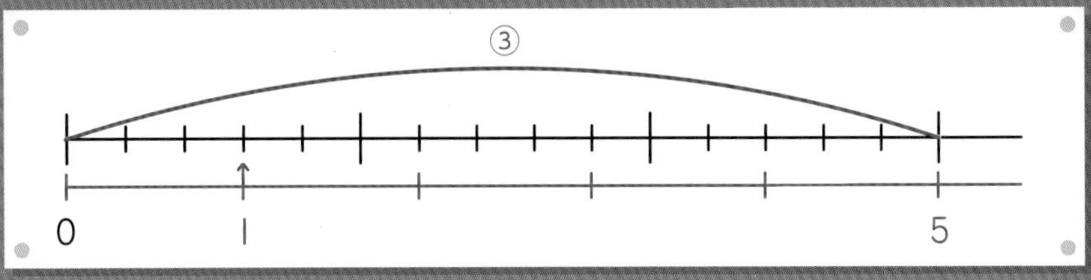

5 ÷ 3 $= \frac{5}{3}$

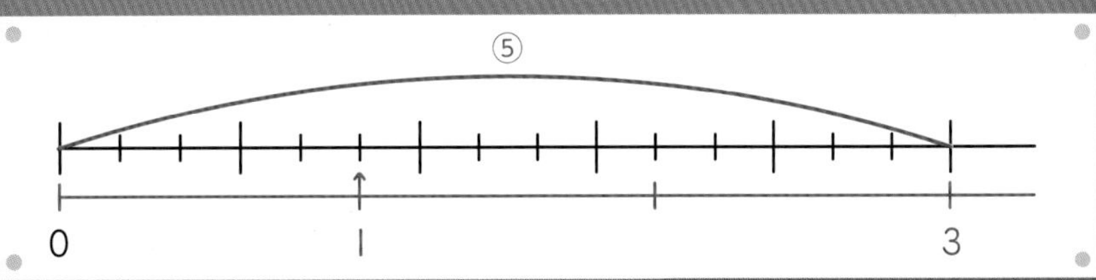

Summary

Answers can be found for the addition, multiplication, and division of whole numbers. However, as for subtraction, there is a case when the answer cannot be found.

The division can be answered by fractions.

Want to Connect

In weather forecasts, the temperatures below 0℃ are represented using a "minus," can I do the same with usual calculations?

Continues at Junior High School!

How many times ~ Times a fraction ~

Softball throw

1

The throwing records for Yukie's softball team were taken and the mean was 18 m. Let's compare the throwing distance and the mean.

① Yukie's record was 24 m. How many times of the mean is 24 m ? Let's represent using fractions.

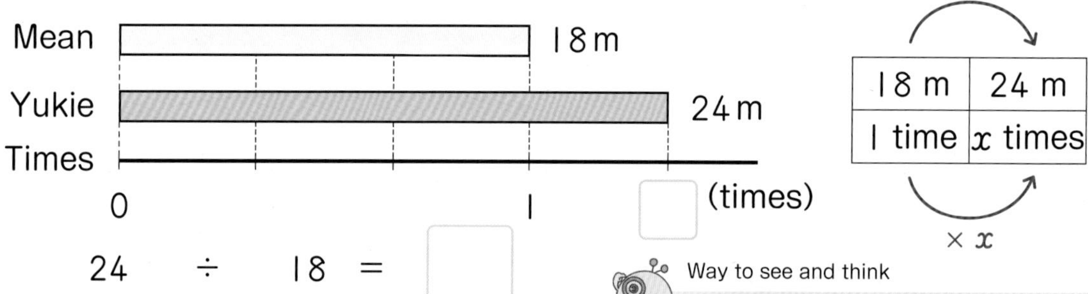

$\times x$

18 m	24 m
1 time	x times

$\times x$

24 ÷ 18 = ☐

Compared quantity　Base quantity　Times

Way to see and think

Can think as compared quantity ÷ base quantity = times

In 5th grade, we learned that times can be represented in fractions.

Want to know

Purpose Let's think about various problems using times a fraction.

② Mari's record was 15 m. How many times of the mean is 15 m?

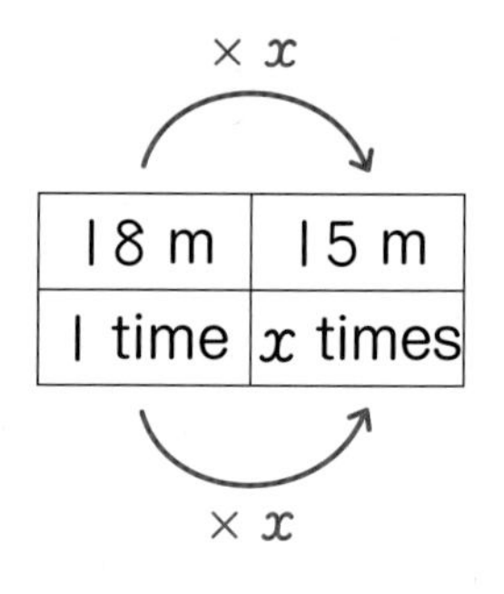

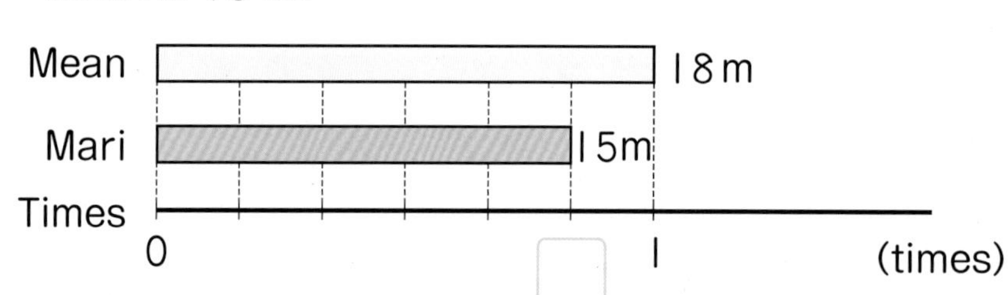

15 ÷ 18 = ☐

Compared quantity　Base quantity　Times

Sometimes the ratio represented by the fraction is smaller than 1.

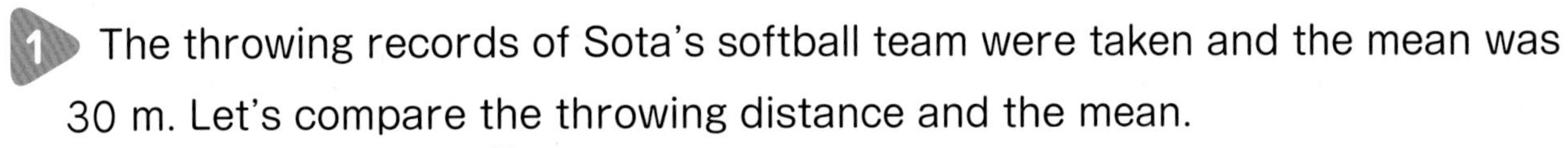

1 The throwing records of Sota's softball team were taken and the mean was 30 m. Let's compare the throwing distance and the mean.

① Sota's record was $\frac{7}{5}$ times of the mean. How many meters was Sota's throwing distance?

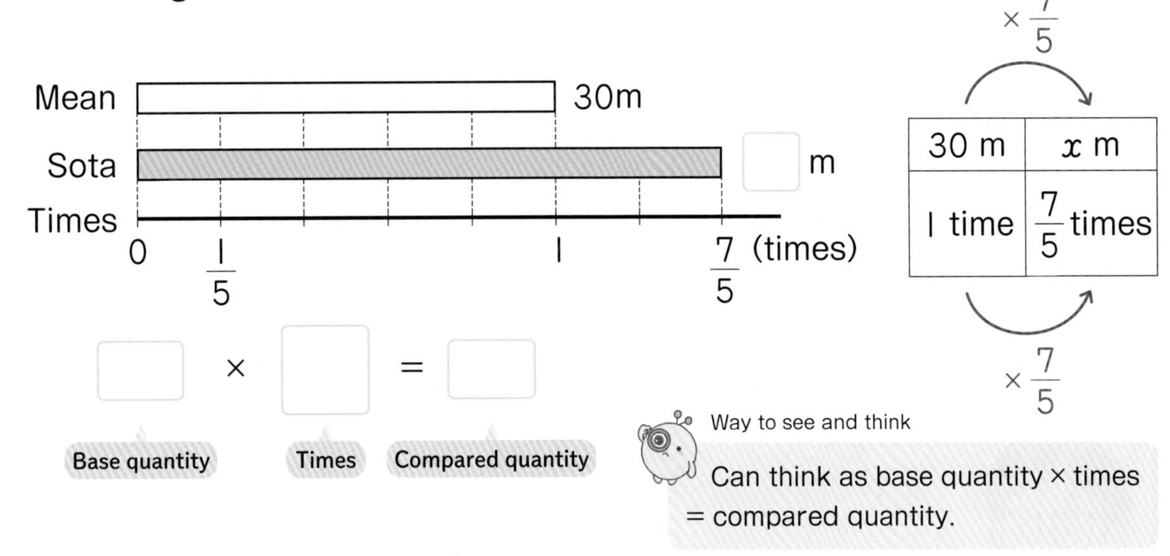

$\square \times \square = \square$

Base quantity　Times　Compared quantity

Way to see and think

Can think as base quantity × times = compared quantity.

② Sota's throwing distance was $\frac{7}{6}$ times of Haruto. How many meters was Haruto's throwing distance? Let's think and represent through a diagram and a table considering Haruto's record as x m.

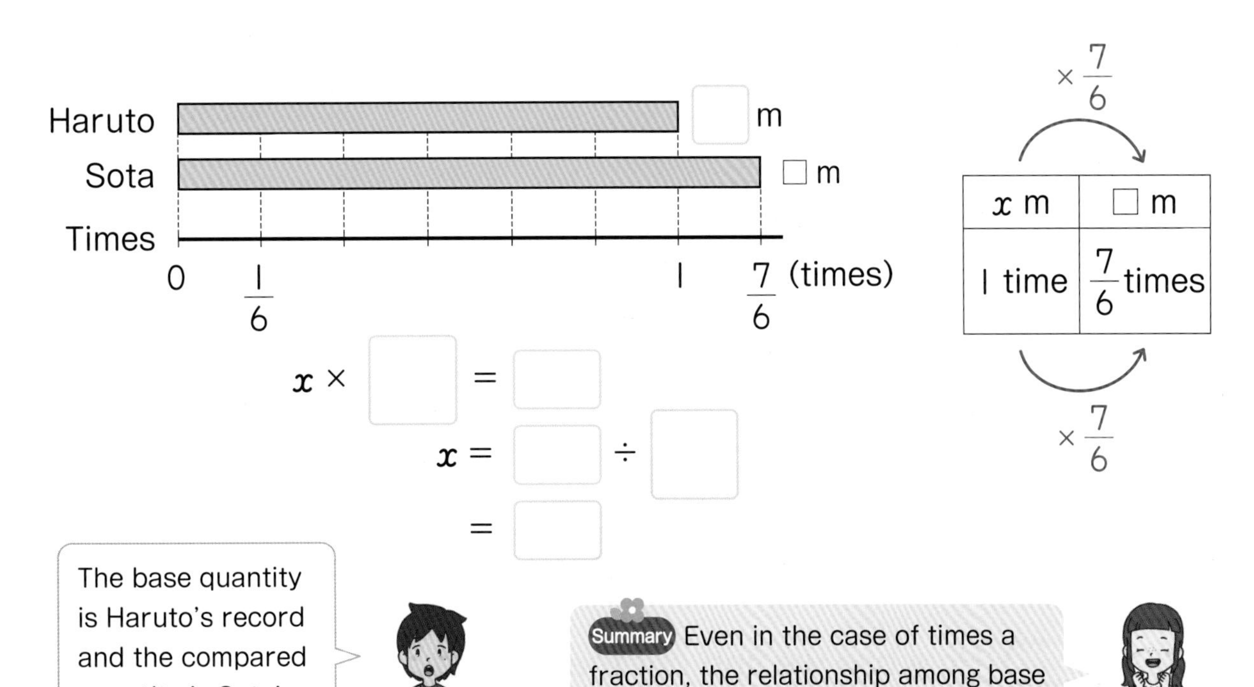

Summary Even in the case of times a fraction, the relationship among base quantity, compared quantity, and times is the same as what we have learned.

Find the ?

What is the area of the circle?

▷

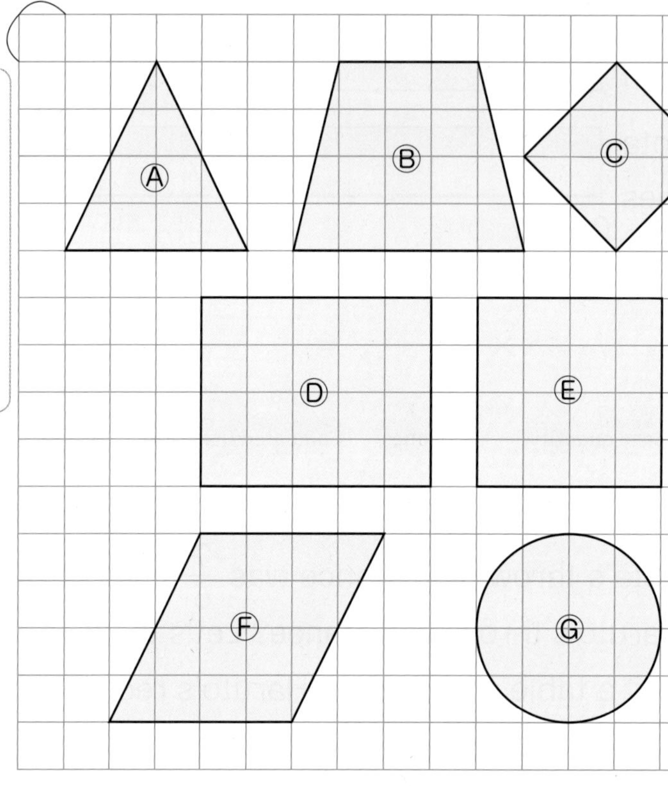

As shown on the right, there are several figures that are similar in size. Let's put them in order from the one with the largest area to the smallest.

1

Let's find out the area.

I know which one has the largest area.

Some figures have the same area.

If I can find out the area of the circle in Ⓖ, I can put them in order...

2

\ Want to know /

Purpose How can we find out the area of a circle?

Area of a Circle

Let's think about how to find out the area of a circle.

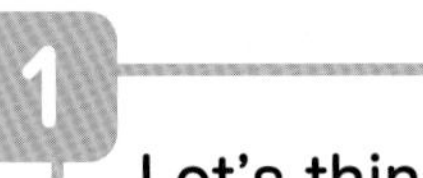

1 Let's think about how to find out the area of a circle with a radius of 10 cm.

When we considered the length of the circumference of a circle, we compared it to the length around a square or a regular hexagon.

Looking at the shapes Ⓐ to Ⓖ, I see that Ⓒ fits perfectly inside Ⓖ and Ⓖ fits perfectly inside Ⓔ.

❶ Let's look at the following shapes and estimate the area of circle Ⓖ.

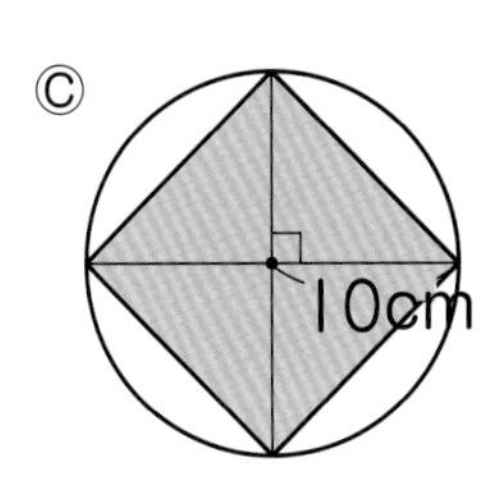

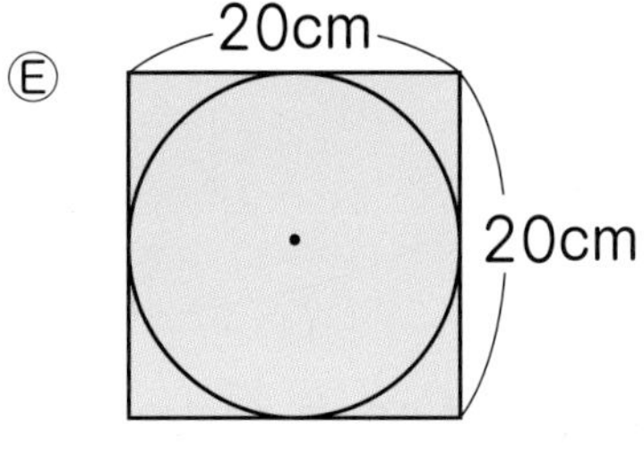

The area of the square Ⓒ is ☐ cm^2. The area of the square Ⓔ is ☐ cm^2. The area of the circle Ⓖ is larger than ☐ cm^2, but smaller than ☐ cm^2.

I wonder if I can count it by making one centimeter for every square of the grid.

❷ Using a 1 cm grid, find the area of the circle Ⓖ.

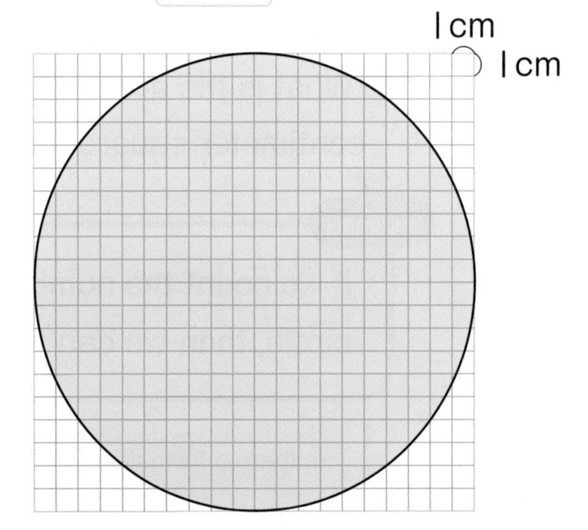

3 Let's separate the circle into 4 equal parts and consider one part.

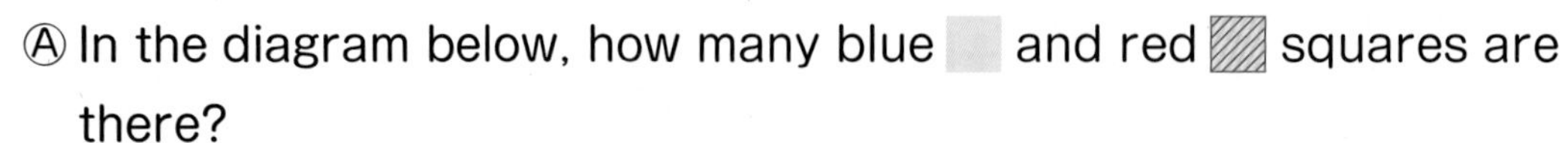

Ⓐ In the diagram below, how many blue and red squares are there?

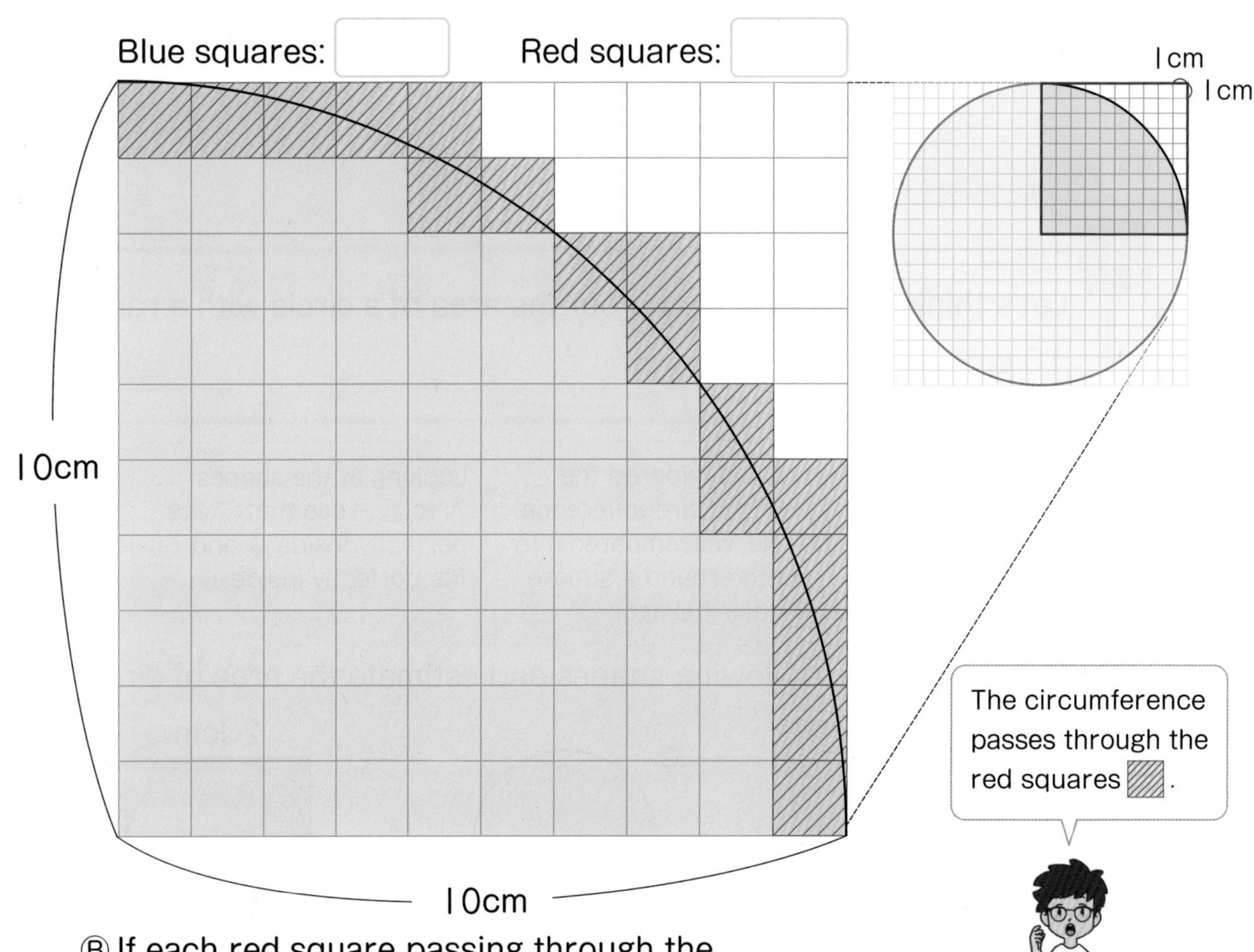

Ⓑ If each red square passing through the circumference has an area of 0.5 cm^2, what is the area of the quarter of this circle in centimeter square (cm^2)?

Blue squares: $1 \times \square$ (cm^2)

Red squares: $0.5 \times \square$ (cm^2)

Area of $\frac{1}{4}$ of the circle: $\square$ (cm^2)

4 What is the area of the entire circle in centimeter square (cm^2)?

! Summary

If we count the number of squares with area of 1 cm^2 that are inside the circle, then we can find a value that is close to the area of the circle.

? Is there a formula to find the area of a circle like for the other figures?

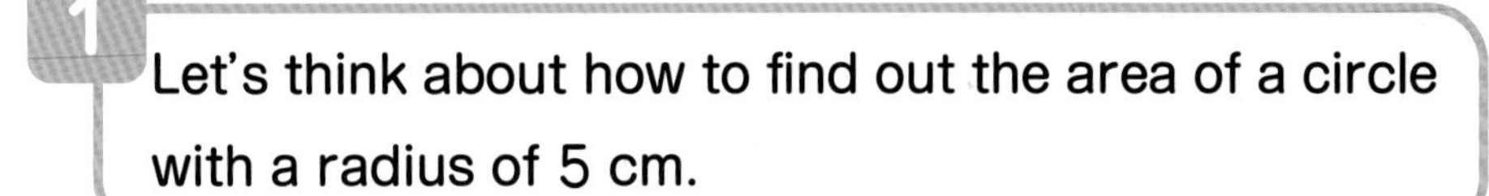

2 Area of circle formula

1 **Let's think about how to find out the area of a circle with a radius of 5 cm.**

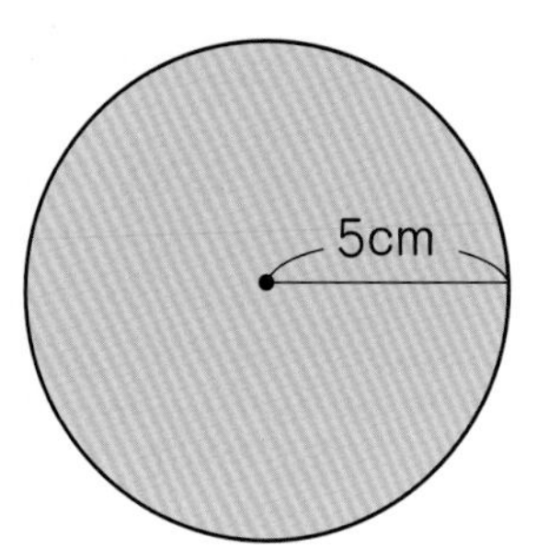

Yu: With parallelograms and triangles, we change the layout of the figure to a known figure.

Akari: We have formulas for rectangles and triangles. I wonder if there is one for circles as well.

\ Want to know /

? **Purpose** **Is there a formula to find the area of a circle?**

① Haruto and Sara rearranged the circle using small equal parts. Let's explain the ideas of the two children.

Haruto's idea

The circle was separated into ☐ parts, and rearranged as a ☐.

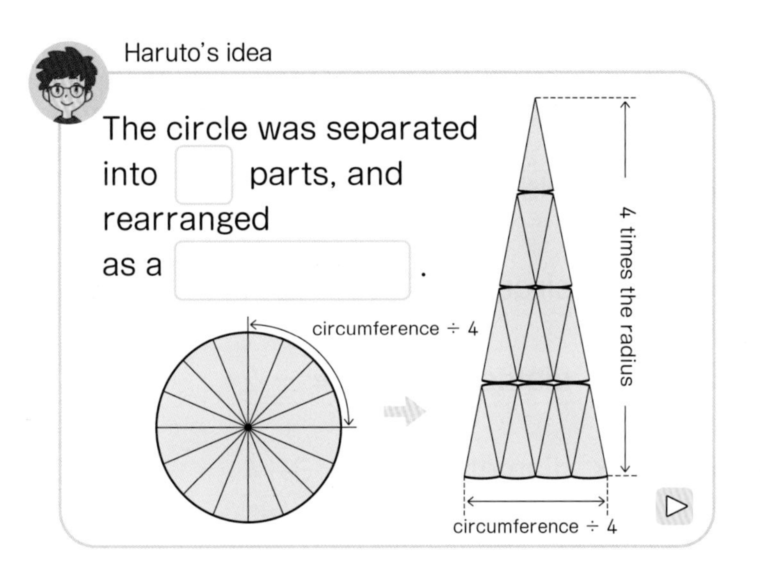

Let's use page 259 to confirm.

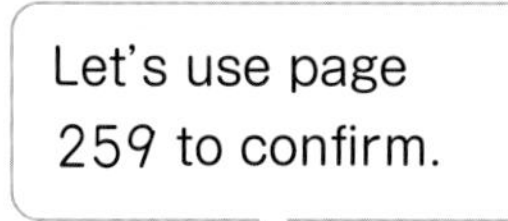

Way to see and think

It is transformed to use the known area of triangle and rectangle formula.

Sara's idea

The circle was separated into ☐ parts and rearranged as a ☐.

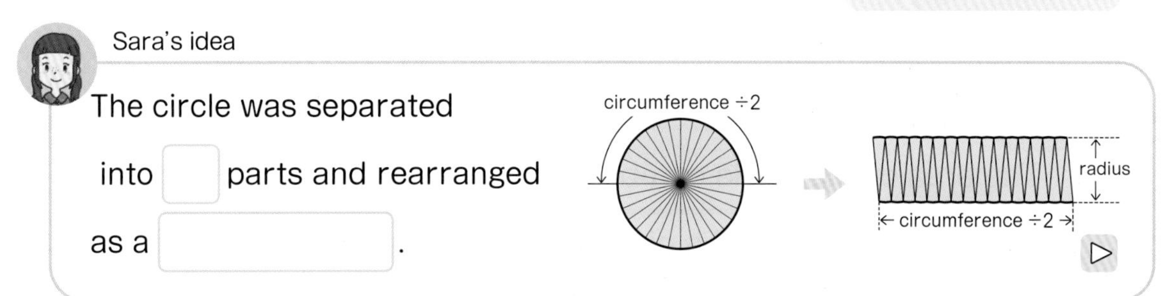

② Let's find out the area of the circle based on the ideas shown on ①.

③ Based on Sara's idea on the previous page, let's explain the formula to find out the area of a circle.

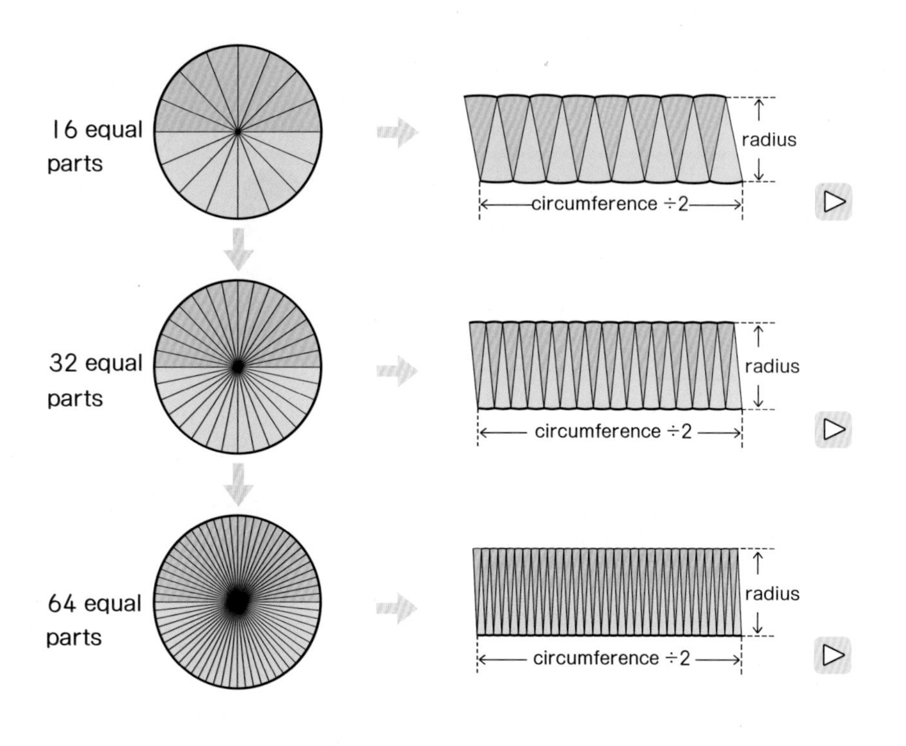

If you divide the circle into small parts, you can get closer to a rectangular figure.

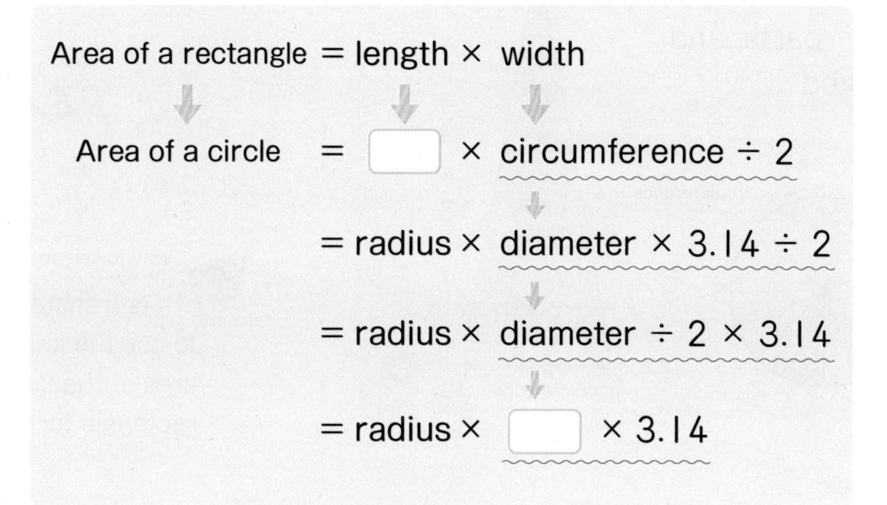

Summary

The area of the circle can be found with the following formula:

Area of a circle = radius × radius × 3.14

? Can we use the formula for the area of a circle to find the area of circles of various sizes?

2

Let's find out the area of a circle with the following radius.

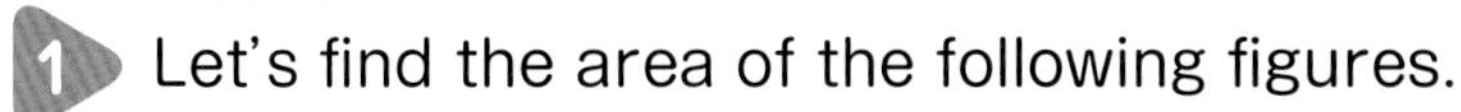

❶ 8 cm ❷ 10 cm ❸ 12 cm

\ Want to try /

Purpose Let's consider various problems using the formula for the area of a circle.

Yu

1 Let's find the area of the following figures.

① 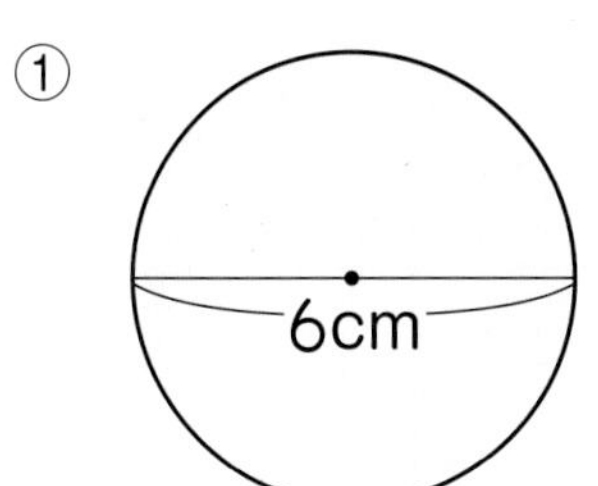

② 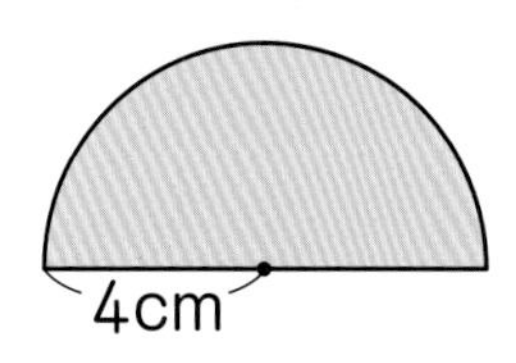

③ 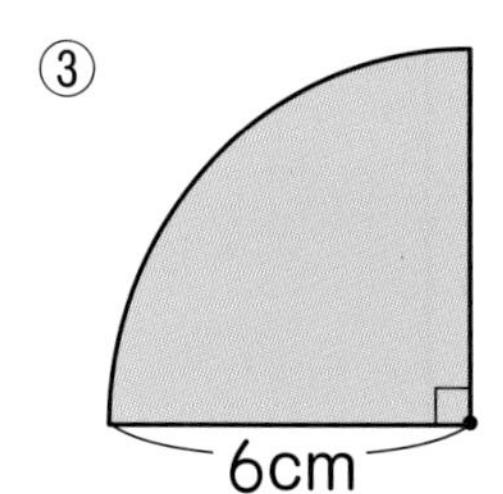

2 The following ①~③ are the circumferences of circles. Let's find the radius and area of each circle.

① 62.8 cm ② 18.84 cm ③ 15.7 cm

3 Circle Ⓐ has a diameter of 4 cm, and Circle Ⓑ has a diameter of 8 cm.

① Let's find the circumference and area of each circle.

② The diameter of Circle Ⓑ is 2 times the diameter of Circle Ⓐ. How many times the circumference and the area of Ⓐ are of Ⓑ?

Ⓐ 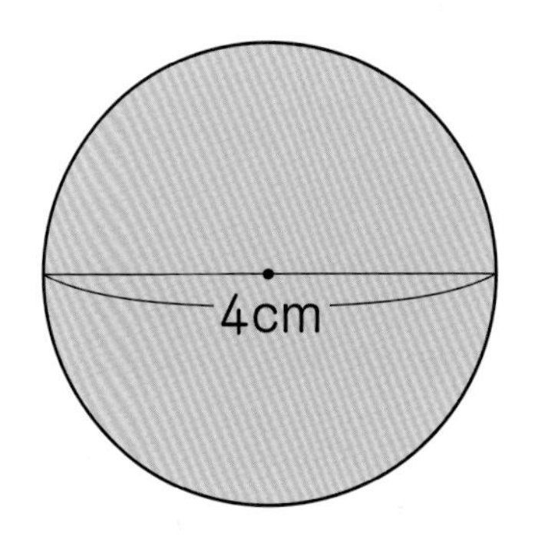

Ⓑ 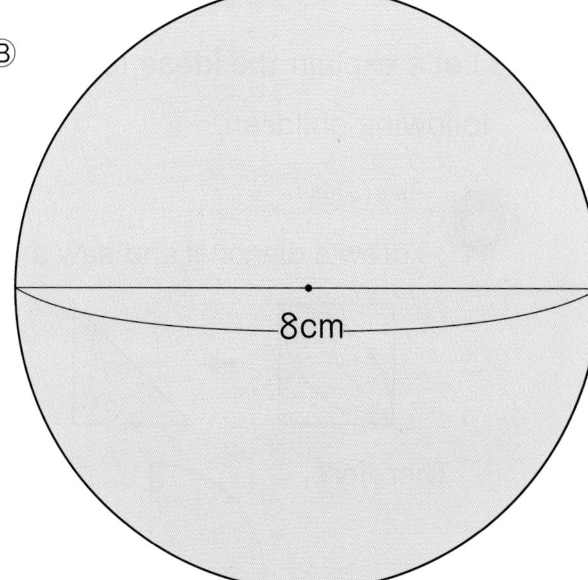

? Can we find the area of other shapes with the formula for the area of a circle?

1 Let's find out the area of the colored parts for the following diagrams.

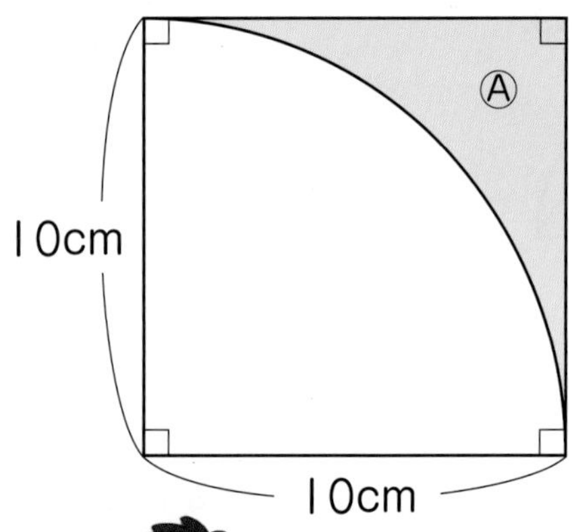

Way to see and think

Search for a familiar figure from which you know how to find the area.

Haruto: It is not a circle nor a part of circle...

Sara: In 4th grade, I thought about dividing and combining them...

Want to think

Purpose Can you find the area of different shapes by applying some creative ideas?

Haruto

① Let's write a math equation to find the area of Ⓐ.

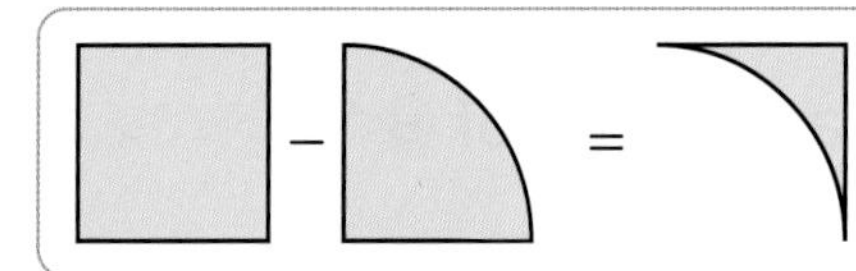

② Let's explain the ideas to find the area of Ⓑ by the following children.

Akari's idea

I drew a diagonal and saw a triangle.

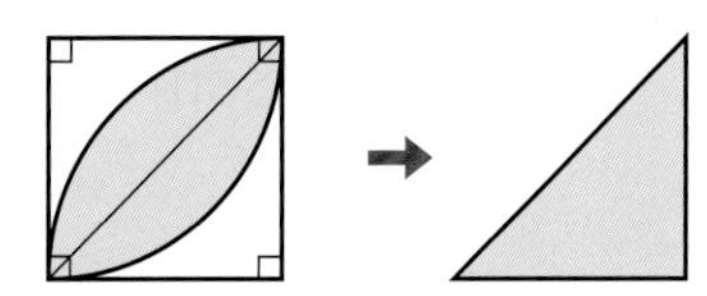

Therefore,

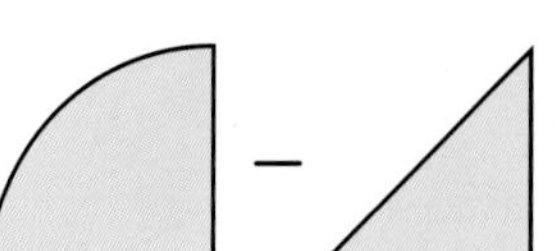

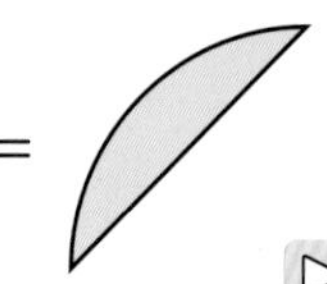

Way to see and think

Using the formula for the area of $\frac{1}{4}$ of a circle and the area of a right triangle.

Haruto's idea

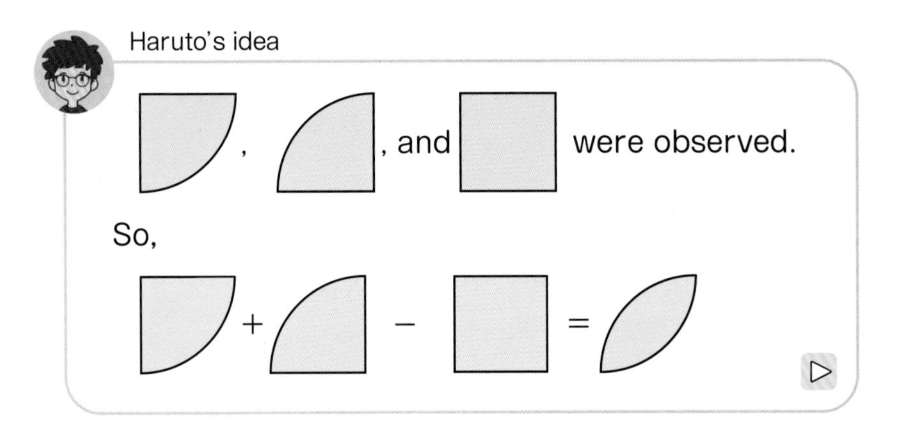

Way to see and think

Using the formula of the area of $\frac{1}{4}$ of a circle and the area of a square.

Sara's idea

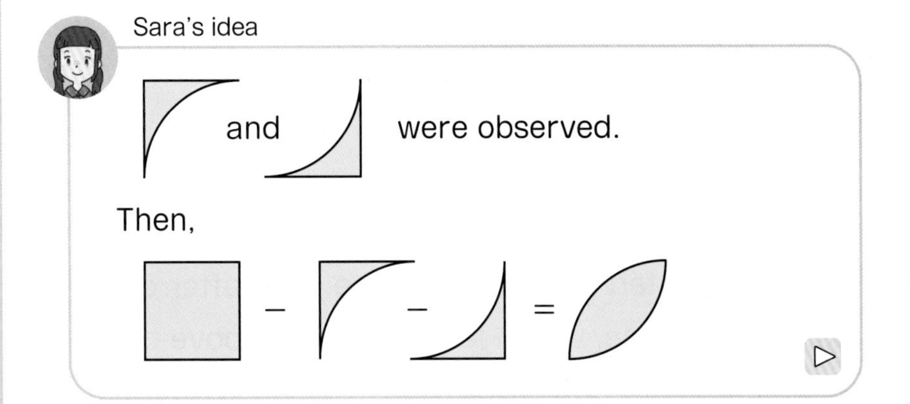

Way to see and think

Using the area of Ⓐ and the formula of the area of a square.

❸ Let's find out the area of Ⓑ using the ideas from ❷ .

Summary We can find out the area using the methods we have learned so far by adding or subtracting.

1 Let's find out the area of the colored parts for the following diagrams.

①

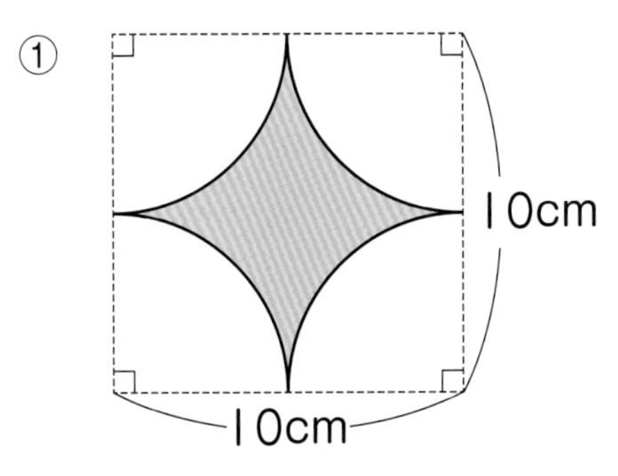

②

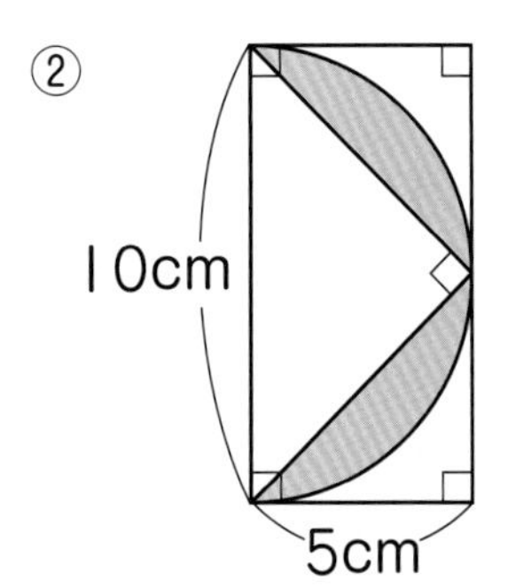

③

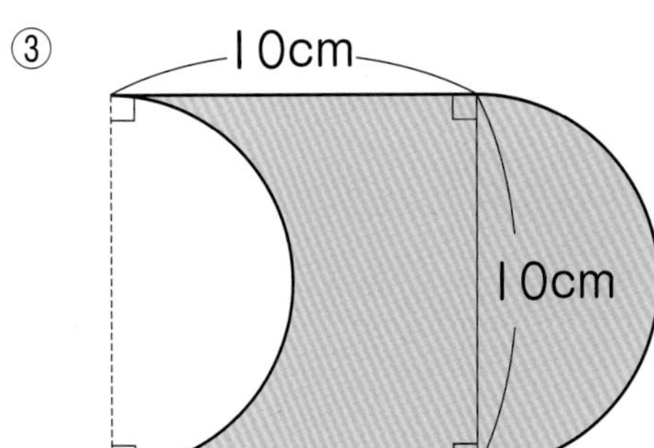

④

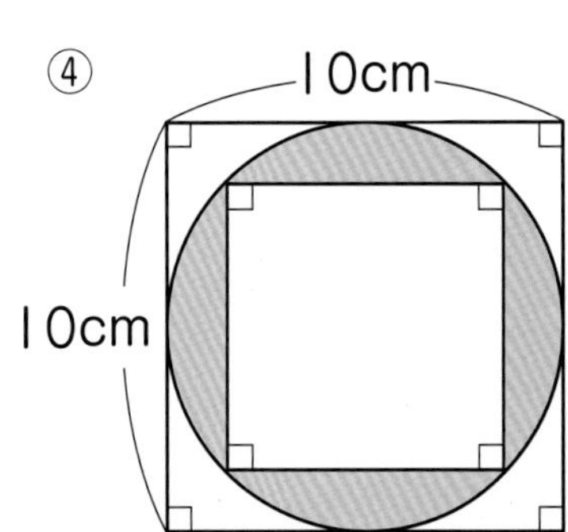

That's it!

Let's confirm the area of a circle using a rope.

A rope is twisted around to make a circle with 5 cm radius, as shown in the picture below.

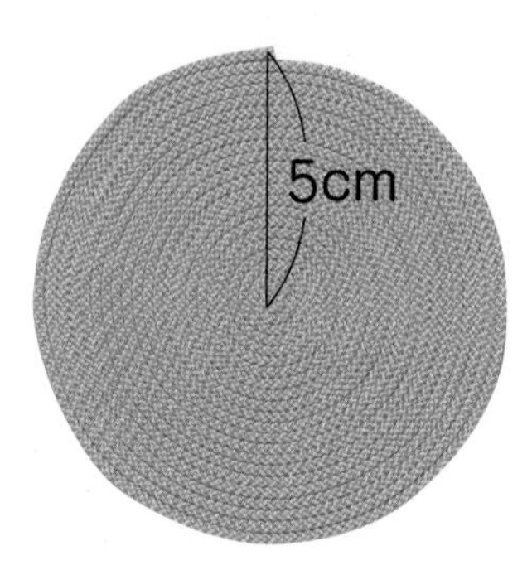

① The following figure was created by unwinding the rope after cutting it through the radius. In this case, which lengths of the above circle correspond to the lengths of AB and CD?

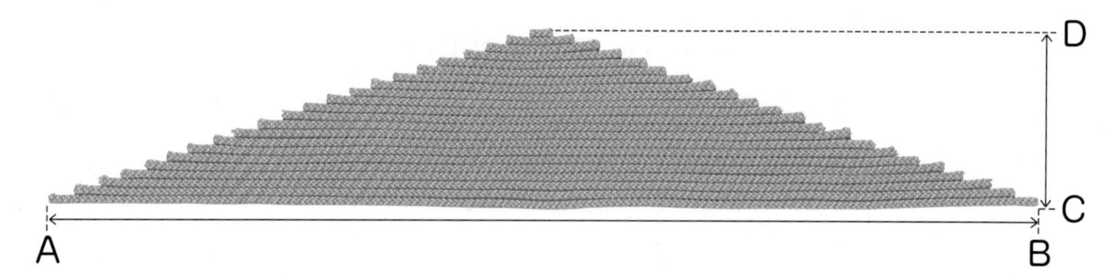

② Manami thought about a formula to find out the area of a circle based on the figure ①. Let's write the appropriate words or numbers on each ☐ to complete her idea. Then, explain it to your classmates.

Area of a triangle = ☐ × ☐ ÷ ☐

Area of a circle = ☐ × 3.14 × ☐ ÷ ☐

= ☐ × 2 × 3.14 × ☐ ÷ ☐

= ☐ × ☐ × 3.14

4 Approximate area

1 A field between rivers is shown on the right. Let's find out the approximate area of the field surrounded by the black line.

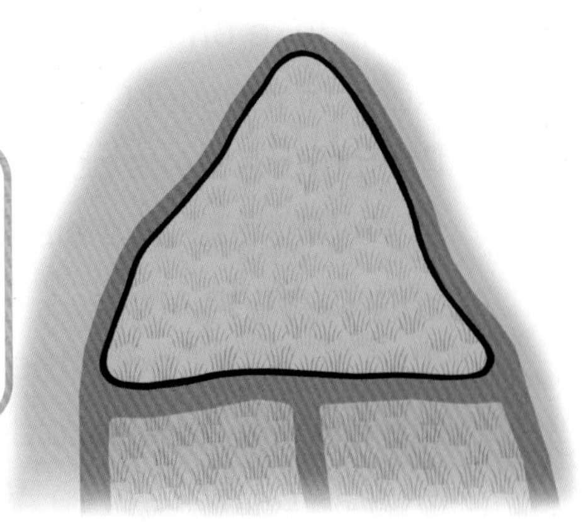

If we sketch it on a grid....
Yu

It looks like a triangle...
Akari

Want to know

? **Purpose** How can we find the approximate area?

1. In the diagram on the right, how many blue and red squares are there?
2. Let's find out the area of the field by considering the area of any 2 red squares where the perimeter passes through as $100\ \text{m}^2$.

Way to see and think: Can we do the same to calculate the approximate area of a circle?

10m
10m

3. If the field shown on the right represents a triangle, let's find out the area of it.

Way to see and think: Look for similar shapes and try to use them to find the area.

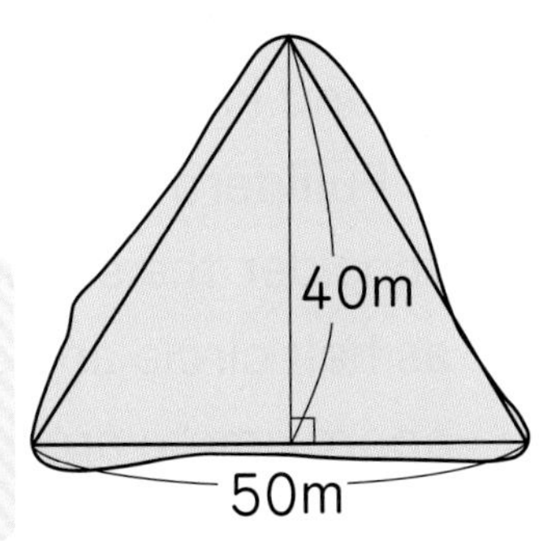

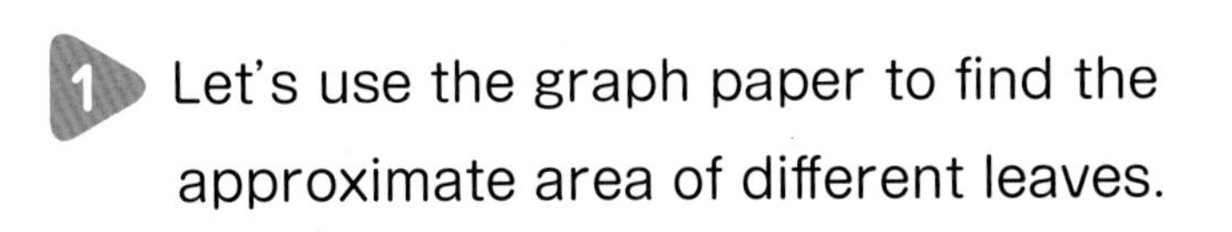

1 Let's use the graph paper to find the approximate area of different leaves.

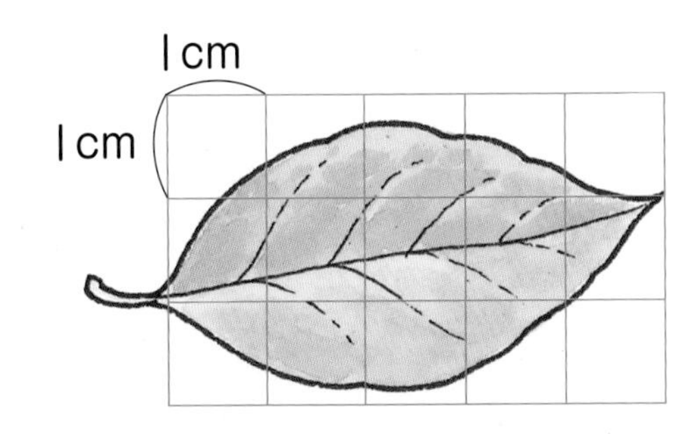

2 The lake on the right is Lake Ikeda in Ibusuki City, Kagoshima Prefecture. Let's consider the shape of Lake Ikeda as a circle to find its approximate area.
Also, let's consider the shape of the lake as a trapezoid to find its approximate area. Which approximation is closer to the actual area?

Lake Ikeda (Ibusuki City, Kagoshima Prefecture)

Let's find the real area of the lake on an encyclopedia or the internet.

2km

5km
3km
2km

Summary

The approximate area of various shapes can also be calculated by considering the shapes we have studied so far.

3 The picture on the right is the burial mound known as Nakahime-no-mikoto-ryo in Fujiidera City, Osaka Prefecture. Let's consider the shape of the burial mound as half circle and a trapezoid to find the approximate area of it, as shown in the following figure.

Nakahime-no-mikoto-ryo(Fujiidera City, Osaka Prefecture)

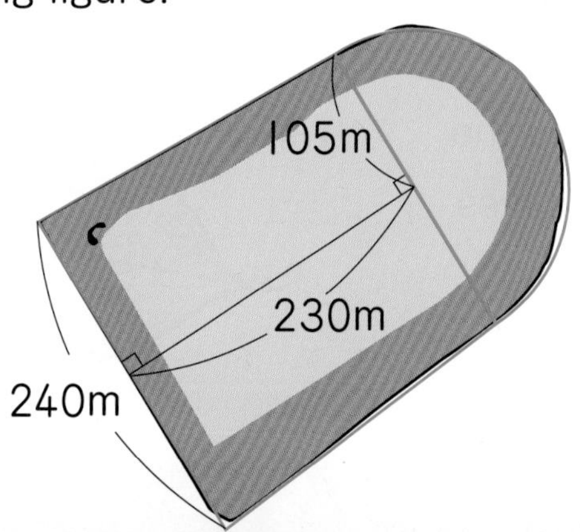

CAN What can you do?

☐ We can use the area of circle formula. → p.133

1 Let's find out the area of the following circles.

①

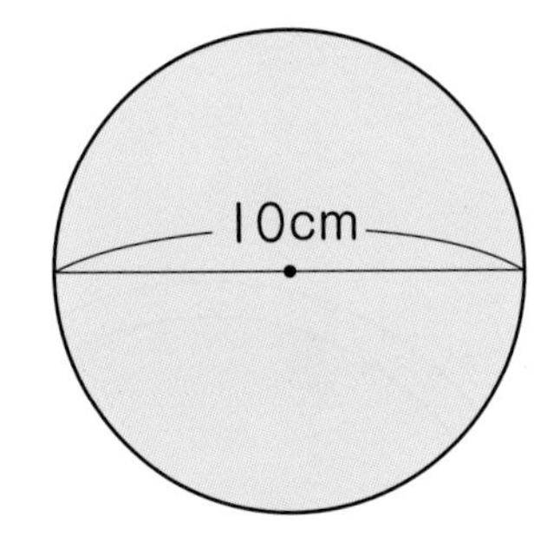

②

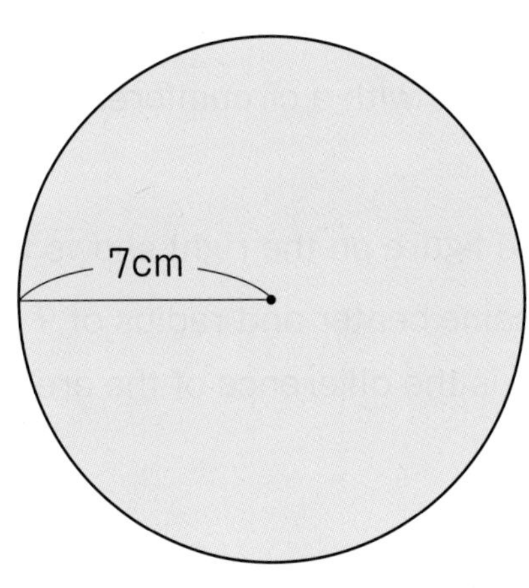

③

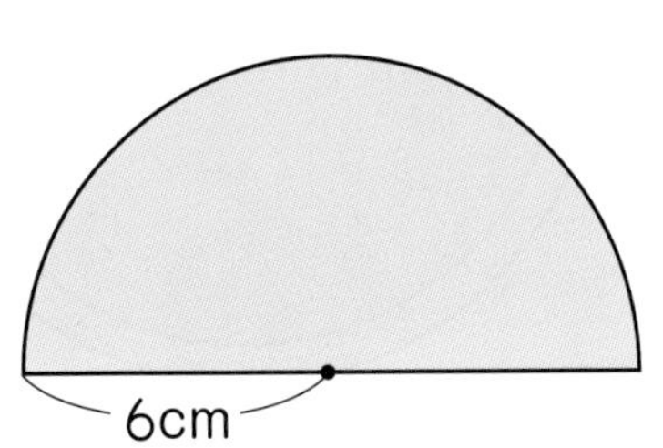

④

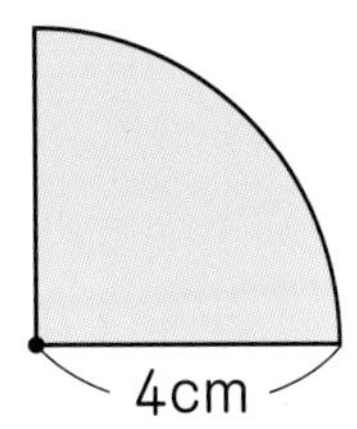

☐ We can find the area of figures using new ideas. → pp.134 ~ 135

2 Let's find out the area of the colored parts for the following diagrams.

①

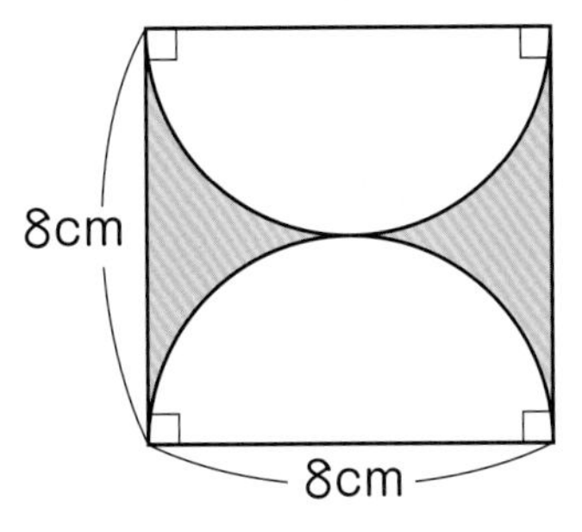

②

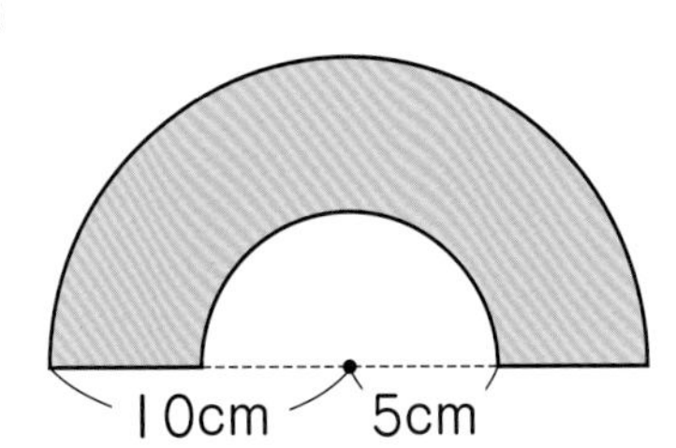

Supplementary Problems → p.239

Which "Way to See and Think Monsters" did you find in "9 Area of a Circle"?

I found "Change" when finding the area of a circle.

Haruto

When finding the area of a circle....

Sara

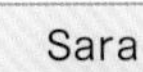

Usefulness and Efficiency of Learning

1 Let's find the diameter and area of the following circles.

① a circle with a circumference of 6.28 cm

② a circle with a circumference of 12.56 cm

2 The figure on the right shows two circles with the same center and radius of 9 cm and 10 cm. What is the difference of the areas in cm^2?

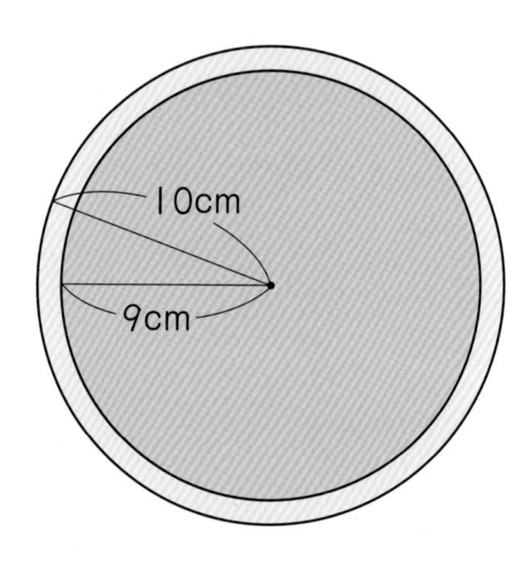

3 Let's find the area of the colored parts for the following diagrams.

①

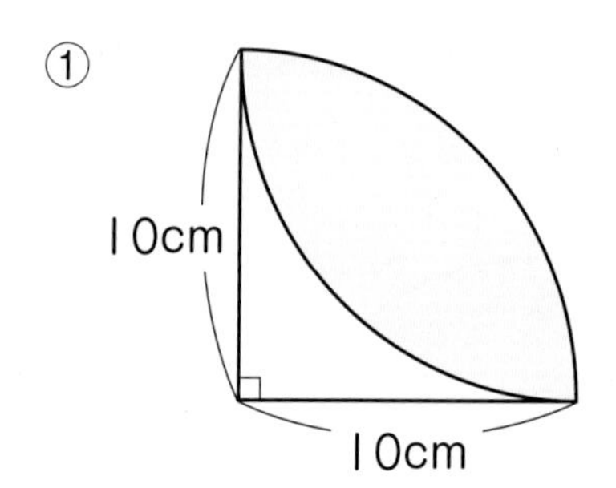

②

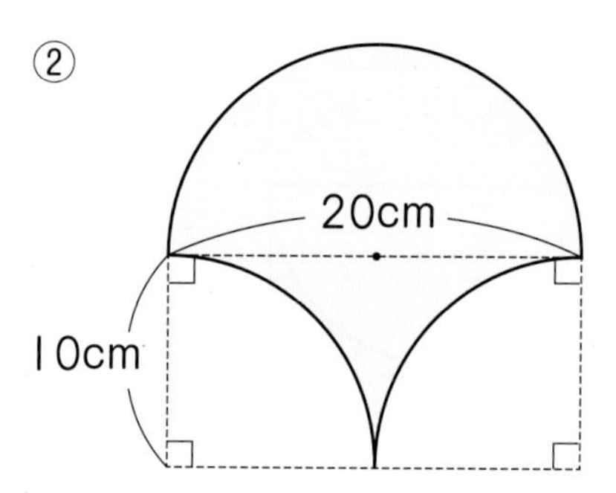

③

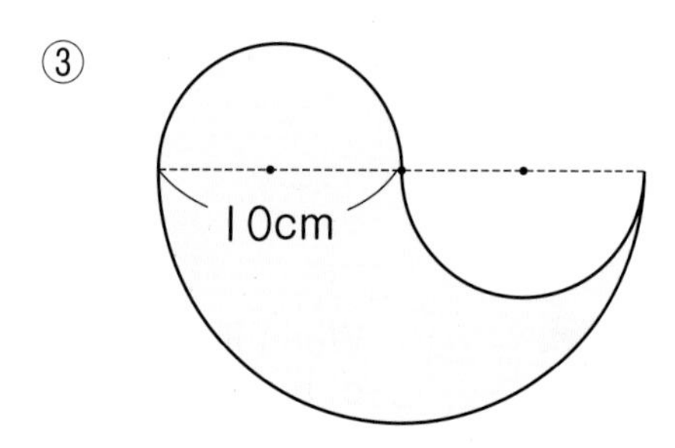

④

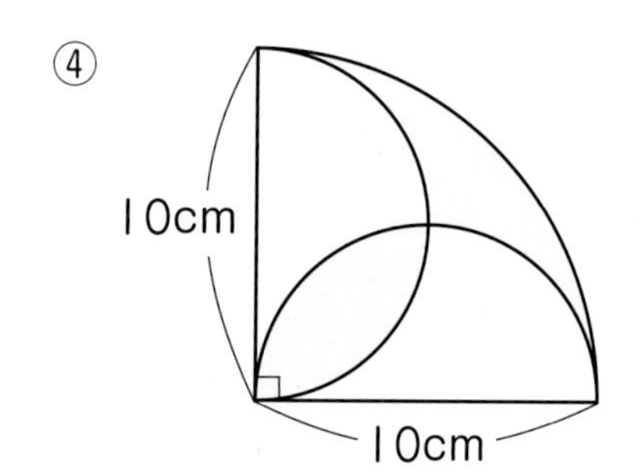

With the Way to See and Think Monsters...

Let's Reflect!

Let's reflect on which monster you used while learning "9 Area of a Circle."

I was able to find the area by dividing the figure and changing it to a figure I knew.

Sara

In the 5th grade, when finding out the area of a parallelogram, we divided the figure and changed it to a ______.

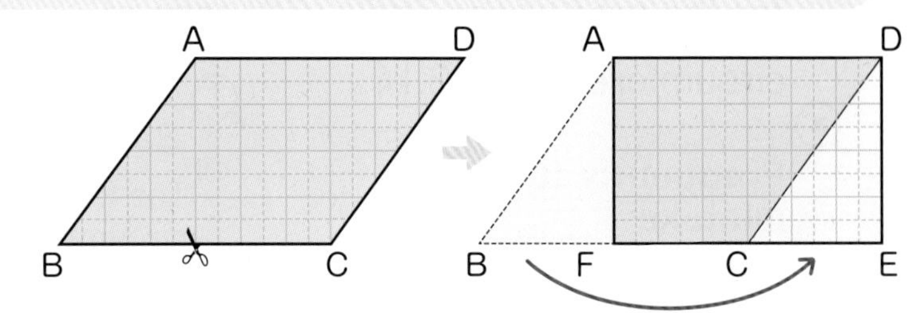

Yu

To find out the area of a triangle, we tried by changing it to a ______ or a ______.

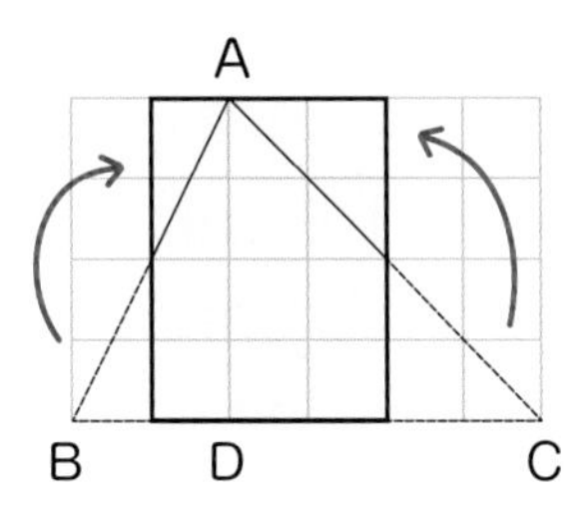

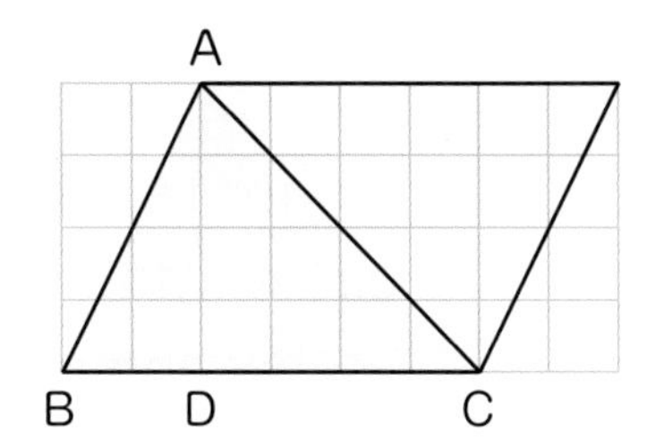

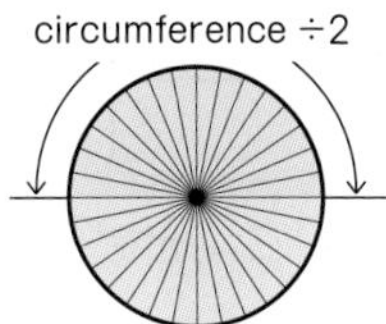

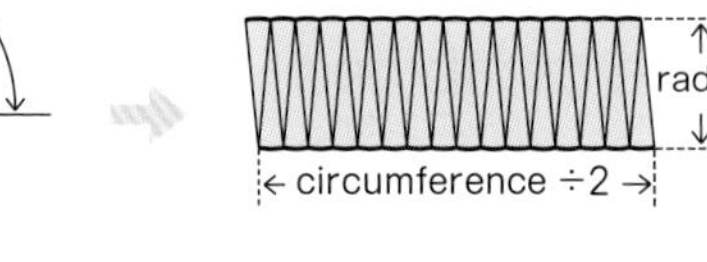
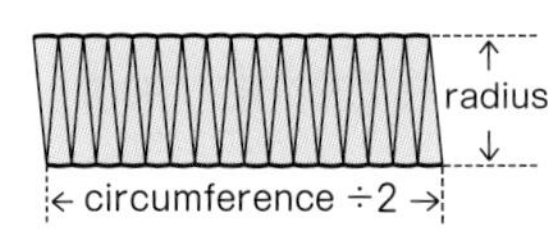

Area of a rectangle = length × width

Area of circle = ______ × ______
= ______ × diameter × 3.14 ÷ 2
= radius × ______ × 3.14

The circle was divided into ______ equal parts and then rearranged as a ______.

Haruto

Let's deepen. → p.245

? Solve the ?

By changing the circle to a known figure, we were able to find the area of the circle.

Sara

→

Want to Connect

We could find out the area of ◺ or ◠, but can we find out the area of figures like ▽?

Yu

The area of a circle could be obtained by "radius x radius x 3.14." While the circumference of a circle, which we studied in 5th grade, could be obtained by "diameter x 3.14." Although these two formulas are similar, let's not lose sight of what they are trying to find.

To find the area of a circle....

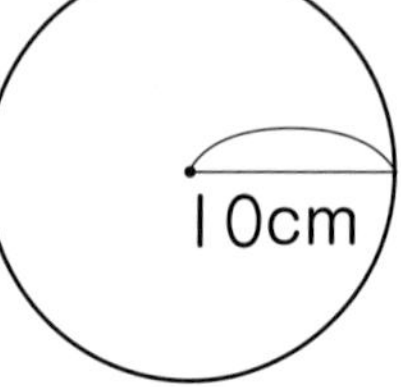

① Let's write the equation and answer for the area of the circle with ratio of circumference as 3.14.

Frequently found mistake
Use the formula to find out the length of the circumference of a circle.

Be careful!
The formula for finding out the length of the circumference of a circle is:
diameter x ratio of circumference.
The formula for finding out the area of a circle is: radius x radius x ratio of circumference.

② As for rectangle Ⓐ and Ⓑ, draw a portion of a circle with a radius of 10 cm centered on the vertex and paint it black, as shown in the figure on the right.
What can be said about the combined area of the four black areas of rectangle Ⓐ and the combined area of the four black areas of rectangle Ⓑ?
Choose the correct one from the following ⓐ to ⓒ.

ⓐ The combined area of the four black areas is larger in rectangle Ⓐ.
ⓑ The combined area of the four black areas is the same.
ⓒ The combined area of the four black areas is larger in rectangle Ⓑ.

Ⓐ 10cm 10cm 10cm 10cm

Ⓑ 10cm 10cm 10cm 10cm

Frequently found mistake
Judging by appearance, consider that one of them is larger than the other.

Be careful!
When the black areas are combined, they both form a circle with a radius of 10 cm.
The fact that the sum of the angles of a rectangle is 360° can be used to explain this.

Since the sum of the four corners of a rectangle is 360°, the black areas together form a circle.
Therefore, the total area of the four black areas of both rectangle Ⓐ and Ⓑ is "the area of a circle with a radius of 10 cm."

What is the volume of the box?

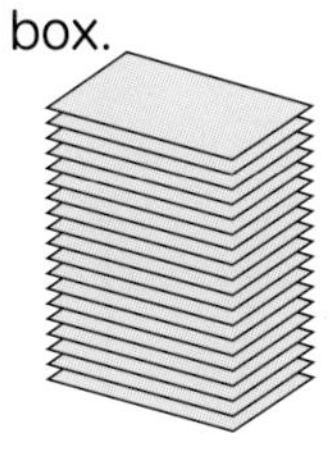

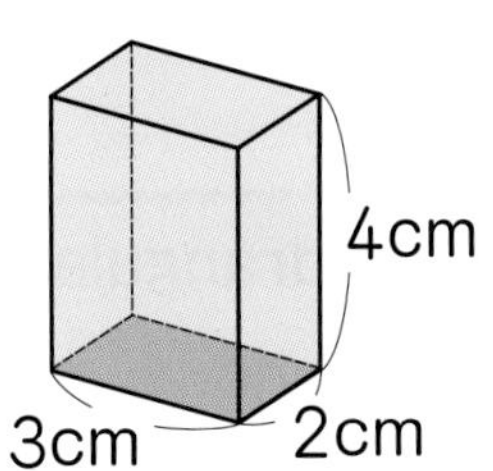

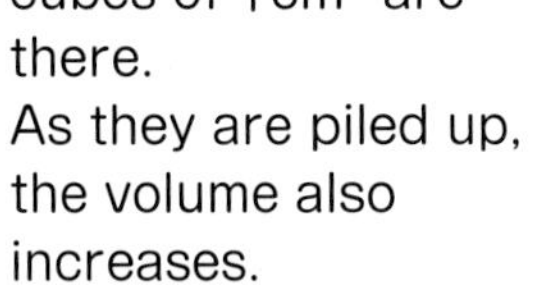

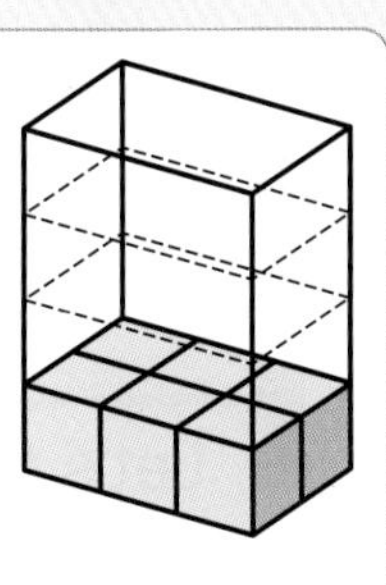

If you stack cards whose length is 2 cm and width is 3 cm, you can make a box.

The volume can be calculated by counting how many cubes of 1cm^3 are there. As they are piled up, the volume also increases.

1

If the length and width were the same, the volume would be proportional to the height.

Does this mean that the volume of a box made with a stack of cards is proportional to the number of cards?

2

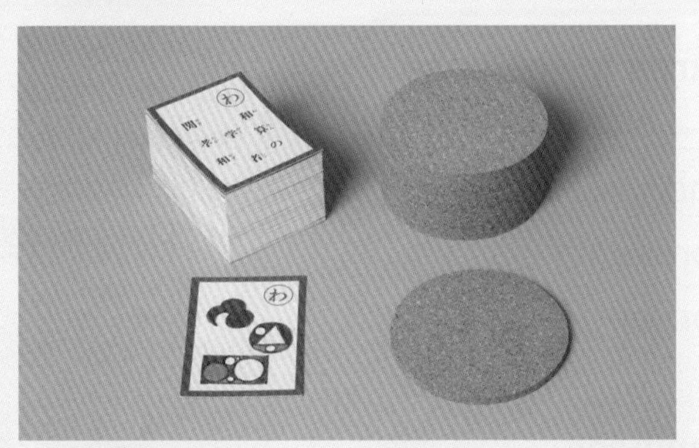

Think of a box as a stack of cards...

I wonder if we can find the volume of a prism or cylinder like the one on the left in the same way.

3

Can we find the volume of a prism or cylinder?

Volume of Solids

Let's think about how to find out the volume of a solid and its formula.

1 Volume of prisms

Volume of quadrangular prisms ↓

1 Let's find out the volume of a quadrangular prism with a rectangular base as shown on the right.

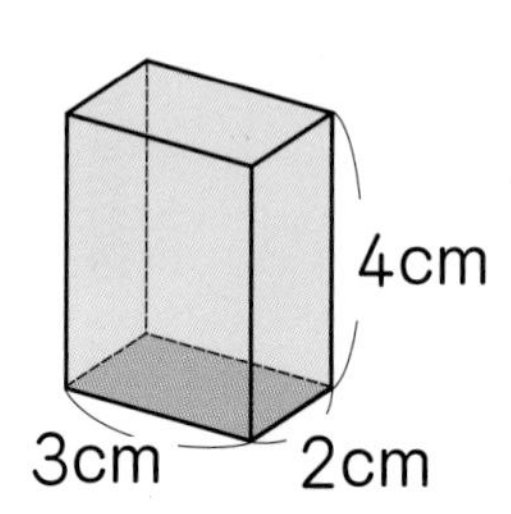

\ Want to know /

? Purpose How can we find out the volume of a prism?

Way to see and think

A quadrangular prism with a rectangular base can be seen as a cuboid.

① Let's find out the volume of the quadrangular prism using the volume formula of a cuboid.

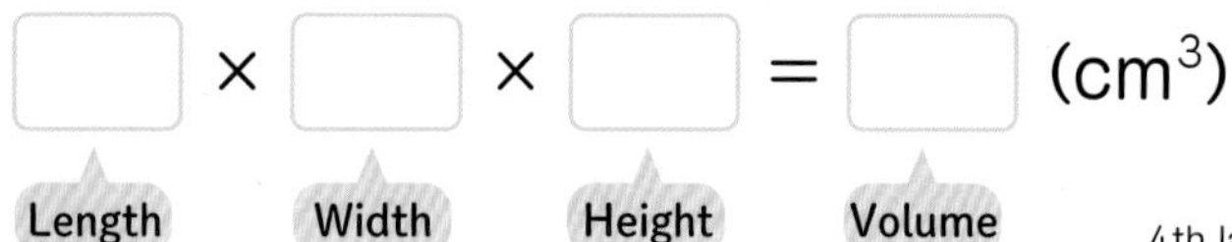

② Let's find the volume of the quadrangular prism with a height of 1 cm.

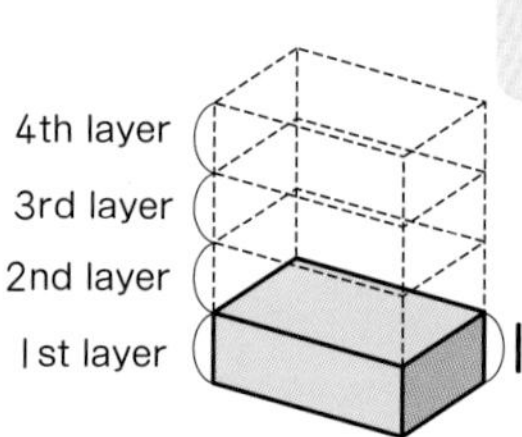

The bottom area is called the **area of the base**.

③ Let's find the area of the base and compare it to the number that represents the volume of the quadrangular prism with a height of 1 cm.

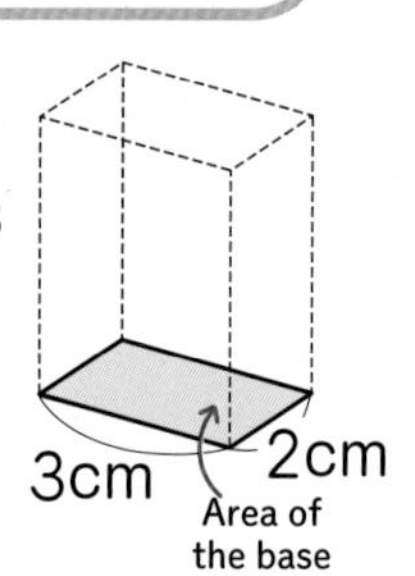

What looks like a quadrangular prism is made by overlapping a lot of bases. ▷

④ Let's find the volume of a quadrangular prism by using the area of the base.

Length Width Height Volume

$\square \times \square \times \square = \square$ (cm³)

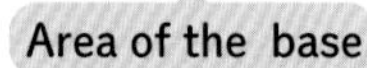

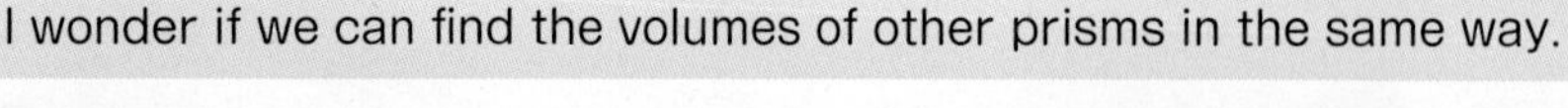

? I wonder if we can find the volumes of other prisms in the same way.

2 Let's think about how to find the volume of a triangular prism as the one shown on the right.

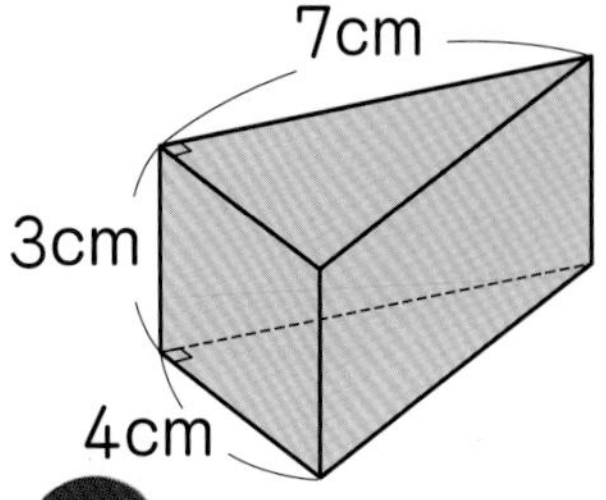

I thought the triangular prism was half of a quadrangular prism.

Can we find the triangular prism volume in the same way as for the quadrangular prism, with (area of the base) × (height)...

If we put two of these triangular prisms together, we get a cuboid, right?

1. Let's find the volume of the triangular prism considering it as the half of a cuboid.

$7 \times 4 \times 3 \div 2 = \square$ (cm^3)

Volume of a cuboid

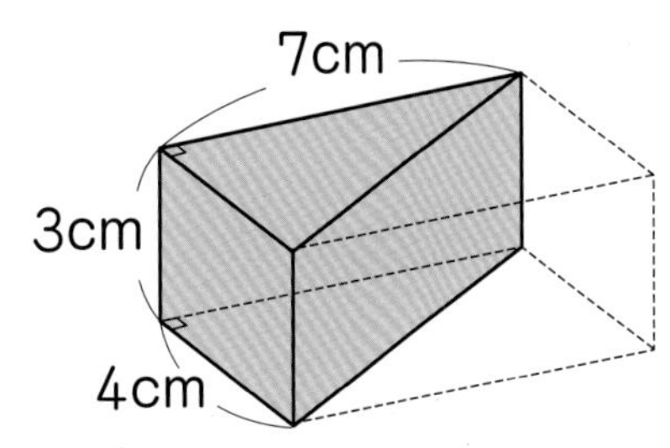

2. Let's find the volume of the triangular prism using the area of the base.

$7 \times 4 \div 2 \times 3 = \square$ (cm^3)

Area of the base

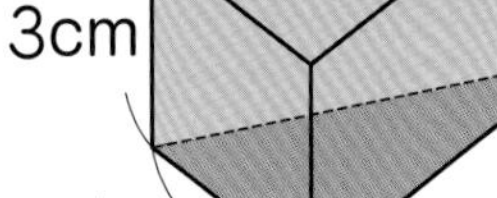

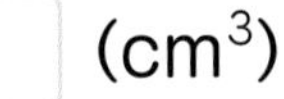

Way to see and think

If the answers are the same with the two methods to find the volume, we can assume that both are correct.

Answer 1 is equivalent to answer 2, so the volume of the triangular prism can also be found by the using the formula (area of the base) × (height).

1 There is a prism with a rhomboidal base, as shown on the right. Let's find its volume using the 2 methods.

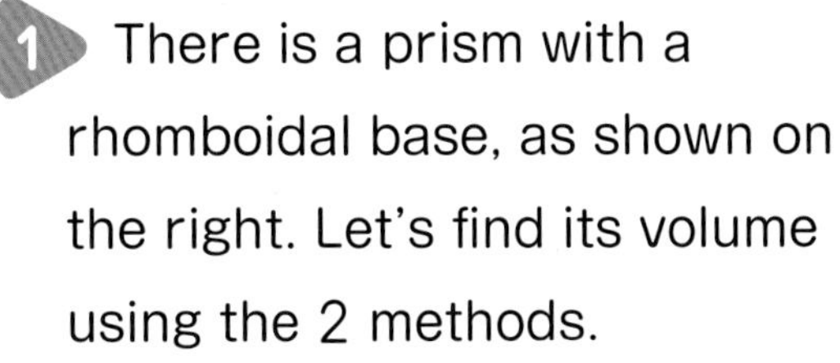

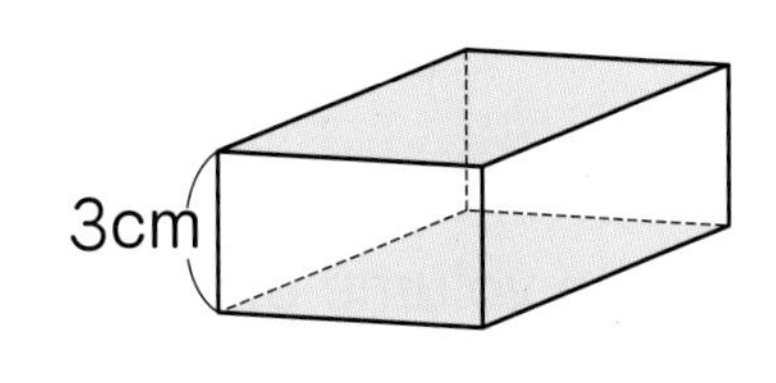

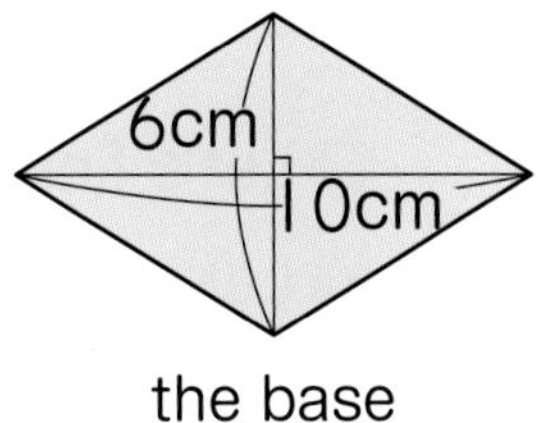

the base

Summary

The volume of all prisms can be calculated with:

Volume a prism = area of the base × height

? Can we find the volume of a cylinder in the same way?

2 Volume of cylinders

1 **Let's think about how to find the volume of a cylinder as the one shown on the right.**

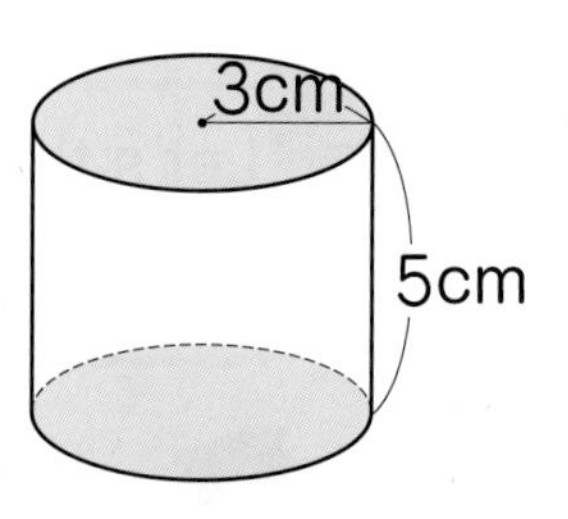

Want to know

Purpose **Can we calculate the volume of a cylinder with the formula (area of the base) × (height)?**

① Let's think about how to find the volume of a cylinder by looking at the following diagrams.

Area of the circle = radius × radius × 3.14

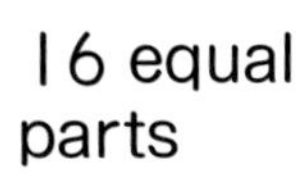
16 equal parts

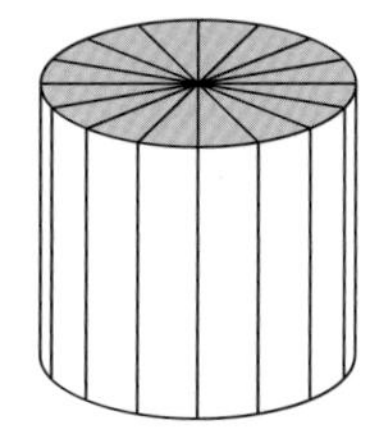

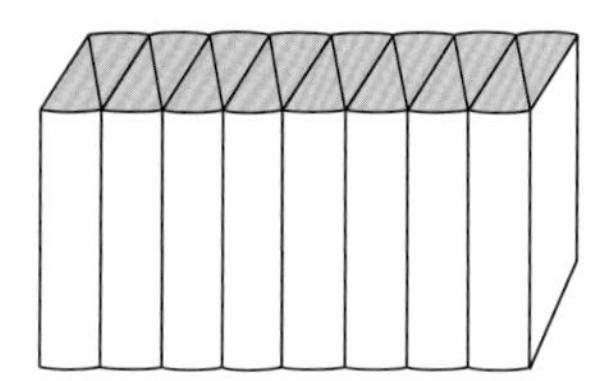

32 equal parts

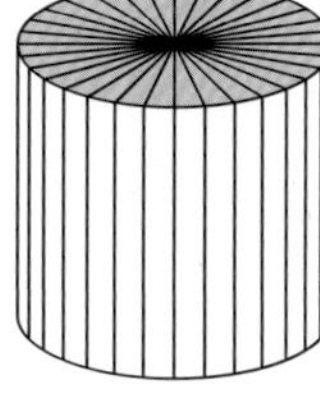

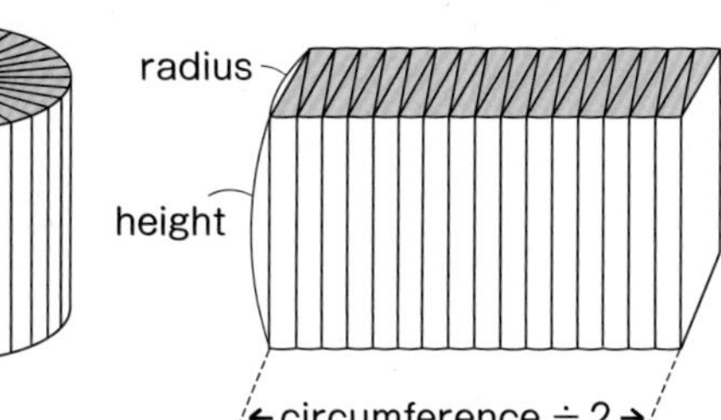

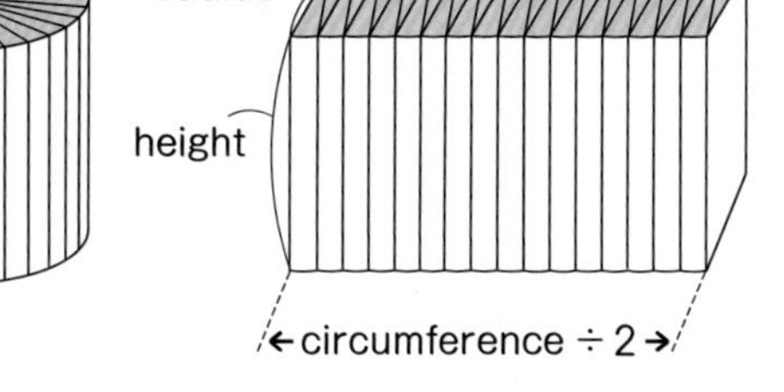

Way to see and think: Think about it in the same way as when you found the area of a circle, you divide it on equal parts and change it to the area of a rectangle.

$(3 \times 3 \times 2 \times 3.14 \div 2) \times 5 = \square$ (cm^3)

Radius / Half of the circumference / Height / Volume

② Let's find out the area of the base of the cylinder.

③ Let's find out the volume of the cylinder in the same way as the prism.

$(3 \times 3 \times 3.14) \times 5 = \square$ (cm^3)

Area of the base / Height / Volume

④ Let's explain what is equal on the math equations ① and ③. Then, make a conclusion about it.

Way to see and think: Try to compare both methods to find the volume.

Summary

The volume of a cylinder can be calculated with:

Volume of a cylinder = area of the base × height

1 Let's find out the volume of the following cylinders.

① 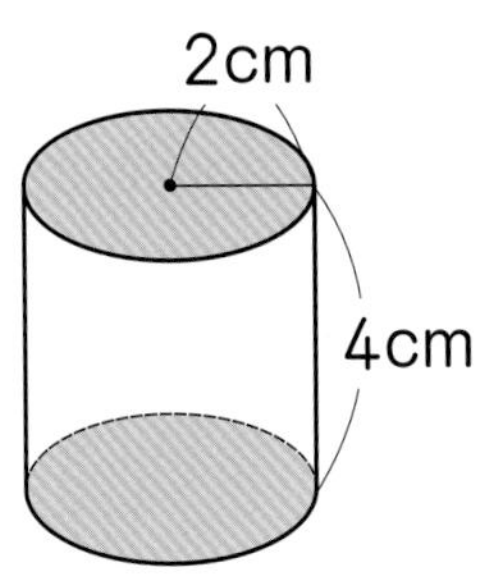

② 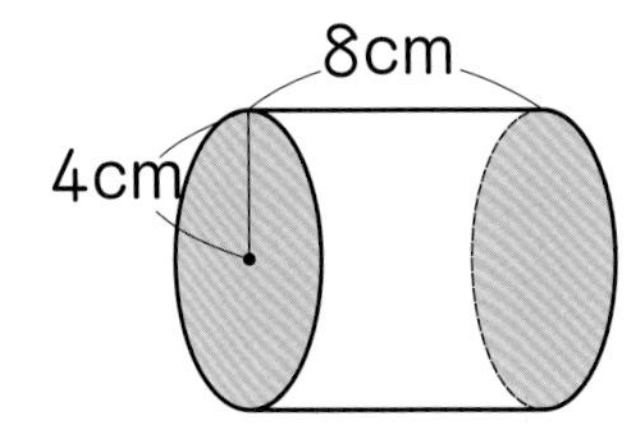

Way to see and think

From what we have learned so far, the volume of a prism or a cylinder can be calculated by (area of the base) × (height).

2 Let's find out the volume of the following solids.

① 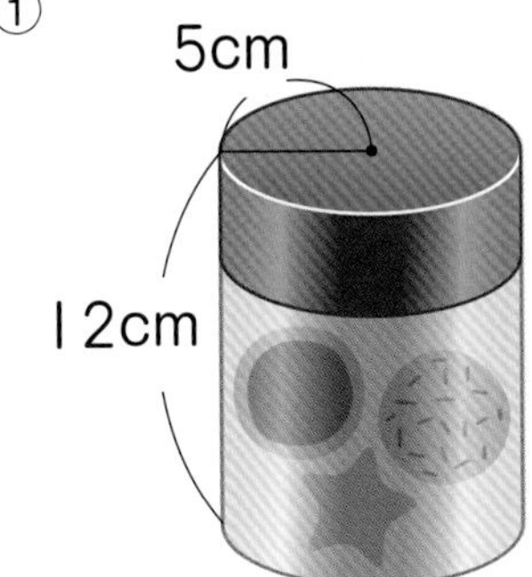

② 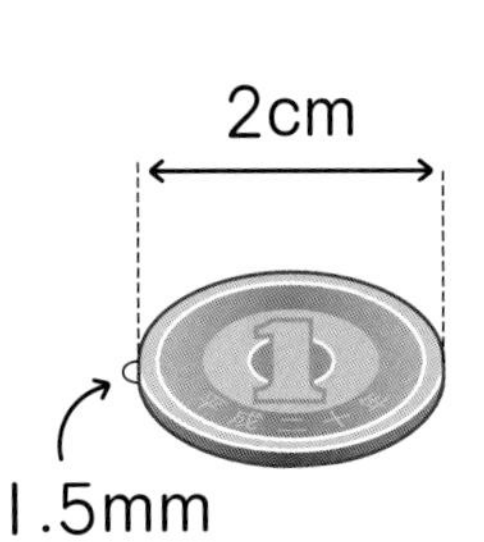

③ 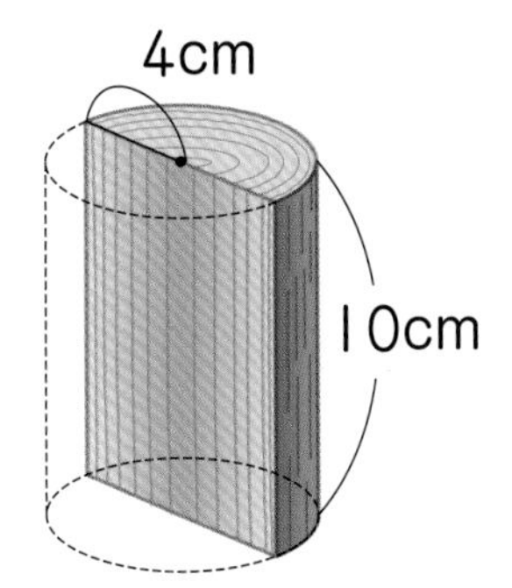

3 There is a circular coaster with a diameter of 10 cm and a thickness of 0.8 cm as shown on the right.
10 of the coasters are stacked to form a cylindrical shape. Let's find out the volume of it.

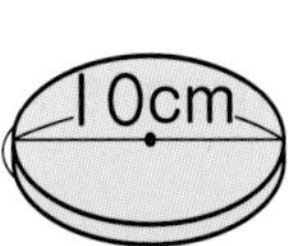

? Can we use the same idea to find the volume of other solids?

That's it!

Compare the volume of different solids.

ADVANCED
Junior High School

The figures on the right are called **pyramids** and **cones**. The pyramids can have different polygonal bases like pentagon or more.

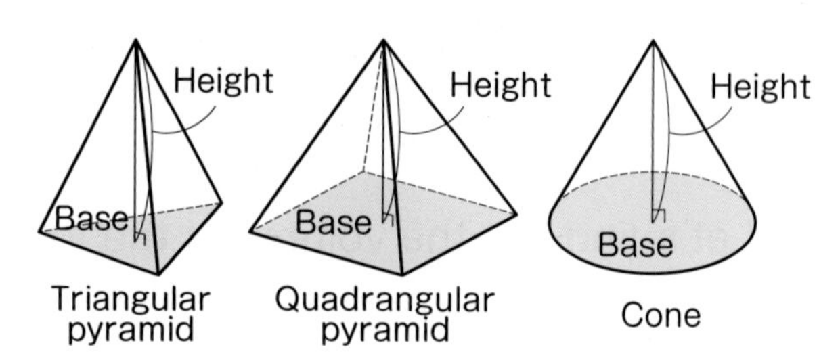

① Let's compare the volume of a quadrangular pyramid and a quadrangular prism when both have the same height and the same base.

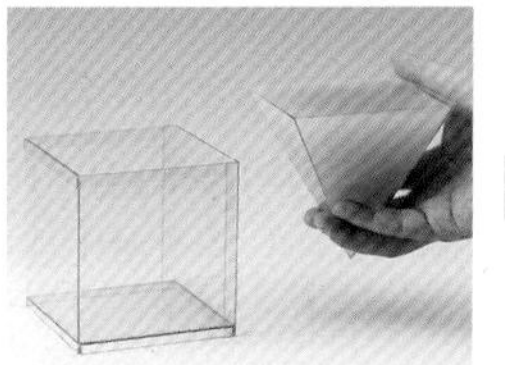
Ⅰ cup is poured...

2 cups are poured...

3 cups are poured...

② Let's compare the volume of a cone and a cylinder when both have the same height and the same base.

Ⅰ cup is poured...

2 cups are poured...

3 cups are poured...

③ What can you conclude from the above experiment?

④ Haruto came up with the following formula for calculating the volume of pyramids and cones.
Let's write the appropriate number in □ . Let's discuss why this math equation is like this.

Volume of pyramid or cone
$= \text{area of the base} \times \text{height} \times \frac{1}{\square}$

You can fill the prism or cylinder with 3 times the water of the pyramid or the cone.

Sara

3 Volume of various shapes

Let's find out the volume of the solid shown on the right.

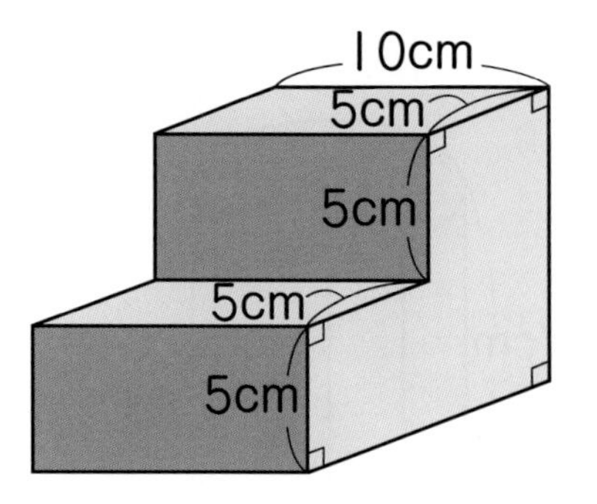

I asked for the volume of a similar shape in 5th grade.

I wonder if it can still be found by area of the base x height.

? Purpose \ Want to think / **Can we also use the volume formula on solids like the one above?**

Yu's idea

Let's do the same we did in 5th grade. Let's separate the solid into 2 cuboids.

10cm, 5cm, 5cm, 5cm, 5cm

Akari's idea

First, let's find the area of the base. Then, let's find the volume.

5cm, 5cm, 5cm, 10cm, 5cm

1. Let's find out the volume using Yu's idea.
2. Let's find out the volume using Akari's idea.

! 

The volume of the solid above, seeing as a prism, can also be found out using the formula (area of the base) × (height).

1 Let's find out the volume of the following solids.

① 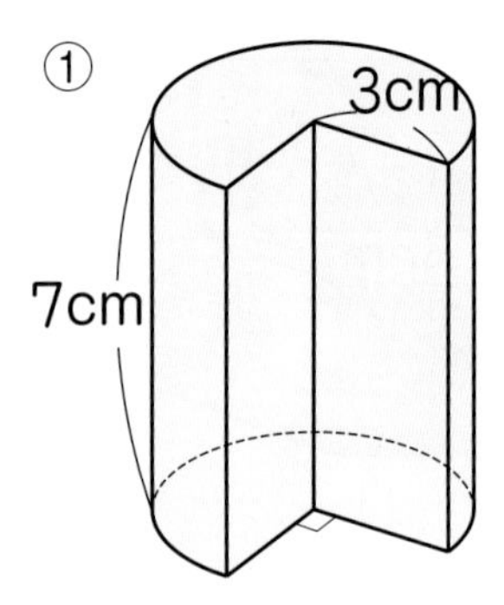

② There is a cylindrical hole in the middle.

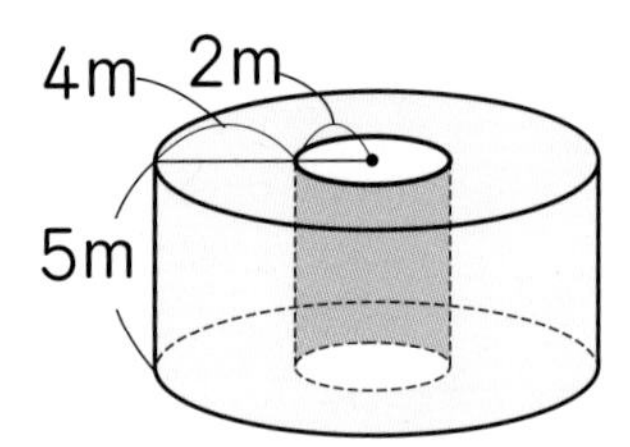

③ 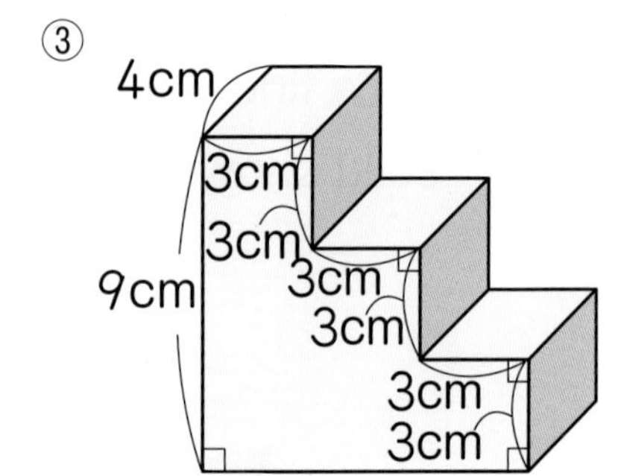

? Can we also find the volume of solids around us?

2 **Let's find the approximate capacity of the storage box shown on the right.**

The inner volume of objects like a storage box can also be called capacity.

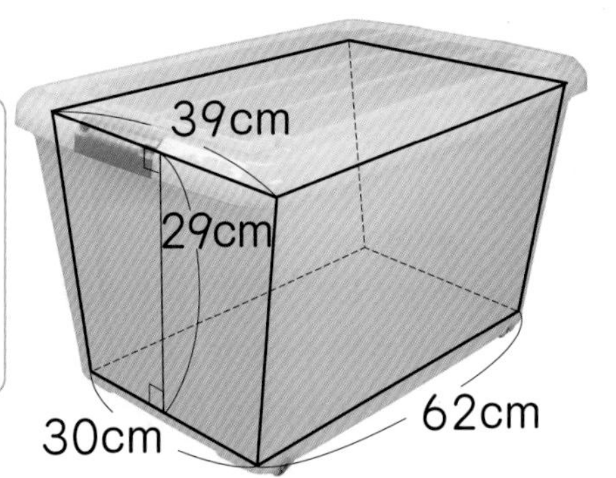

❶ What is the shape of the storage box?

❷ Let's consider the storage box as a quadrangular prism with a trapezoidal base to calculate the approximate volume.

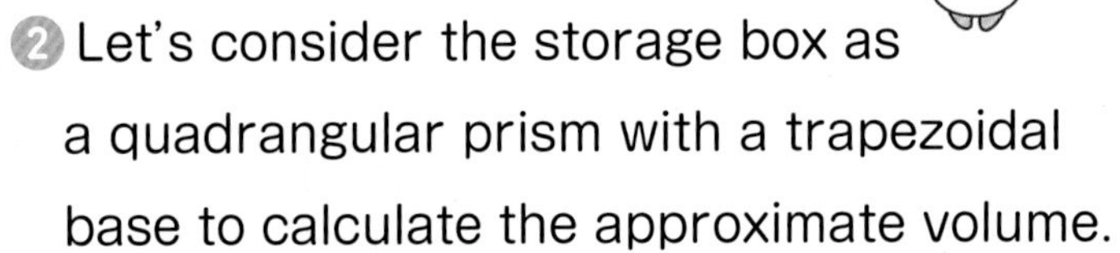

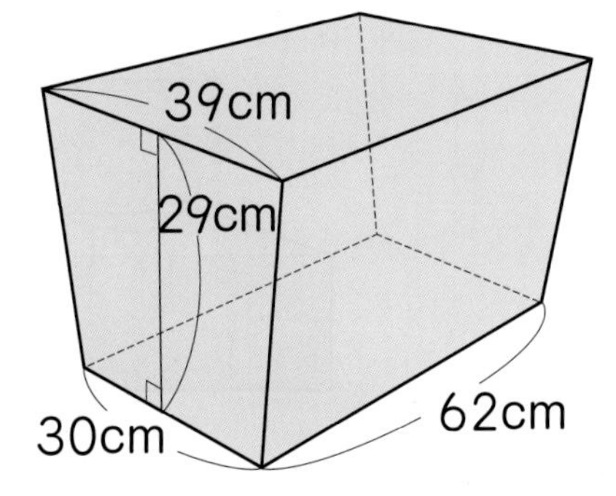

1 There is a cup like the one on the right. Let's look at it as a cylinder to find the approximate volume.

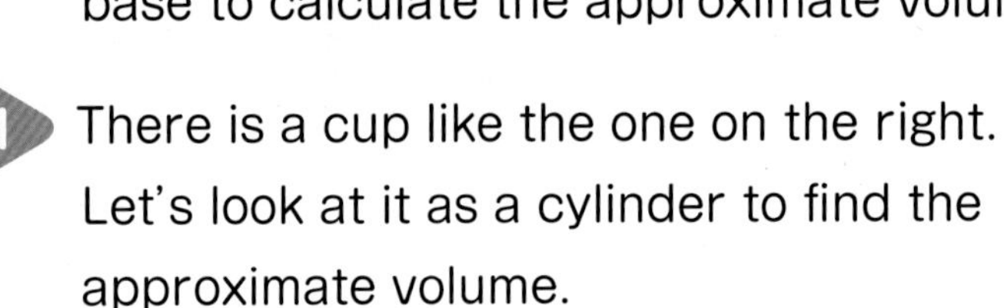

\ Want to try /

Purpose Let's find the approximate volume of various objects.

Haruto

2 Let's find out the approximate volume or capacity of objects around you.

CAN What can you do? 

□ We can find the volume of a prism and a cylinder. → pp.144 ~ 147

1 Let's find the volume of the following prisms and cylinder.

①

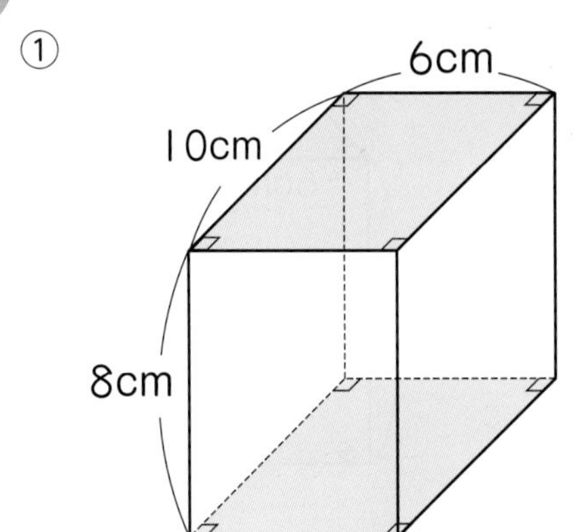

②

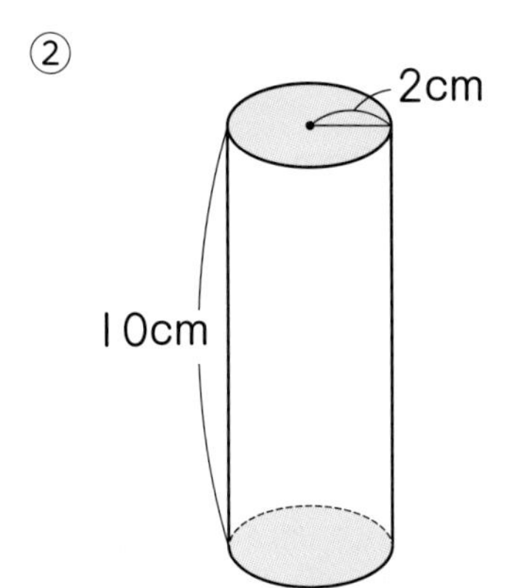

③

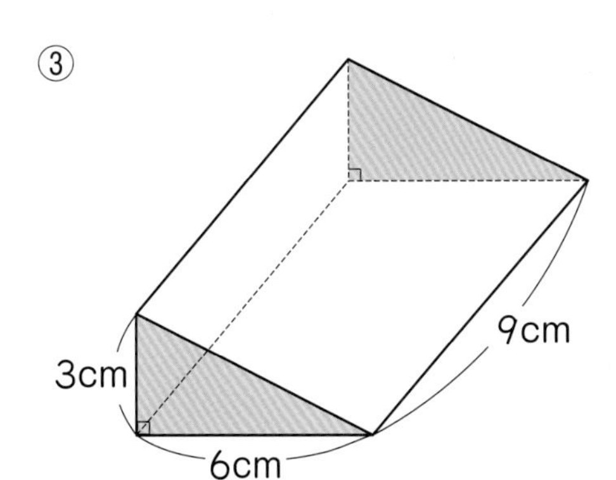

④

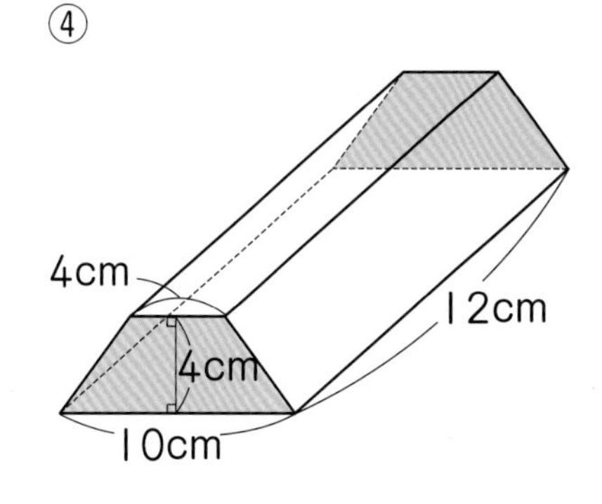

□ We can use the formula to find the volume of different solids. → p.149

2 The figure on the right shows a solid with faces intersecting perpendicularly.
Let's find out the volume of this solid.

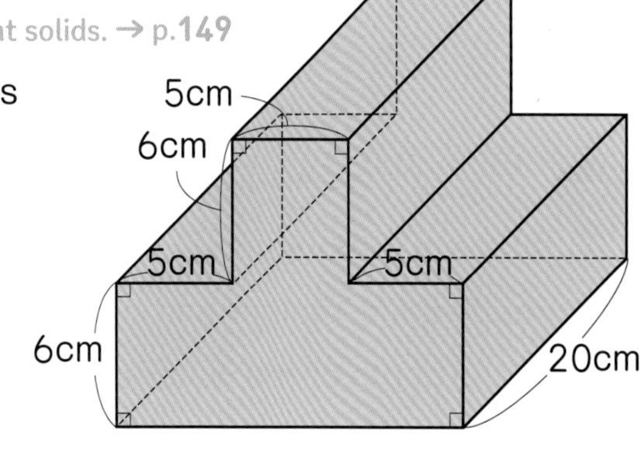

Supplementary Problems → p.240

Which "Way to See and Think Monsters" did you find in " 10 Volume of Solids"?

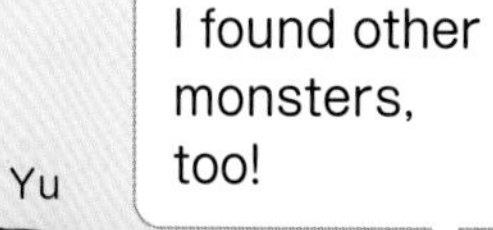

When thinking about how to find volume, I found "Summarize."

Yu
I found other monsters, too!
Akari

Utilize Usefulness and Efficiency of Learning

1 Let's find out the volume of the following solids.

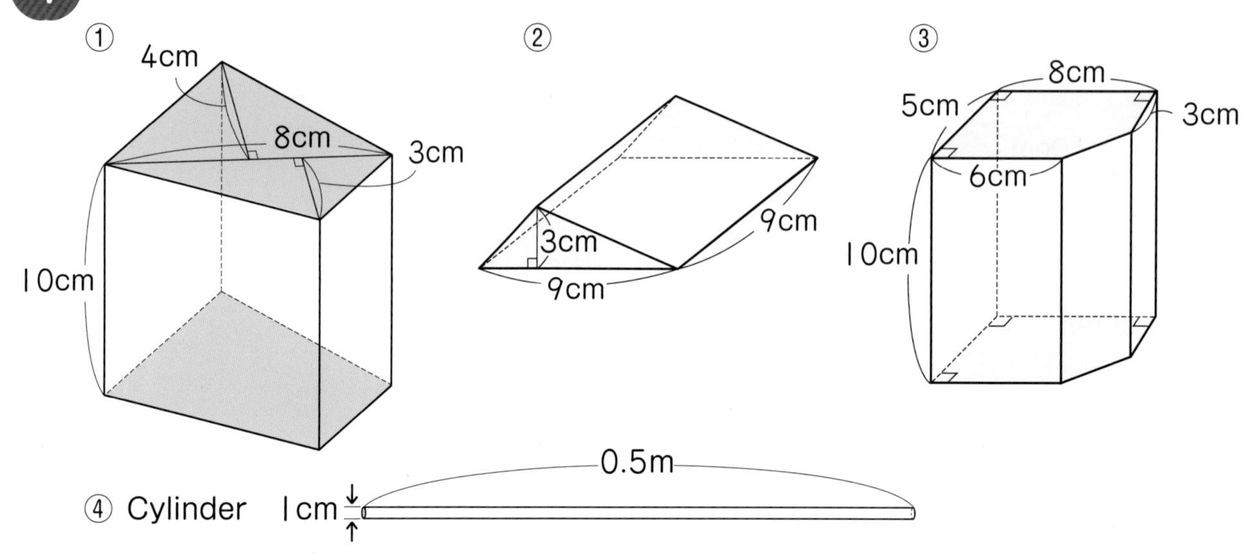

2 The figure on the right shows a solid with faces intersecting perpendicularly. Let's find out the volume of the solid.

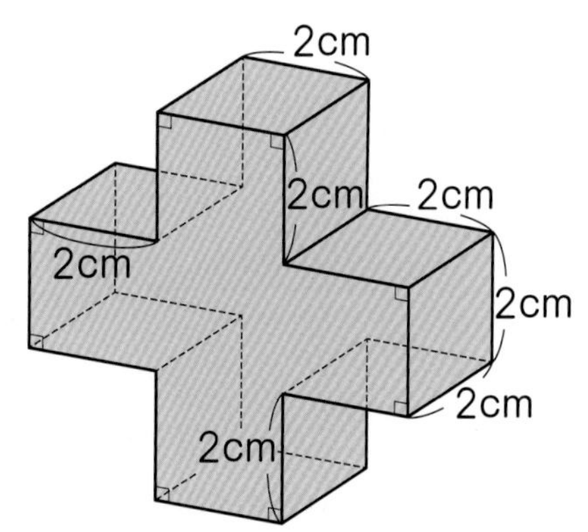

3 Figure 1 on the right shows a container in the shape of a triangular prism filled with water. If we rotate the container and place it as shown in Figure 2, the depth of the water changes to 5 cm. With this information, answer the following questions.

① Find out the volume of water inside the container. Do not consider the thickness of the container.

② What is the depth of the water on Figure 1 in centimeters (cm)?

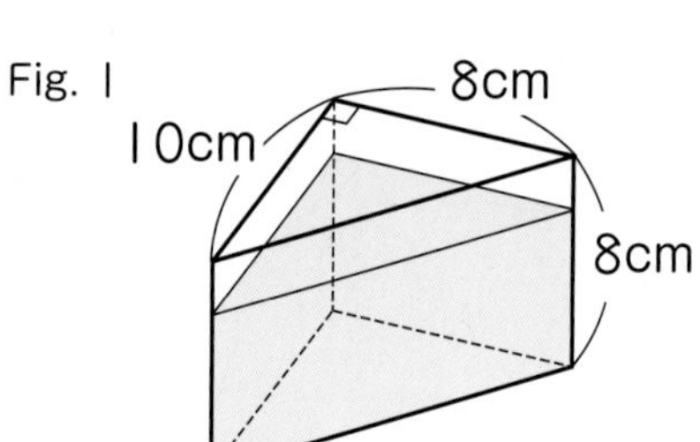

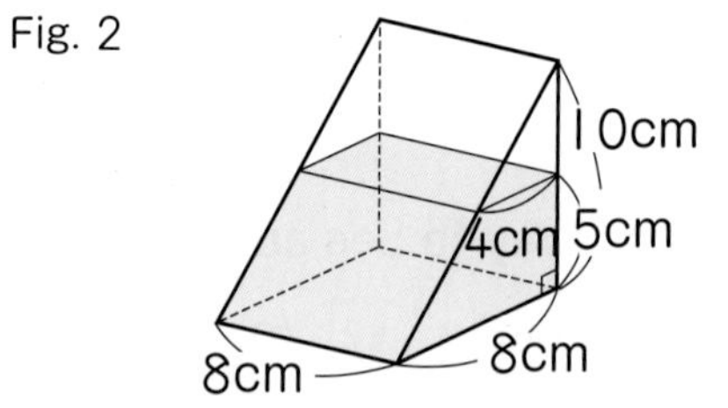

With the Way to See and Think Monsters...

Let's Reflect!

Let's reflect on which monster you used while learning "10 Volume of Solids."

Summarize

When we summarized how to find the volume of a prism or a cylinder, we noticed that they can all be obtained by (area of the base) x (height).

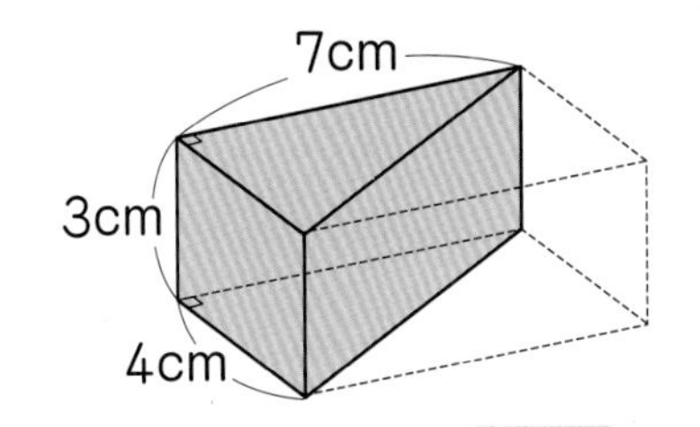

$7 \times 4 \times 3 \div 2 =$ ☐ (cm^3)

Volume of cuboid

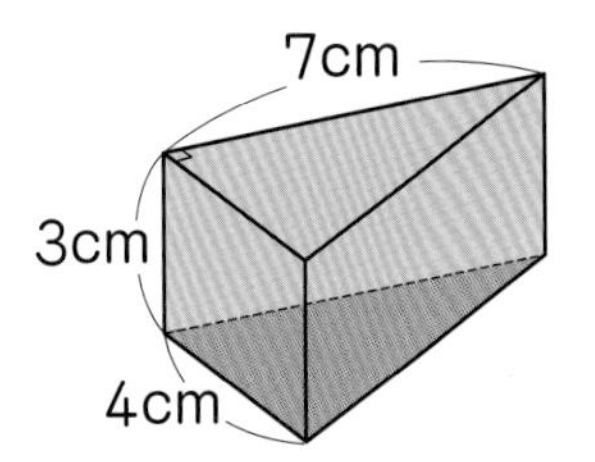

$7 \times 4 \div 2 \times 3 =$ ☐ (cm^3)

Area of the base

The volume of the triangular prism was the same whether we considered it as half of volume of the cuboid or using the area of the base.

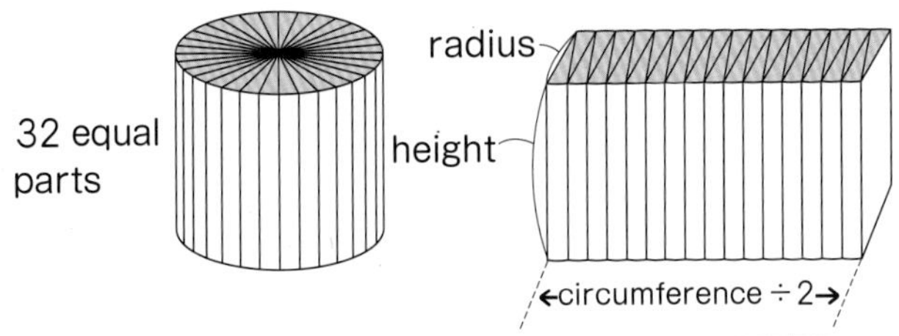

$(3 \times 3 \times 2 \times 3.14 \div 2) \times 5 =$ ☐ (cm^3)

Radius | Half of the circumference | Height | Volume

$(3 \times 3 \times 3.14) \times 5 =$ ☐ (cm^3)

Area of the base | Height | Volume

The volume of the cylinder was the same whether it was divided to make it a cuboid or considered using the base area.

Solve the ?

For both prisms and cylinders, if the base area and height are known, we could find out the volume.

Want to Connect

Are there any other solids that we can find out the volume of?

Reflect

Connect

Problem Let's draw a regular dodecagon.

How should we draw it?

Use a circle to divide the center angle into 12 equal parts.

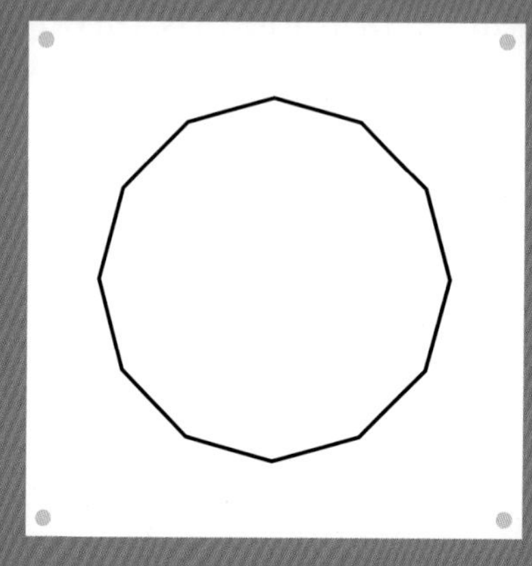

You can think based on one angle of a regular dodecagon.

If you divide the center angle of a circle into 12 equal parts ...

$360 \div 12 = 30$ 30°

I could draw!

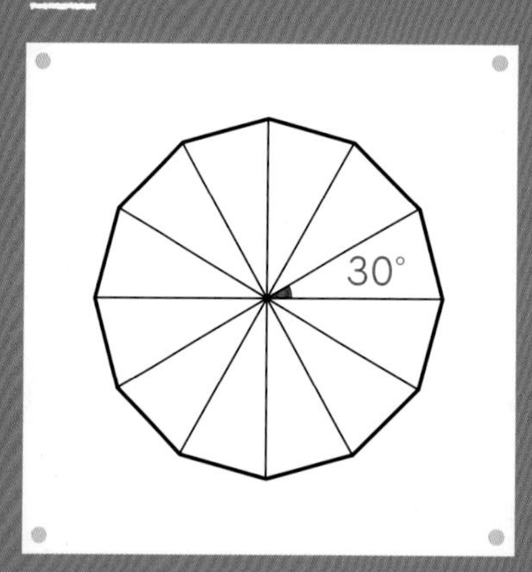

There are a lot of congruent figures.

How many degrees is one angle of the regular dodecagon?

The area is?

How should we draw a regular dodecagon? After drawing it, let's try to explore various things about the regular dodecagon.

Yu

Using a circle, the center angle can be divided into 12 equal parts.

Sara

We can also draw by considering the size of one angle.

Haruto

The number is a divisor of 12.

Can it be divided into congruent figures?

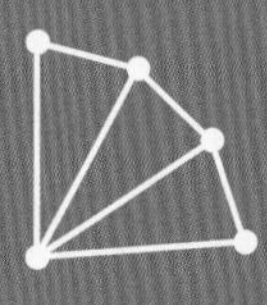

There are (12) isosceles triangles.

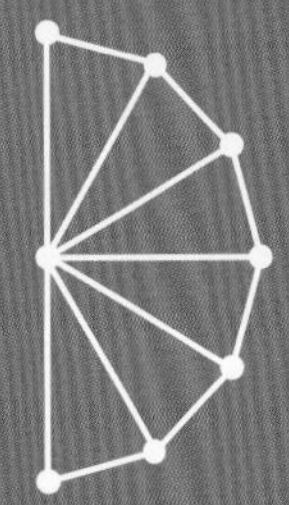

There are (6) quadrilaterals.

There are (4) pentagons.

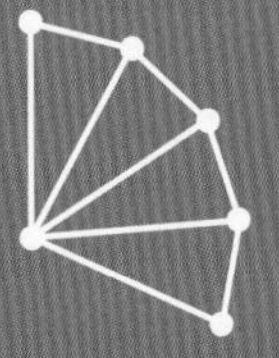

There are (3) hexagons.

There are (2) heptagons.

One angle is?

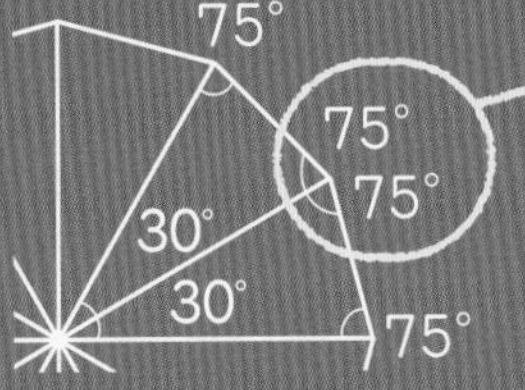

In total, 150°

What about the area?

If you gather a regular hexagon as shown on the right, you can see the quadrilateral ABCD. Let's try to think about this quadrilateral ABCD.

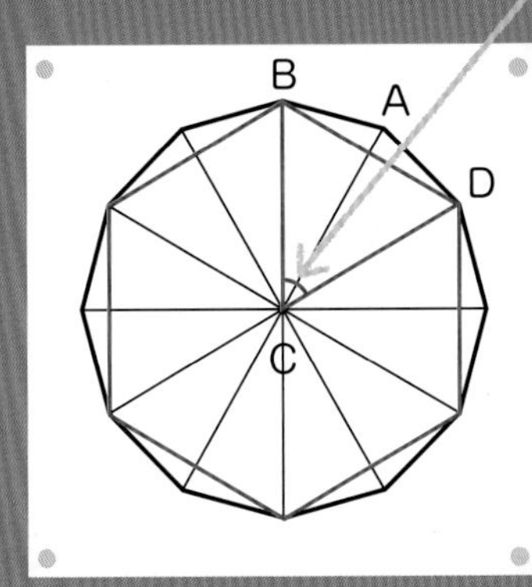

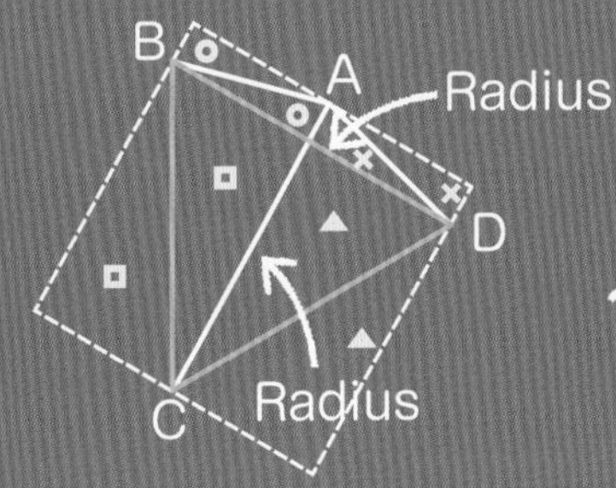

A square can be made using two times the quadrilateral ABCD.

A hexagon is two times the quadrilateral ABCD.

The area of two times the quadrilateral ABCD is (radius) × radius.

→ The area of the (hexagon) is radius × radius.

↓

Since there are three hexagons, radius × radius × 3.

Since it's an equilateral triangle, the length of BD is the same as the radius.

What kind of congruent figures can it be divided into? What is the size of each angle?

If we know the radius of the original circle, we can easily find out the area of the regular dodecagon.

Want to Connect

Can we draw a regular dodecagon using a circle, triangle ruler, and a compass?

Continue at junior high school!

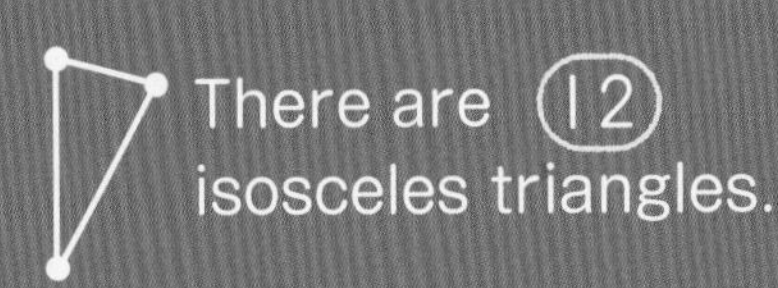

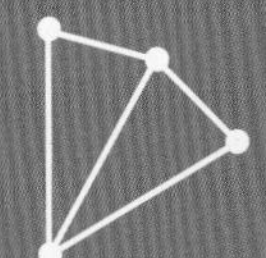

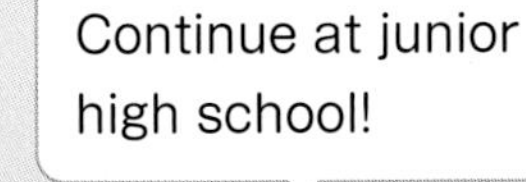

Reflect Connect

Problem

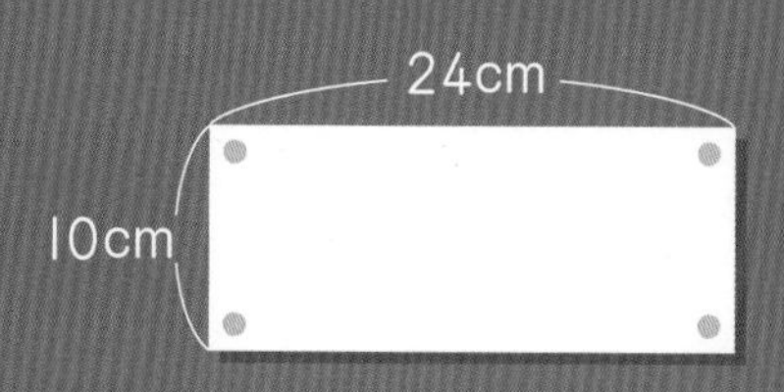

With the rectangle on the left, a prism or cylinder can be made by folding it or rolling it by the 10 cm sides until they overlap. Let's make the largest volume prism or cylinder following this method.

What kind of solid can you make?

Triangular prism, quadrangular prism, cylinder, ...

Is the volume the same if the lateral area is the same?

Quadrangular prism

1cm, 11cm, 1cm, 11cm, 10cm

$1 \times 11 \times 10 = 110$

110 cm³

The volume increases as the shape of the base approaches a square.

Change

2cm, 10cm, 10cm	3cm, 9cm, 10cm	4cm, 8cm, 10cm	5cm, 7cm, 10cm	6cm, 6cm, 10cm
$2 \times 10 \times 10 = 200$	$3 \times 9 \times 10 = 270$	$4 \times 8 \times 10 = 320$	$5 \times 7 \times 10 = 350$	$6 \times 6 \times 10 = 360$
Area of the base	Area of the base	Area of the base	Area of the base	Area of the base
200 cm³	270 cm³	320 cm³	350 cm³	360 cm³

Which solid has the largest volume?

Akari

I think the volume is the same because the lateral area is the same.

Let's examine by changing the length of the sides of the quadrangular prism by 1 cm.

Yu

Triangular prism I can do a lot, but...

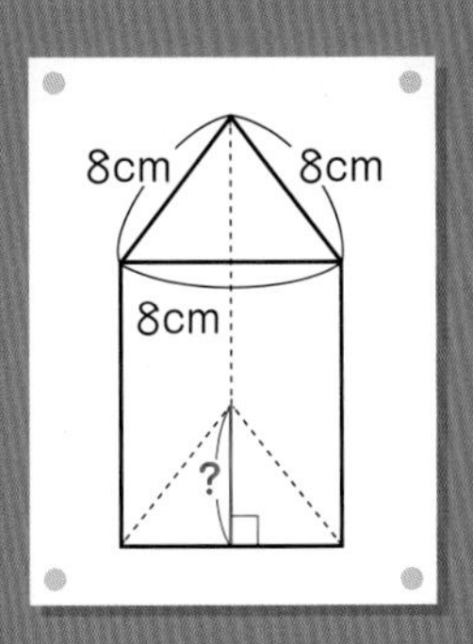

For example, if the shape of the base is an equilateral triangle, the height of the base is unknown so the area cannot be found.

I can find the area if it is a right triangle.

Will the volume be the largest if the shape of the base is a square?

↓

What happens if the shape of the base is a circle?

・Let's make the base triangle with the sides of 6 cm, 8 cm, and 10 cm.

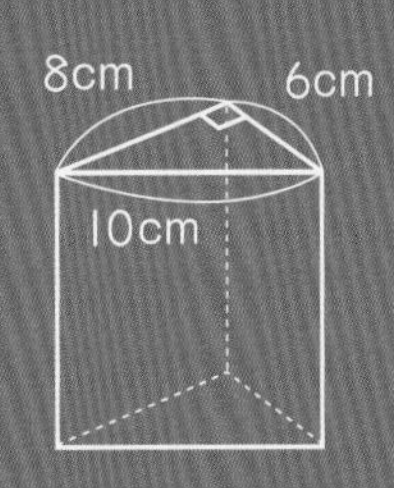

It will be a right triangle.

$6 \times 8 \div 2 \times 10 = 240$

Area of the base

240 cm³

Cylinder

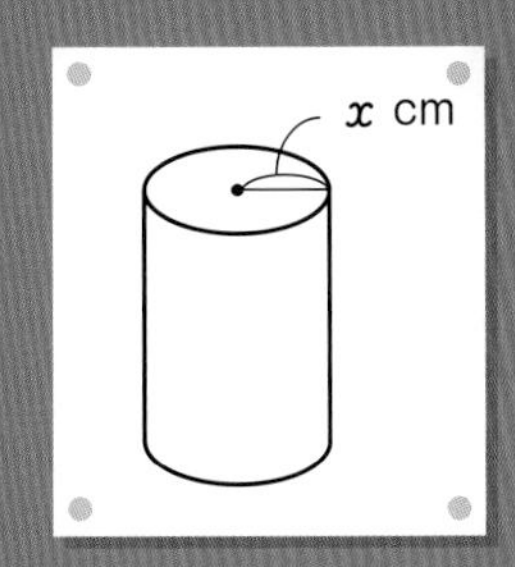

Width of the rectangle = Circumference = 24 cm

If the radius is x cm

$x \times 2 \times 3.14 = 24$

$x = 24 \div 6.28 = 3.82\cdots$

Considering a radius of about 3.8 cm,

$3.8 \times 3.8 \times 3.14 \times 10 = 453.416$ About 453.416 cm³

Area of the base

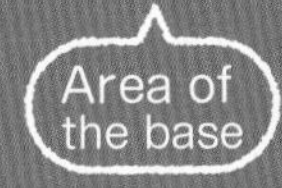

Just think in terms of the area of the base.

If the length around the base is the same, the area of the circle will be the largest.

Summary

- Even if the lateral area is the same, the volume is different.
- When the height is the same, the volume of the solid with the largest area of the base is the biggest.

Haruto: The volume increases as the shape of the base gets closer to a square. But, how about a triangular prism?

Sara: Is the volume the biggest when the shape of the base is a square?

Akari: Can the area of the base be bigger than the area of the circle?

Can we make the same taste?

We are cooking at Haruto's house.

French dressing (for 2 servings)

Vinegar: 4 teaspoons
Salad oil: 6 teaspoons

First, let's make the French dressing.
I want to make it for 4 people, but it only lists the ingredients for 2.
1

[For 2 servings]

Vinegar: 4 teaspoons
Salad oil: 6 teaspoons

2 times ➡

[For 4 servings]

Vinegar: 8 teaspoons
Salad oil: 12 teaspoons

For 4 people, we should consider twice as much as for 2.
Can we really say that the taste will be the same with that idea?
2

What methods can we use to make it taste the same?

11 Ratio and its Applications

Let's think about how to express by ratio and how to use it.

1 Ratio and ratio value

\ Want to think /

Purpose What are the proportions of each ingredient?

Haruto

Proportions ↓

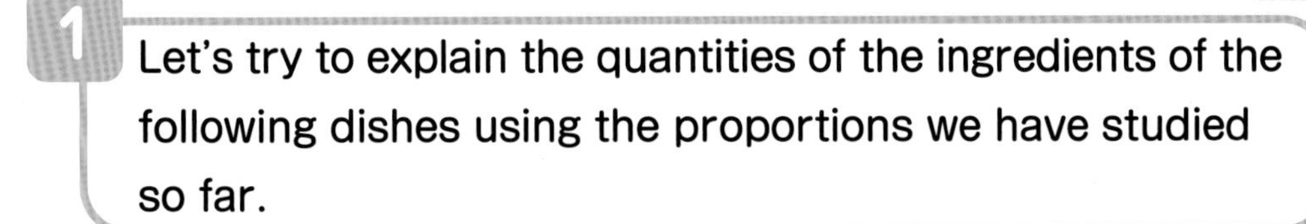

1 Let's try to explain the quantities of the ingredients of the following dishes using the proportions we have studied so far.

French dressing (for 2 servings)	Rice (for 3 servings)	Miso soup (for 3 servings)
Vinegar··· 4 teaspoons	Rice···300 mL	Water···450 g
Salad oil···6 teaspoons	Water···360 mL	Miso···50 g

In miso soup, the amount of water needed is 9 times as much as miso.

If rice is 1, then the water is ...

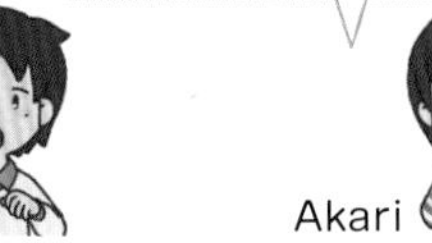

Yu

Akari

We have learned how to express proportions:
- A is □ times B.
- If A is 1, then B is □.
- If A is 100%, then B is □%.

? Are there other ways to express proportions?

Ratio and ratio value ↓

2 Let's consider the ratio of vinegar to salad oil when making the french dressing.

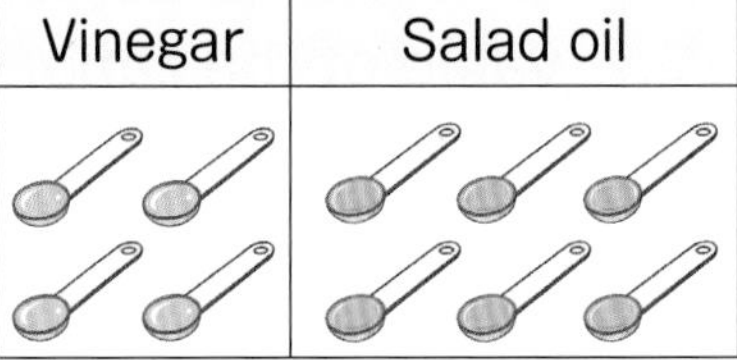

① How can the ratio of the amount of vinegar to the amount of salad oil be represented using two numbers?

Sara

We could use multiples and percentages to express proportions....

Haruto

The amount of vinegar and salad oil...

\ Want to know /

? Purpose Can we express the ratio of two quantities in a different way than with multiples and percentages?

"When the quantity of vinegar is 4, then the quantity of salad oil is 6" can be represented by using the " : " symbol as follows: **4 : 6** and it can be read as **"four to six"**. This way of the representation is called **ratio**.

4 : 6 is also read "ratio of 4 to 6".

❷ How many times the quantity of vinegar is of salad oil? Let's represent it with a fraction.

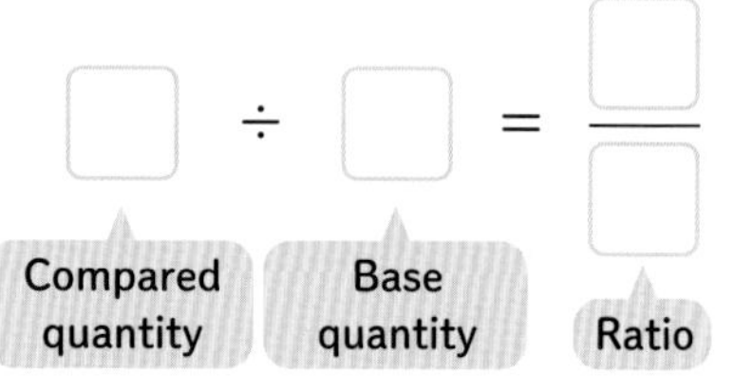

When a ratio is represented as $a : b$, the quotient of a divided by b is called the **value of ratio**. The value of ratio represents how many times b is a.

Value of ratio $a : b$ is the quotient of $a \div b$.

In particular, if a and b are whole numbers, the value of ratio $a : b$ can be represented as $\frac{a}{b}$.

The ratio of two quantities can be expressed as ratios and with the value of the ratio.

1 Let's represent the following expressions as ratios and let's find the value of the ratio.

① 450 g water and 50 g miso

② 300 mL rice and 360 mL water

③ 3 tablespoons of vinegar and 5 tablespoons of salad oil

④ 36 g of tomato sauce and 43 g of mayonnaise

? After all, can we say that the French dressing for 2 and 4 servings that Haruto came up with on page 158 tastes the same?

2 Equal ratios

1 On page 158, Haruto thought that if the ingredients for French dressing for 2 servings are "4 teaspoons of vinegar and 6 teaspoons of salad oil," then the ingredients for 4 servings are "8 teaspoons of vinegar and 12 teaspoons of salad oil." Let's see if this idea is correct.

① Let's find the value of the ratio of the amounts of vinegar and salad oil for 2 and 4 servings.

\ Want to know /

Purpose I wonder if we can say that they are the same ratio or not.

Akari

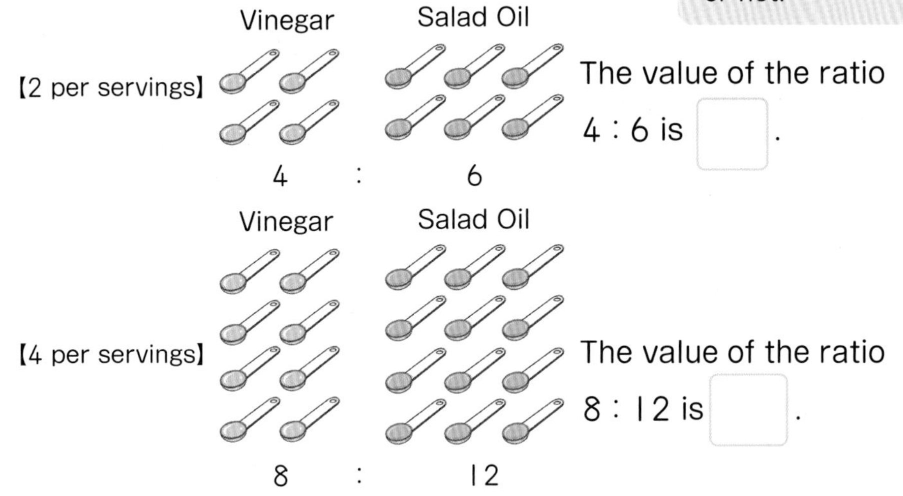

The value of the ratio 4 : 6 is ☐.

The value of the ratio 8 : 12 is ☐.

Way to see and think

The ratio of vinegar to salad oil is represented with equal-sized teaspoon.

② Is the consistency of the French dressing the same for 2 and 4 servings considered in ① ?

When the value of the ratio is equivalent, like 4:6 and 8 : 12, the **two ratios** are considered **equal**, and can be written as follows: $4 : 6 = 8 : 12$

Summary If the ratio values are equal, we can say that the proportions are the same.

Yu

1 Which of the following coffee milk has the same concentration?

Ⓐ 90 mL of coffee and 60 mL of milk

Ⓑ 600 mL of coffee and 200 mL of milk

Ⓒ 3 cups of coffee and 2 cups of milk

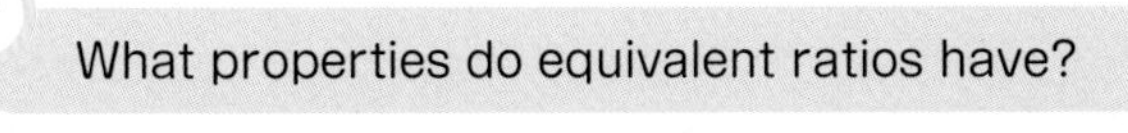

2 **There are 2 combinations of rice and water on the right. Let's examine the relationship between the ratios of rice to water.**

Ⓐ
Rice...60 mL
Water...72 mL

Ⓑ
Rice...300 mL
Water...360 mL

❶ From the ratios of rice to water in Ⓐ and Ⓑ. Let's find the values of ratio based on the water.

Value of ratio Ⓐ is $\frac{\square}{\square}$ Value of ratio Ⓑ is $\frac{\square}{\square}$

\ Want to know /

? Purpose **What is the relationship between the 2 equal ratios?**

❷ Let's explore the relationship between the 2 equal ratios.

60 : 72 = 300 : 360
60 : 72 = (60 × □) : (72 × □)
= 300 : 360

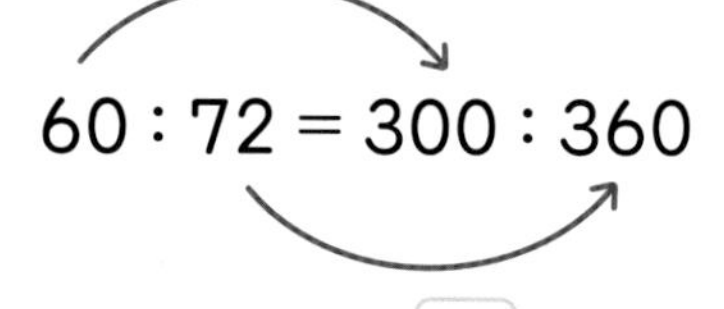

On the other hand,
300 : 360 = 60 : 72
300 : 360 = (300 ÷ □) : (360 ÷ □)
= 60 : 72

÷ □
300 : 360 = 60 : 72
÷ □

! Summary

The ratio a : b can be divided or multiplied by the same number to create a new equal ratio.

Way to see and think

1 From the following ratios choose the equal ratio to 3 : 1.

① 6 : 3 ② 6 : 2 ③ 1 : 3 ④ 13 : 10 ⑤ 9 : 3

2 Let's write 3 ratios that are equal to 6 : 9.

? How can we find equal ratios?

3

A drink is made by mixing 120 mL of water and 30 mL of syrup. If you have 60 mL of syrup, how much water in mL you need to make the drink with equal concentration?

If you need x mL of water,

$120 : 30 = x : 60$

$x = 120 \times$ ☐

$=$ ☐

× ☐

$120 : 30 = x : 60$

× ☐

Way to see and think: To make the same concentration, we need to make equivalent ratios.

\ Want to think /

Purpose Can we find the appropriate value of x to make the equivalent ratios?

Sara

1 Pancakes are made with 200 g of pancake mix and 160 g of milk. If you have 100 g of pancake mix, how much milk in g you need to make the recipe with equal concentration?

If you need x g of milk,

$200 : 160 = 100 : x$

$x = 160 \div$ ☐

$=$ ☐

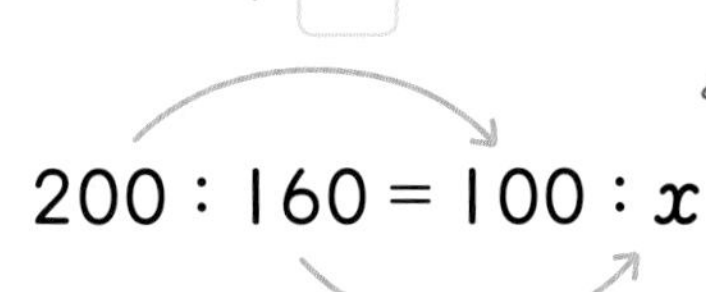

Way to see and think: Make equivalent ratios of pancake mix to milk.

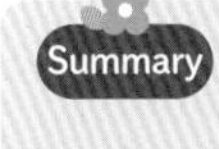

If we multiply or divide by the same number, we can find x.

Yu

2 Let's find out the appropriate value of x.

① $2 : 5 = x : 10$　　② $4 : 5 = 100 : x$

③ $12 : x = 3 : 5$　　④ $x : 20 = 5 : 4$

? There might be many equal ratios, but can we find the ratio with smaller whole numbers?

4 Let's find out a ratio that is equal to 12 : 18 and write it with the smallest whole numbers.

Akari's idea

$12 : 18 = (12 \div 2) : (18 \div 2)$
$= 6 : 9$
$= (6 \div 3) : (9 \div 3)$
$= 2 : 3$

Haruto's idea

$12 : 18 = (12 \div 6) : (18 \div 6)$
$= 2 : 3$

Both of them are using the equivalent ratio property.

Reducing a ratio into the smallest whole numbers without changing the value of the ratio is called **simplifying a ratio**.

1 Let's simplify the following ratios,

① $1.2 : 3.2 = (1.2 \times 10) : (3.2 \times 10)$

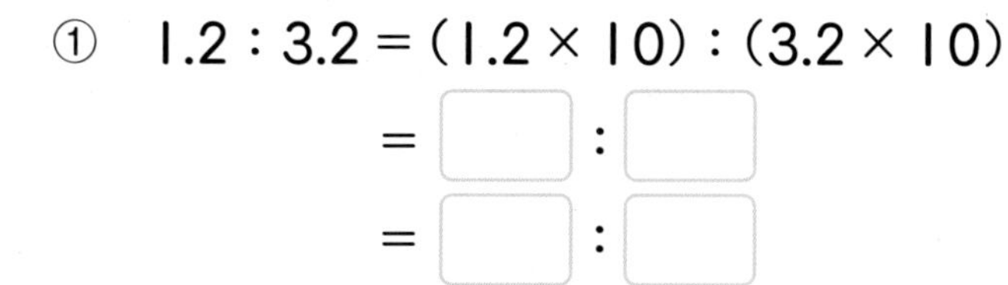

$= \square : \square$
$= \square : \square$

② $\frac{2}{5} : \frac{3}{8} = \frac{16}{40} : \frac{15}{40}$

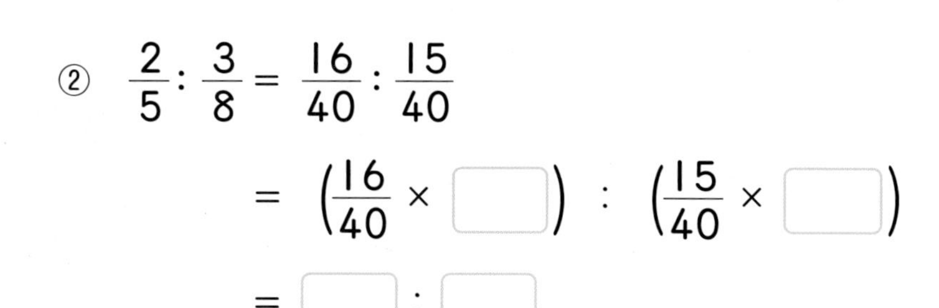

$= \left(\frac{16}{40} \times \square\right) : \left(\frac{15}{40} \times \square\right)$
$= \square : \square$

\ Want to think /

Akari

Purpose I wonder if we can simplify ratios that are decimals or fractions.

Summary Even if it is represented as a fraction or a decimal, it can be easily represented as a ratio of whole numbers.

Yu

2 Let's simplify the following ratios.

① 25 : 35 ② 7 : 28 ③ 180 : 120 ④ 0.6 : 2.9 ⑤ $\frac{3}{4} : \frac{2}{3}$

3 Let's measure the length and width of your desk and represent them as a ratio. Then, let's simplify.

? Can we use ratios to think about problems around us?

3 Applications of ratios

1 Let's find out the height of a tree from its shadow length.

① Pole A with a length of 0.8 m and pole B with a length of 2 m were placed vertically in the school yard. The lengths of their shadows were 1.2 m and 3 m, respectively. For pole A and pole B, let's compare the ratios of the pole length to the pole shadow.

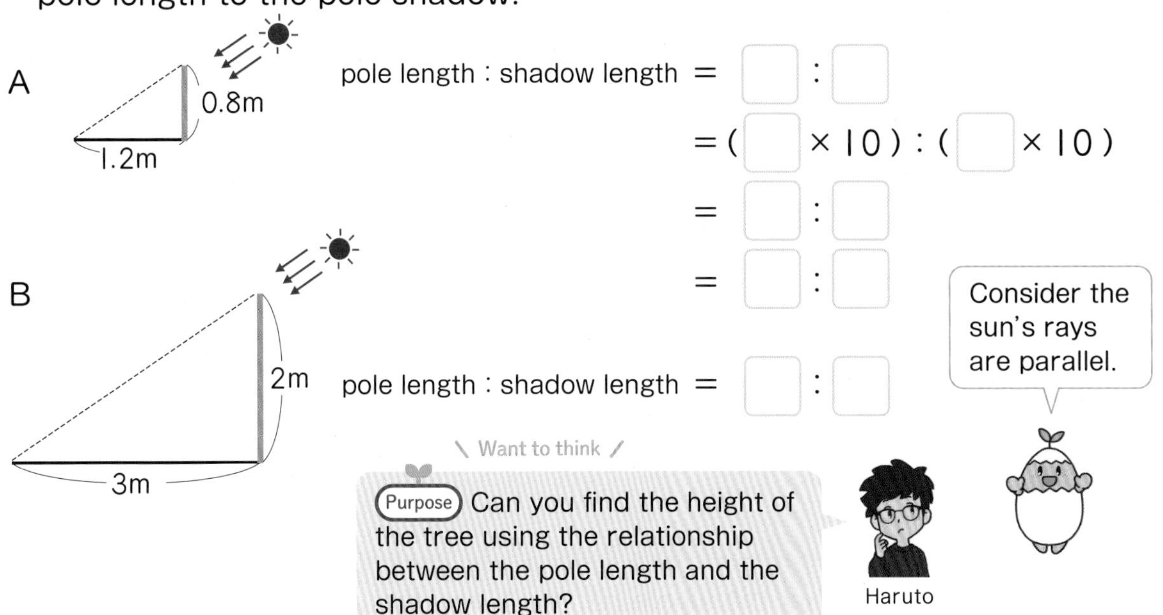

\ Want to think /

Purpose Can you find the height of the tree using the relationship between the pole length and the shadow length?

Haruto

② In this situation, what is the tree's height if the shadow is 12 m?
If the height of the tree is x m, let's write an equivalent ratio equation. Then, let's find the tree's height.

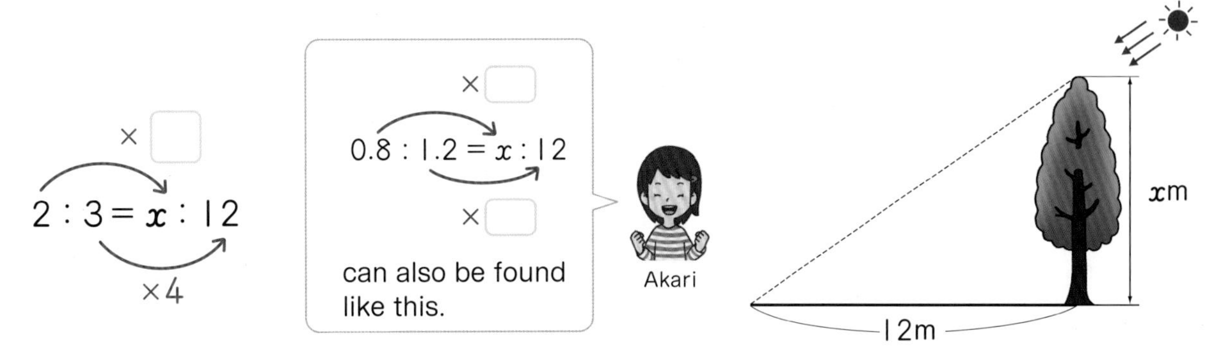

③ Let's find the height of the tree if the shadow length of the tree is 15 m.

2

A ribbon of 72 cm long should be divided between two sisters in the ratio 5 : 4. How many centimeters of the ribbon will each sister receive?

Want to know

? Purpose How can we divide the whole quantity into part-to-part ratios?

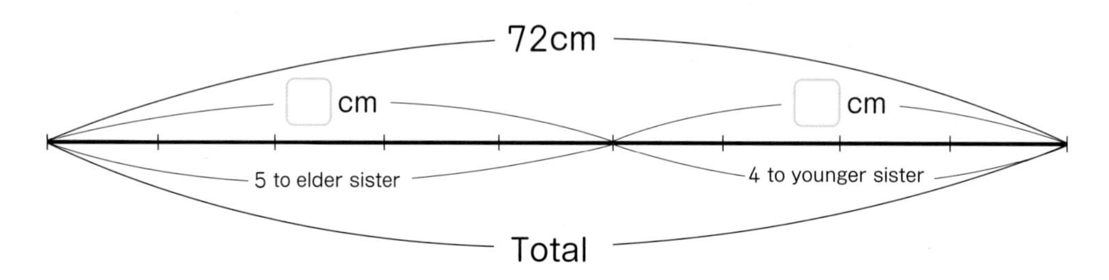

① Let's explain Haruto's and Sara's ideas to find out the answer.

Haruto's idea

To find the measure of the elder sister's ribbon, I used the ratio between her part and the whole. If the length of her ribbon is represented with x, then $5 : 9 = x : 72$

Way to see and think

If you have 5:4, then the total quantity is 5+4.

Sara's idea

I assumed that the whole part is 1 and thought about what part of 1 corresponds to the elder sister's ribbon.

The elder sister's part ... $\frac{5}{9}$ of the total $72 \times \frac{5}{9} = \square$

! Summary

We can divide the whole into ratios by using the ratio of part to whole, or by considering the whole as 1.

② Now, let's find the length of the younger sister's ribbon using the ratio.

1 Let's separate 400 mL of milk between Aoi and her father in the ratio 2 : 3. What is the amount of milk will Aoi receive in mL?

CAN What can you do?

☐ We can represent the ratio between two quantities as ratio and the value of ratio. → pp.159 ~ 160

1 Let's represent the following ratios as ratio and value of ratio.

① The quantity of vinegar and salad oil. ② The length of AB and AC in the triangle ruler.

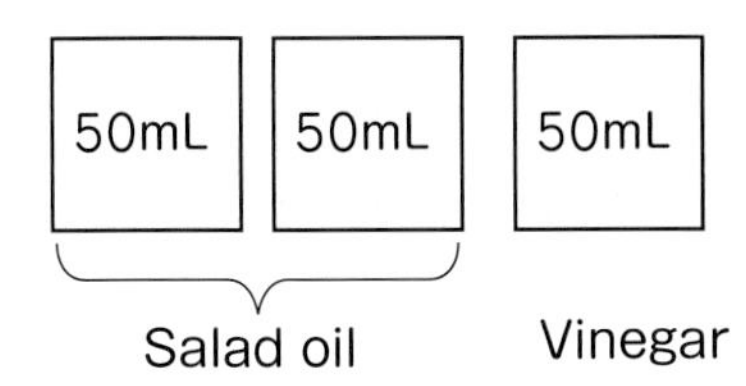

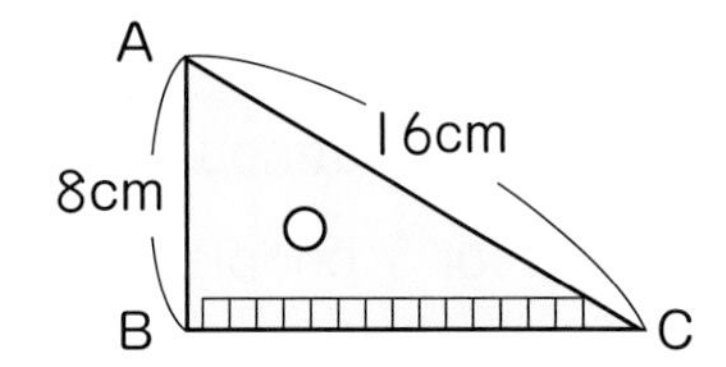

☐ We understand the properties of equivalent ratios. → p.163

2 Let's find the appropriate value for x.

① $3:5=x:10$ ② $7:4=35:x$

③ $80:x=5:8$ ④ $x:125=3:5$

☐ We can use properties of equivalent ratios. → p.163

3 The ratio between the length and width of a rectangle is 4 : 7.

① If the length is 28 cm, what is the width in cm?

② If the width is 28 cm, what is the length in cm?

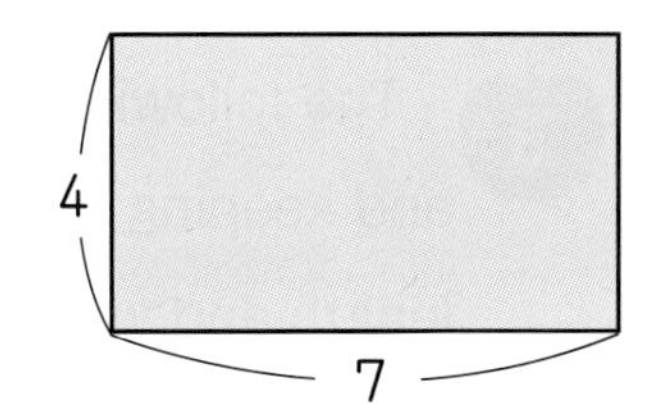

☐ We can simplify ratios. → p.164

4 Let's simplify the following ratios.

① $36:48$ ② $800:1400$ ③ $2.4:0.8$ ④ $\frac{1}{2}:\frac{2}{3}$

☐ We can divide by a ratio. → p.166

5 Let's cut a ribbon that is 2 m long by the ratio 2 : 3. What is the length of each section in cm?

Supplementary Problems → p.241

Which "Way to See and Think Monsters" did you find in "11 Ratio and its Applications"?

I found "Unit" when expressing equivalent ratios.

When you simplify the ratio....

Usefulness and Efficiency of Learning

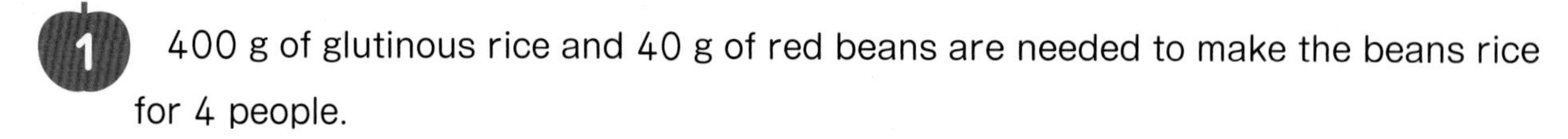

1 400 g of glutinous rice and 40 g of red beans are needed to make the beans rice for 4 people.

① How many grams of glutinous rice and red beans do you need respectively to make beans rice for 1 person?

② How many grams of glutinous rice and red beans do you need respectively to make beans rice for 7 people?

③ There are 600 g of glutinous rice. If you make the beans rice with the same ratio as for 4 people, how many grams of red beans would you need?

2 A lottery box is made so that the ratio of red balls to white balls is 3 : 4. If there are 28 white balls, how many red balls are there?

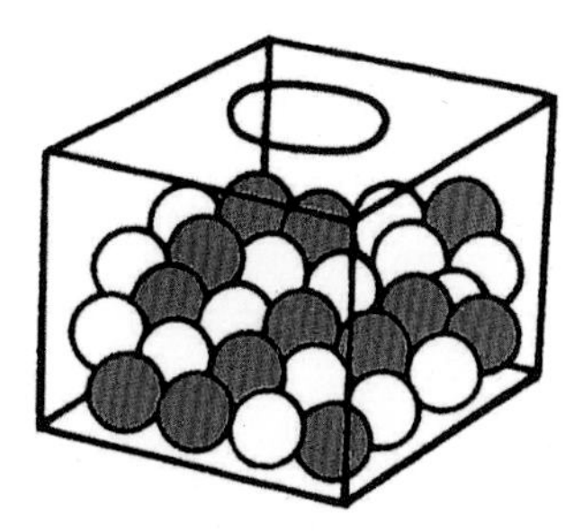

3 The following rectangles were made with a wire that is 60 cm long and keeping the ratios shown on each diagram. What should be the length and width in cm for each rectangle?

①

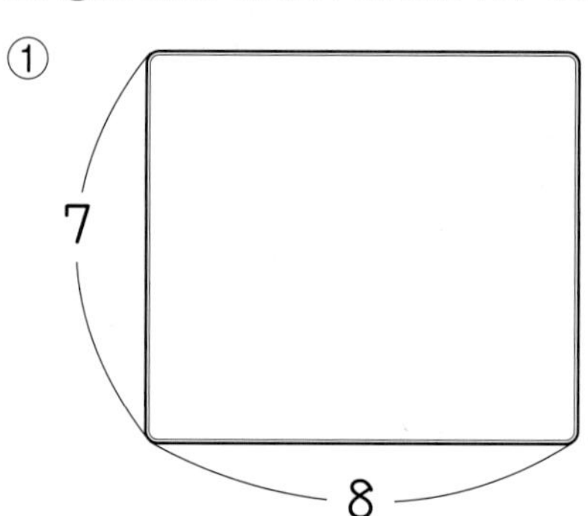

②

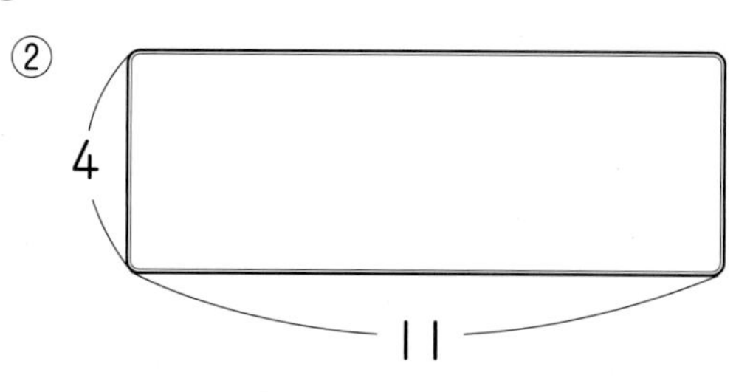

4 Two triangular rulers of different sizes overlap at the corner of a right angle, as shown on the right. Let's find the length of DE.

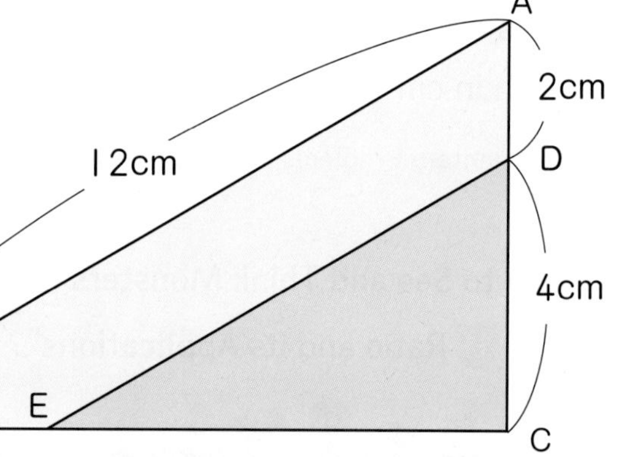

With the Way to See and Think Monsters...

Let's Reflect!

Let's reflect on which monster you used while learning "11 Ratio and its Applications."

Unit

After setting one unit and considering how many of each of the two quantities are equivalent to each other, we were able to express the relationship between the two quantities in terms of ratio.

French dressing (for 2 servings)
Vinegar···4 teaspoons
Salad oil···6 teaspoons

Teaspoons

Vinegar	Salad oil

Considering one teaspoon as one unit, the ratio of vinegar to salad oil can be represented as $\square : \square$.

The value of ratio is $\frac{\square}{\square}$.

Sara

Rule

Using the ratio properties, we are able to make large equivalent ratios and ratios of decimals into simple numbers.

$$12 : 18 = (12 \div 6) : (18 \div 6)$$
$$= 2 : 3$$

$$1.2 : 3.2 = (1.2 \times 10) : (3.2 \times 10)$$
$$= 12 : 32$$
$$= 3 : 8$$

To make the ratios equal, we used the fact that the ratio a:b is also equal if a and b are divided by the same number.

Haruto

Even if a ratio was expressed in decimal form, it could be converted to a whole number ratio using the same rule.

Yu

Let's deepen. → p.246

? Solve the ?

There are two ways of representing the ratio of two quantities: ratio and the value of ratio.

Yu

Want to Connect

When is it useful to use ratios?

Haruto

Find the

Do they have the same shape?

Let's summarize what we learned about life in the Jomon Period in social studies.
1

縄文時代のくらし
It would be easier to understand if you include a picture.
2

When I was trying to make the picture bigger, the shape changed. Which one has the same shape?
3

A man-made house looks like a trapezoid, so we can consider it as a trapezoid....
4

\ Want to know /

Purpose **Where can we check if they have the same shape?**

Let's explore the properties of figures with the same shape and how to draw them.

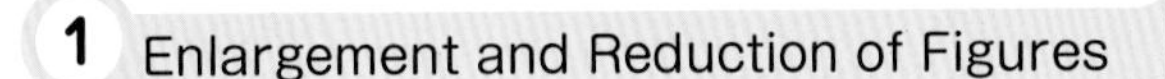

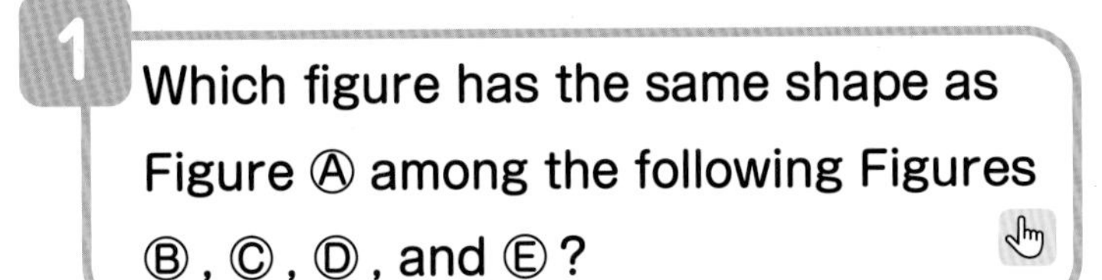

1
Which figure has the same shape as Figure Ⓐ among the following Figures Ⓑ, Ⓒ, Ⓓ, and Ⓔ?

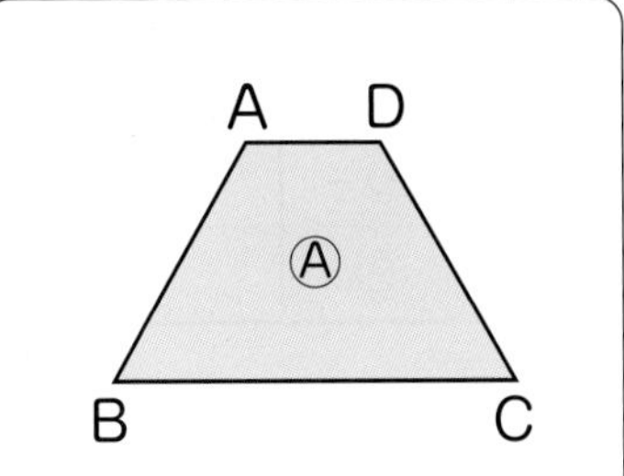

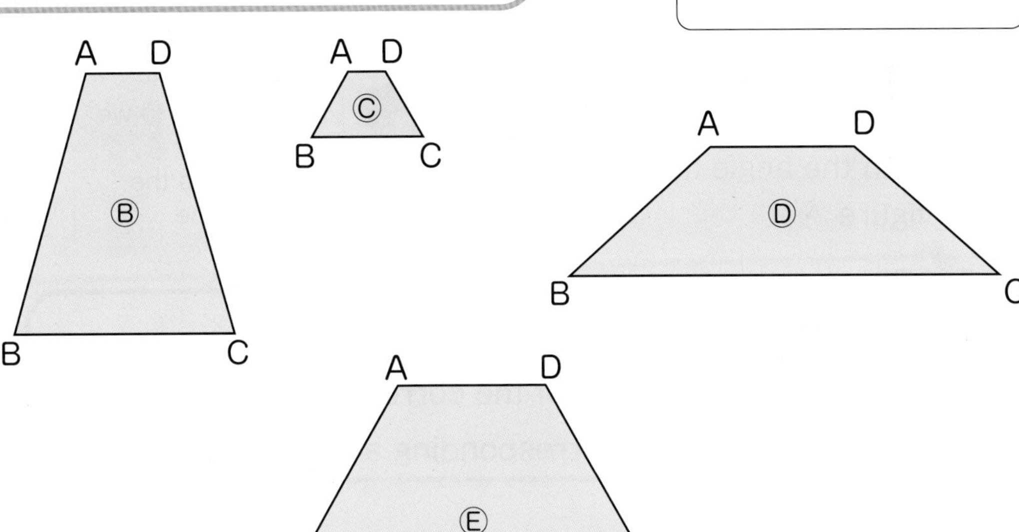

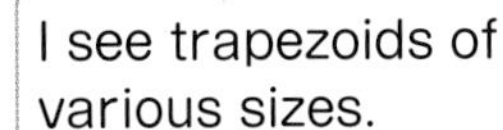

I see trapezoids of various sizes.
What do you mean by the same shape?
Some of them look like they have the same shape though with different sizes.
What are the lengths of the sides and the measurements of the angles?

① Let's measure the side lengths and the angles of the 4 figures on previous page and summarize your findings on the table below.

	Side length (cm)				Angle measure (°)	
	Side AB	Side BC	Side CD	Side DA	Angle A	Angle B
Ⓐ	2	3	2	1	120	60
Ⓑ						
Ⓒ						
Ⓓ						
Ⓔ						

Way to see and think

You have organized the figures by "side length" and "angle measure" and made a table with it.

② Let's compare the side length of figures Ⓑ~Ⓔ with the side length of Ⓐ.

③ Let's compare the angle measure of the figures Ⓑ~Ⓔ with the angle measure of figure Ⓐ.

Haruto: From ② and ③ we can see that Ⓐ, Ⓒ, and Ⓔ have the same shape.

Way to see and think

If you compare the sides length using a ratio, you can understand the property better.

Summary

You can check if they have the same shape by comparing the lengths of the corresponding sides and the measure of the corresponding angles.

? Do the same shapes have any properties?

2 **Figure Ⓐ and Ⓔ from the previous page are shown on the right. Let's investigate the length of the sides and the measure of the angles in detail.**

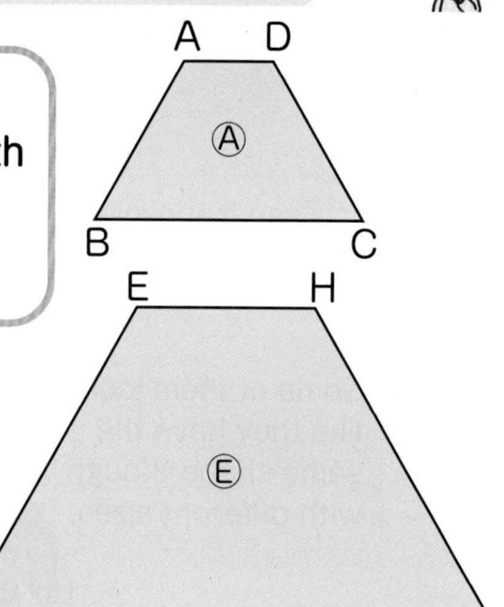

Akari: The side lengths are expressed in ratios that we have studied before...

Want to know

Purpose **What properties do the same shape with different sizes have on its sides and angles?**

① Let's represent the length of the corresponding sides of Figure Ⓐ and Figure Ⓔ with the simplest ratio.

Side BC : Side FG = ☐ : ☐

Side AB : Side EF = ☐ : ☐

Side DA : Side HE = ☐ : ☐

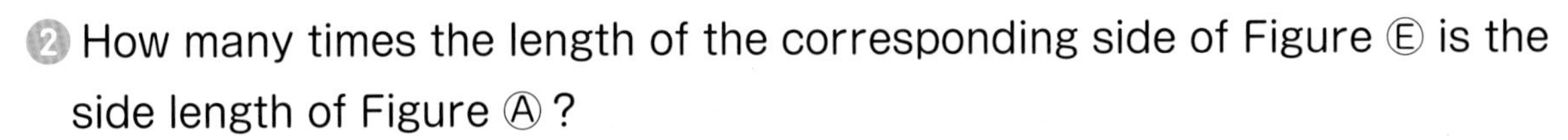

② How many times the length of the corresponding side of Figure Ⓔ is the side length of Figure Ⓐ ?

③ The diagonals AC and EG are corresponding straight lines. Let' s express the lengths of the two lines as the simplest ratio. How many times longer is the diagonal AC than the diagonal EG?

④ Let's compare the measure of the corresponding angles.

When the corresponding angles are congruent and all the lengths of the corresponding sides share the same ratio then it is called an **enlargement**. If the ratio is simplified, then it is called a **reduction**.

Summary

When we do enlargements or reductions, the length of corresponding sides are the same ratio and all the corresponding angles are equal.

Figure Ⓔ is **2 times** figure Ⓐ, so it is an **enlargement**.

Figure Ⓐ is $\frac{1}{2}$ of figure Ⓔ, so it is a **reduction**.

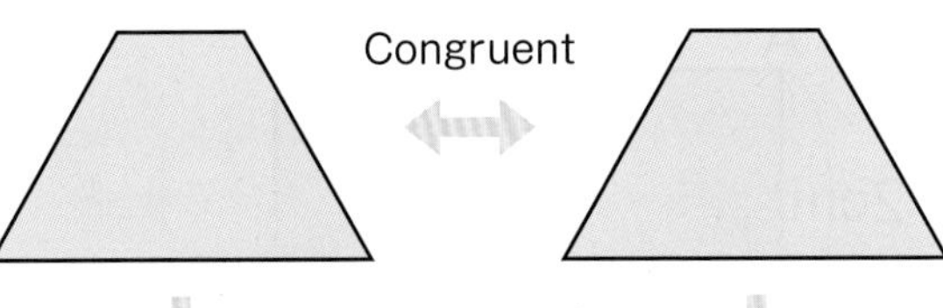

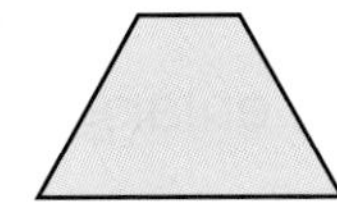

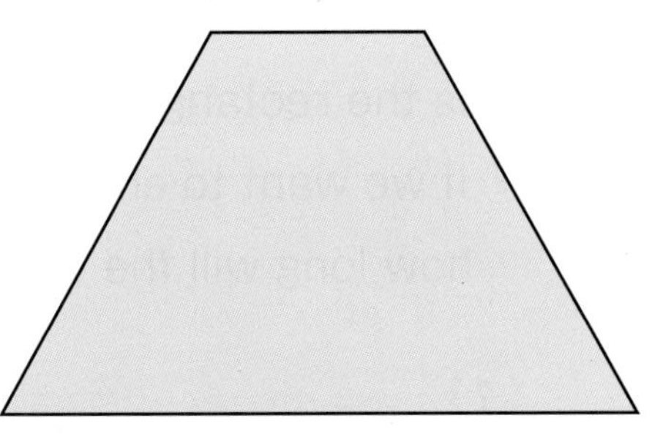

When two figures are congruent, the ratio of the lengths of the two corresponding sides is 1 : 1.

1 Which of the following figures is an enlargement or reduction of Figure Ⓕ? Also, how many times is the enlargement or reduction?

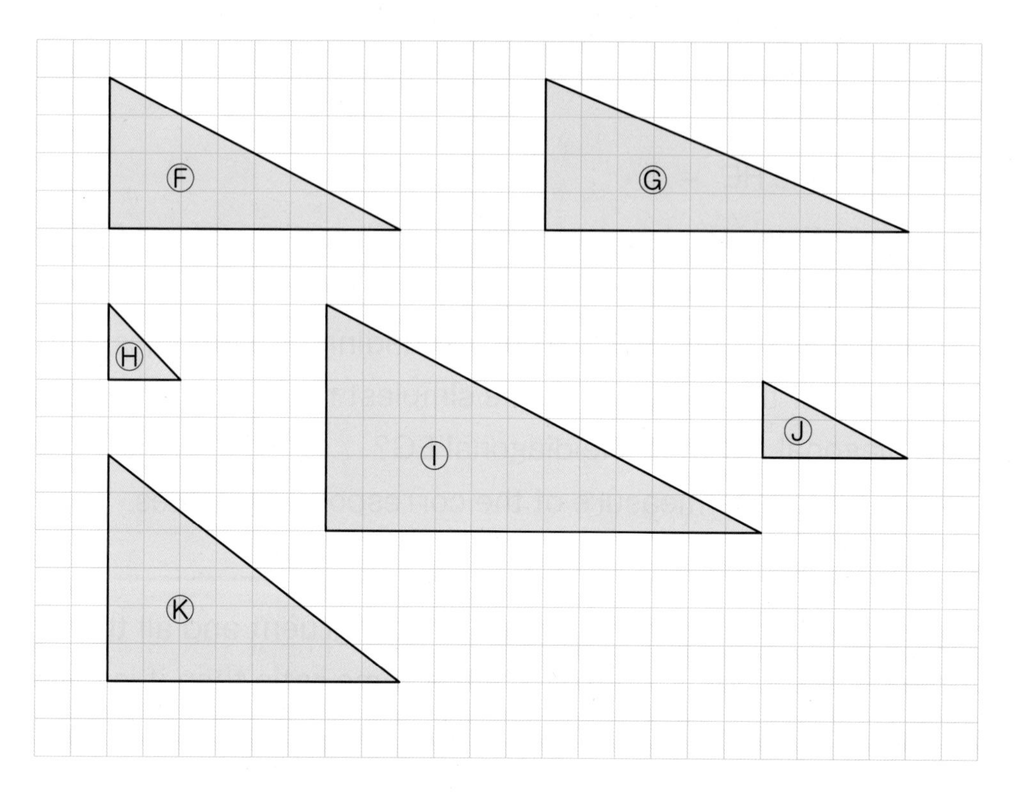

2 Rectangle EFGH was drawn by adding 1 cm to the length and also to the width of rectangle ABCD. Let's answer the following questions about the figures.

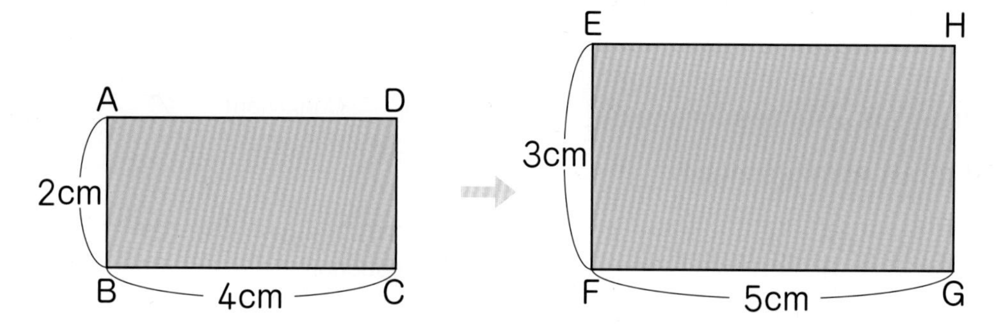

① Is the rectangle EFGH an enlargement of rectangle ABCD?

② If we want to enlarge rectangle EFGH 1.5 times of the rectangle ABCD, how long will the width be?

? Can we use what we have learned so far to draw an enlargement or reduction figure?

1 How to draw enlargement or reduction of figures

\ Want to know /

Purpose Can you use the grid to make enlargement or reduction of the figures?

1 Consider triangle DEF, which is an enlargement by 2 times of triangle ABC. The point E corresponding to point B is determined as shown in the following grid.

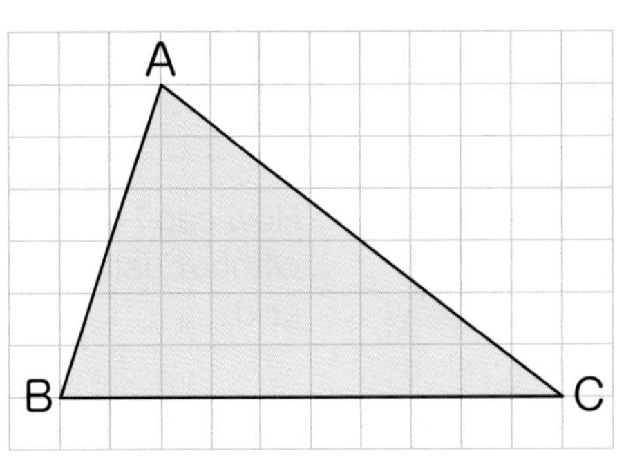

E

After drawing an enlargement or a reduction, confirm the length of the corresponding sides and the measure of the corresponding angles.

Summary Point A is 2 grid points right and 6 grid points up from point B. If you double to enlarge it, these points should also be double.

1 On the following grid, draw triangle DEF, which is $\frac{1}{2}$ of triangle ABC in **1**.

What lengths should be used for the height and the base, respectively?

? Is it possible to draw an enlargement or reduction without using a grid?

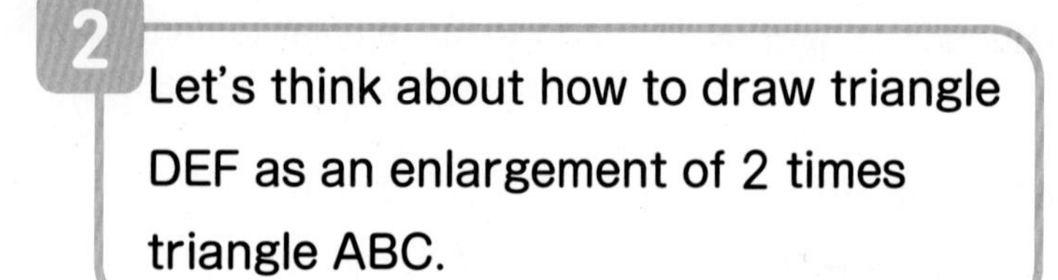

2 Let's think about how to draw triangle DEF as an enlargement of 2 times triangle ABC.

A

B C

How can I draw without using a grid?

How can we draw a congruent triangle?

\ Want to think /

Purpose How can I draw an enlargement of a figure?

① Line EF is twice line BC and it has been drawn below. Let's think about where should we place point D, corresponding to point A. Then, continue the drawing.

Which sides should I measure?

Which angles should I measure?

E F

❷ Let's explain the drawing ideas of the 3 children.

Sara's idea

The length of each of the three sides is doubled.

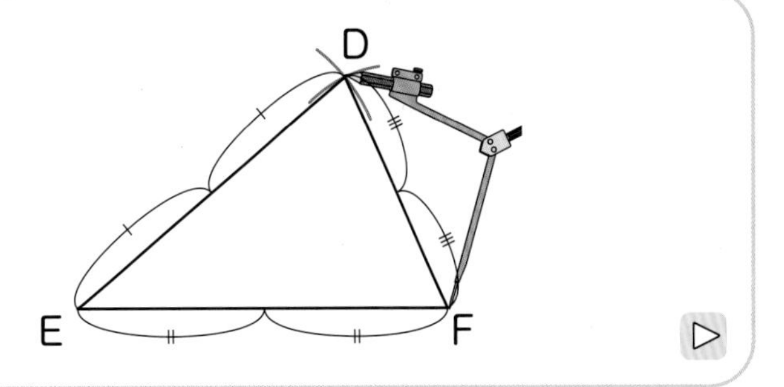

Yu's idea

I used the measure of the angle between them and doubled the lengths of the 2 sides to draw it.

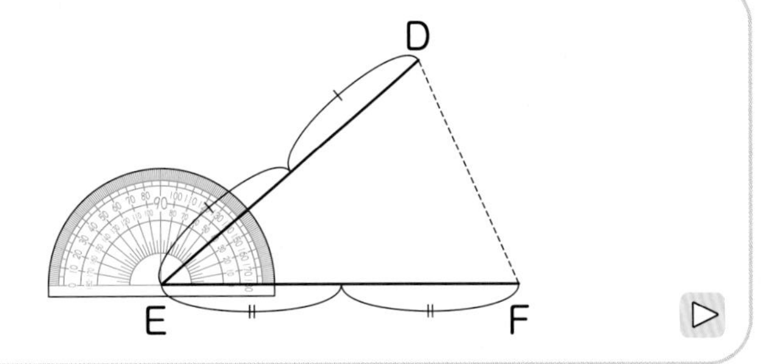

Akari's idea

I doubled the length of one side and used the measure of the two angles on either side of it.

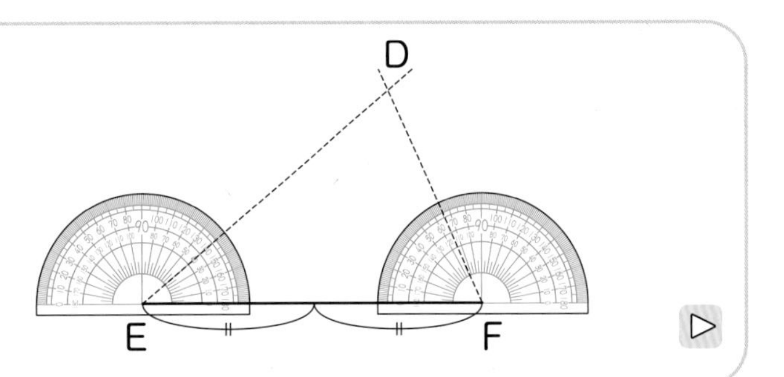

I can use the congruent triangle drawing.

The enlarged figure can be drawn using the congruent figure drawing method, multiplying the lengths of the sides by a factor and keeping the angles with the same size.

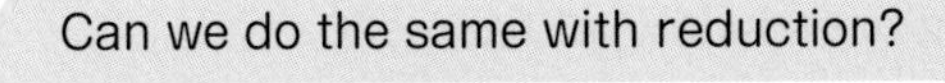

? Can we do the same with reduction?

3

Let's think about how to draw reduced triangle DEF, which is $\frac{1}{3}$ of the triangle ABC shown on the right.

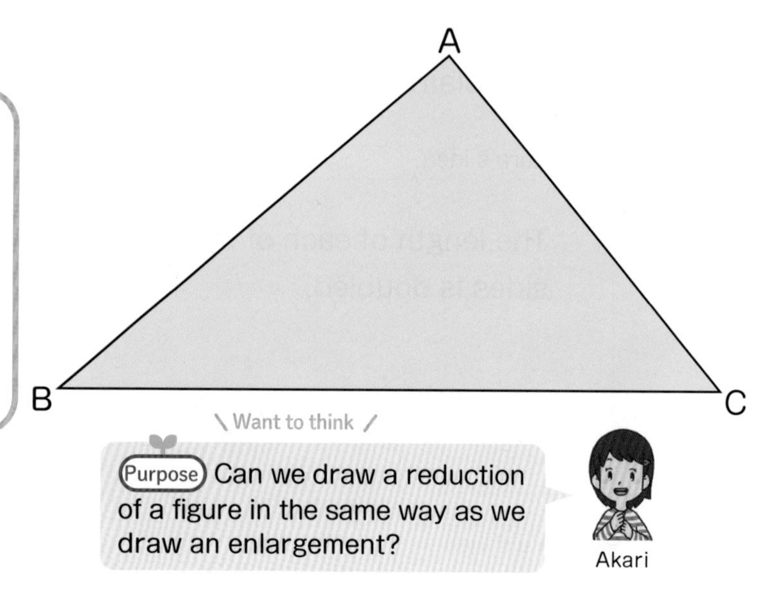

\ Want to think /

Purpose Can we draw a reduction of a figure in the same way as we draw an enlargement?

Akari

① Let's try to draw triangle DEF with your own method. Explain how you did it to a classmate.

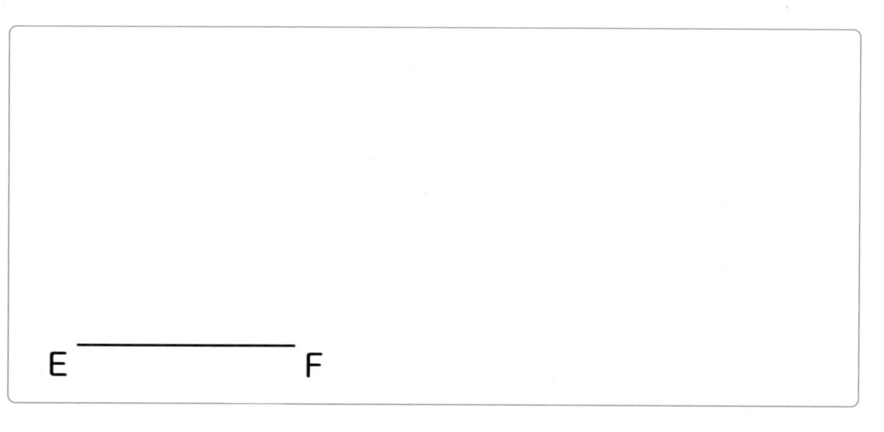

② Is your drawing method similar to those on the previous page?

1 Consider quadrilateral ABCD shown on the right. Let's draw an enlargement of 2 times and a reduction of $\frac{1}{2}$ to it in your notebook.

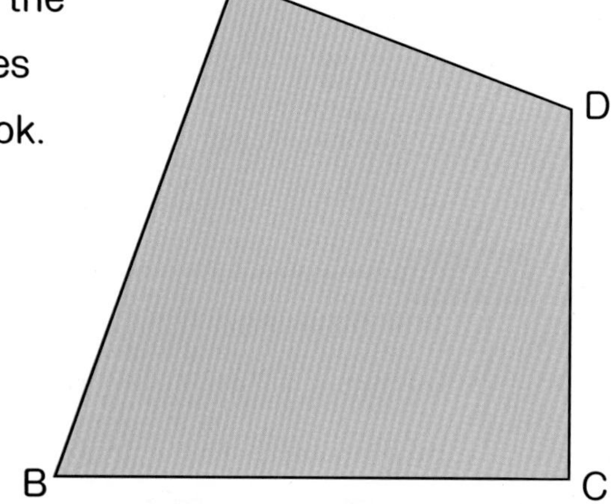

Summary We can make enlargement and reduction using the same method.

? Are there other ways to draw enlargement or reduction of figures?

4 Let's draw triangle DBE that is 3 times an enlargement of the triangle ABC shown on the right.

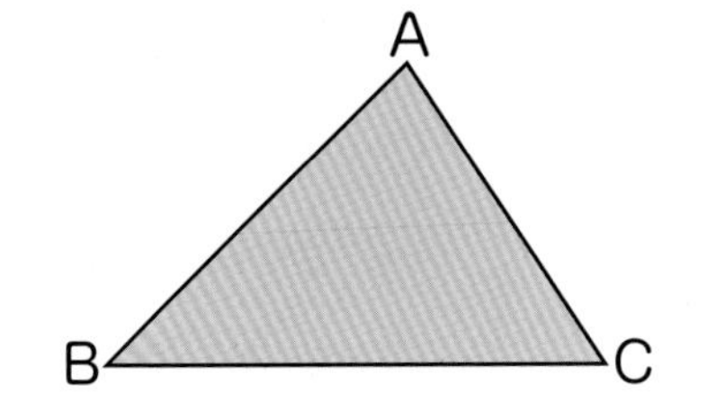

① First, let's extend side BA and draw point D, which is corresponding with point A. Now, let's extend side BC and draw point E, which is corresponding with point C.

\ Want to try /

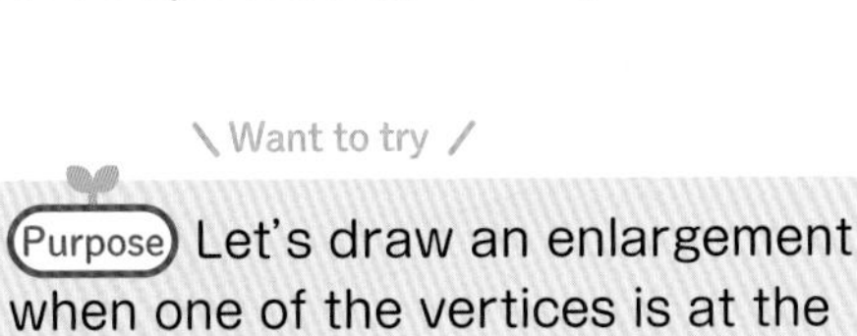

Purpose Let's draw an enlargement when one of the vertices is at the same position.

Sara

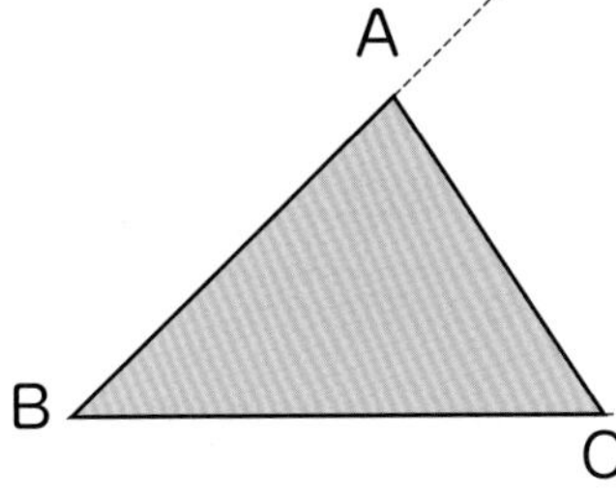

② Let's confirm that the enlarged triangle DBE is 3 times triangle ABC.

As shown above, a line connecting one vertex to another vertex can be used to draw an enlargement or reduction. This reference point is called **the center point** (**center of enlargement** or **reduction**).

1 Let's use the diagram in ① of 4 to draw a reduction of $\frac{1}{2}$ triangle ABC. Use point B as the center point.

2 Let' s use point B as the center point to draw an enlargement of 2 times and a reduction of $\frac{1}{2}$ quadrilateral ABCD.

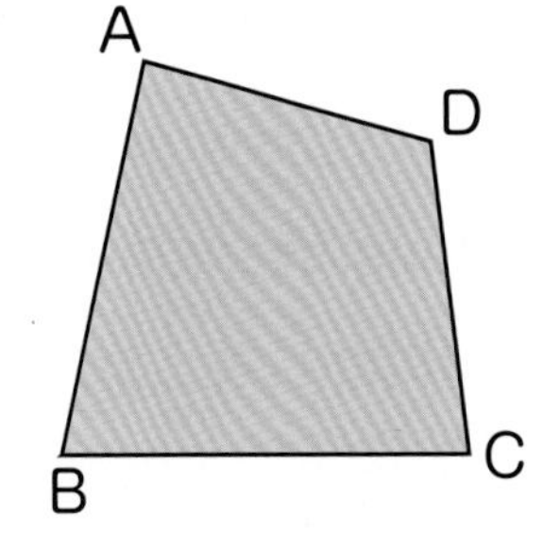

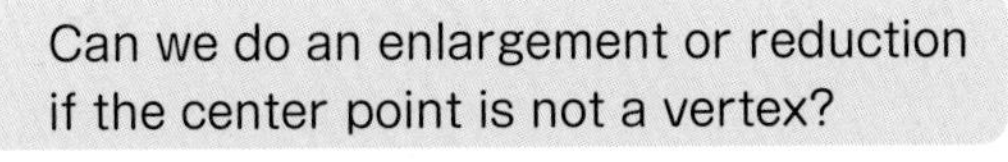

?
Can we do an enlargement or reduction if the center point is not a vertex?

Quadrilateral FGHJ is an enlargement of 2 times quadrilateral ABCD. Let's consider point E as the center point and think about how to draw an enlargement of it.

\ Want to think /

? Purpose I wonder if I can also do an enlargement or reduction centered on a point.

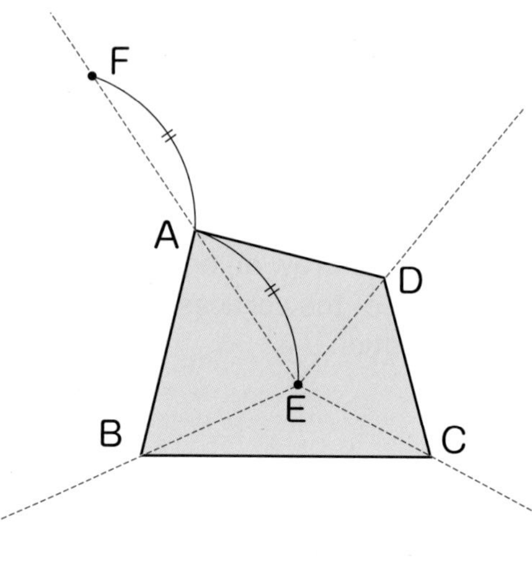

1 Straight line EA was extended and point F, corresponding to point A, was located. Let's continue and finish the drawing.

2 Why can we draw the enlargement using this method? Let's explain the reasons.

1 In the diagram from 5, let's use point E as the center point and draw a reduction of $\frac{1}{2}$ quadrilateral ABCD.

! Summary

No matter where you decide the center point, you can use the straight line to connect the point and each vertex to draw the enlargement or reduction.

2 Let's use point D as the center point to draw an enlargement of 2 times triangle ABC and a reduction of $\frac{1}{2}$ triangle ABC.

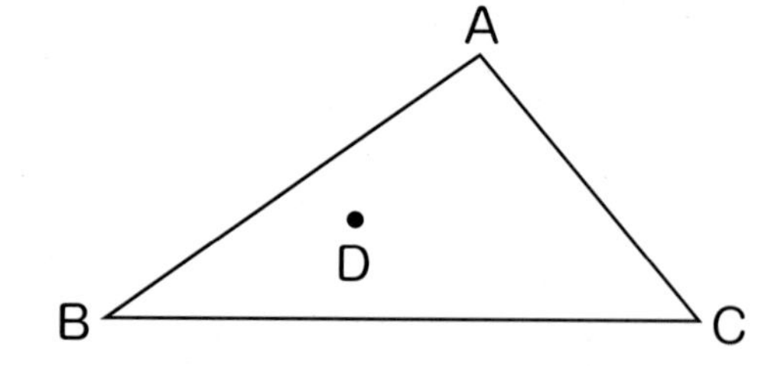

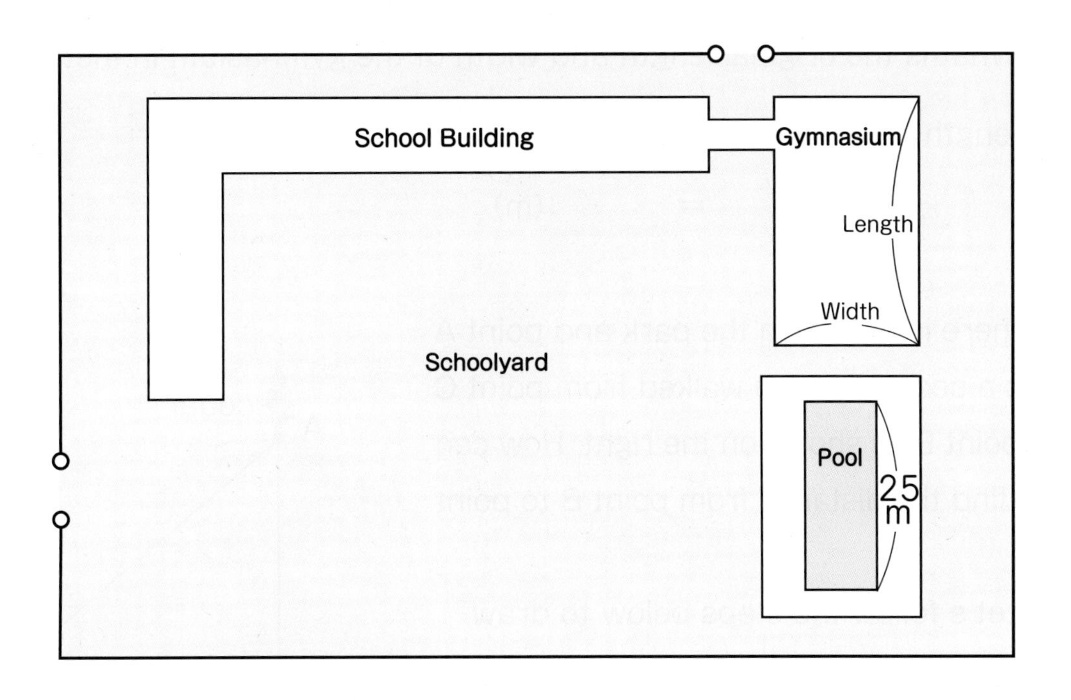

3 Applications of reductions

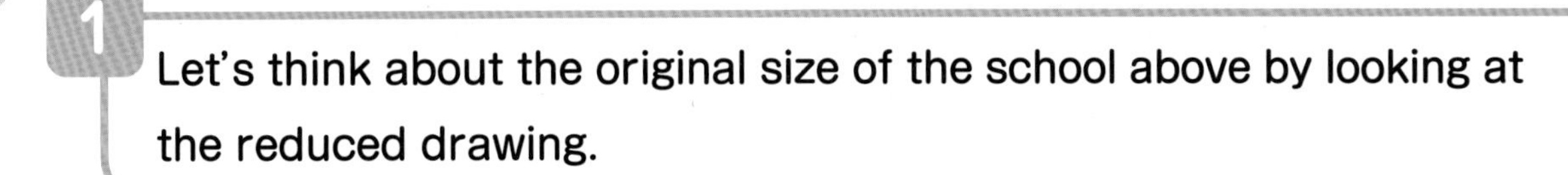

1 **Let's think about the original size of the school above by looking at the reduced drawing.**

① The original length of the pool is 25 m. What is the length of the reduced drawing in cm and mm? By how much was it reduced?

② How much is the original length of 1 cm of the reduced drawing?

The ratio that represents how much is reduced from the original length is called **reduced scale**. The picture above is using a $\frac{1}{1000}$ reduced scale. There are 3 ways to show a reduced scale:

Ⓐ $\frac{1}{1000}$　　Ⓑ 1 : 1000　　Ⓒ 0　10　20　30m

\ Want to know /

? Purpose **How can we use a reduced drawing?**

③ What is the original length and width of the gymnasium in meters?

Length$\cdots 3.3 \times 1000 = \square$ (cm)　Width$\cdots 2 \times 1000 = \square$ (cm)

$= \square$ (m)　　$= \square$ (m)

1 There is a pond in the park and point A has a cedar. Sakura walked from point C to point B as shown on the right. How can we find the distance from point B to point A?

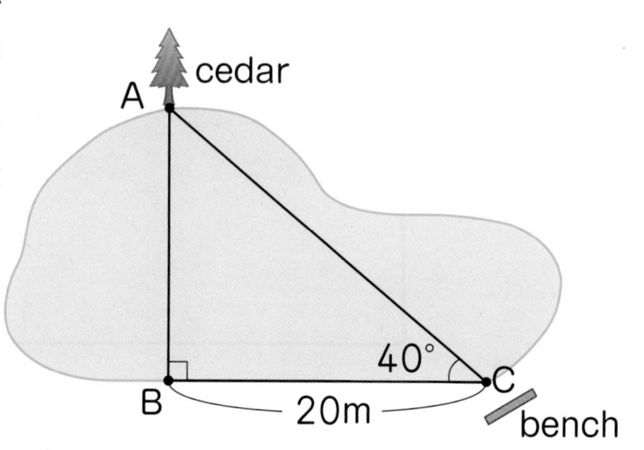

① Let's follow the steps below to draw a reduction of $\frac{1}{500}$ the right triangle ABC.

(1) Find the length of line BC with the new scale and draw it.
(2) Draw a perpendicular line to straight line BC passing through point B.
(3) Make angle C of 40° and locate point A.
(4) Draw the right triangle ABC.

② Let's measure straight line AB in the reduction of ① and find the original distance to the cedar.

Summary

Even if it is difficult to measure the original length of a drawing, it is possible to find the length by using a reduction of it.

2 In the figure on the right, what is the height of the tree in meters? Let's explain how to find out the answer.

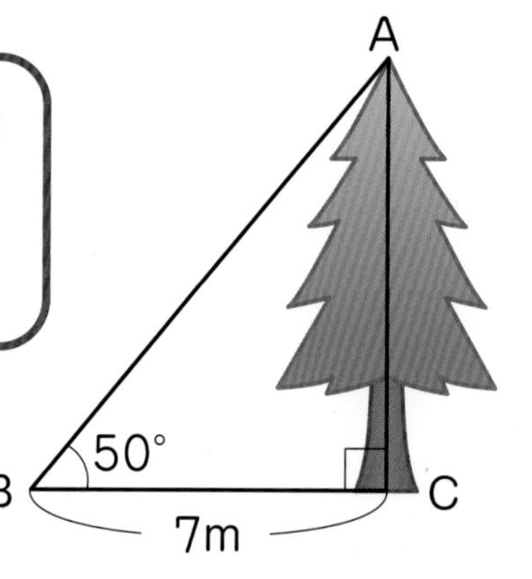

CAN What can you do? 

☐ We understand the properties of enlargements and reductions. → pp.171 ~ 174

1 From the following diagram identify which figures are an enlargement or a reduction of each other? Let's also explain the reason.

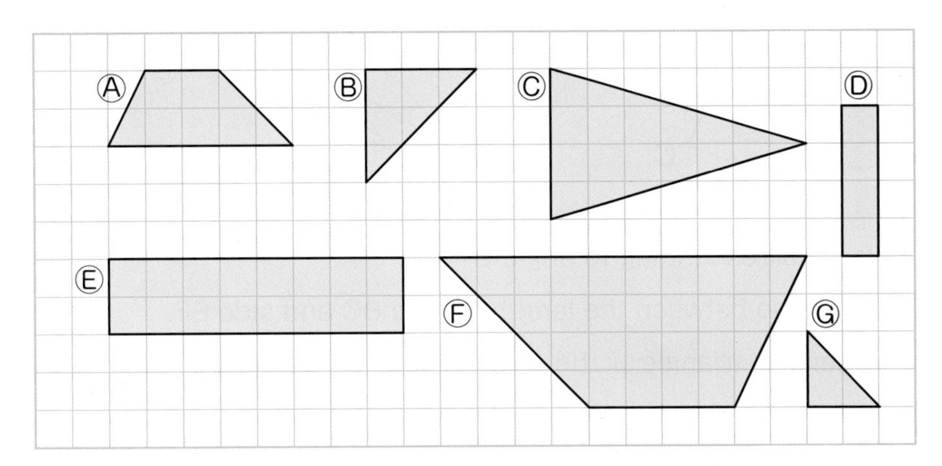

☐ We can draw an enlargement or a reduction of figures. → pp.178 ~ 180

2 Let's draw in your notebook an enlargement of 2 times triangle ABC and a reduction of $\frac{1}{2}$ triangle ABC.

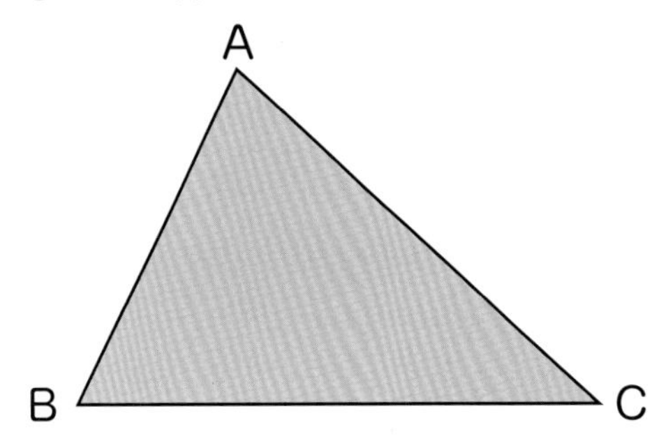

☐ We can use a reduction. → pp.181 ~ 182

3 There is a map of a school made with $\frac{1}{500}$ reduced scale. In the reduction, the gymnasium has a rectangular shape with a length of 6 cm and a width of 3.2 cm. What is the original length and width of the gymnasium?

Supplementary Problems → p. 242

Which "Way to See and Think Monsters" did you find in "12 Enlargement and Reduction of Figures"?

I found "Why" when I tried to explain.

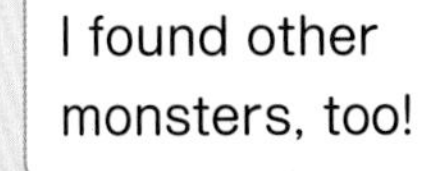

I found other monsters, too!

Usefulness and Efficiency of Learning

Triangle Ⓑ is an enlargement of triangle Ⓐ. Let's answer the following questions.

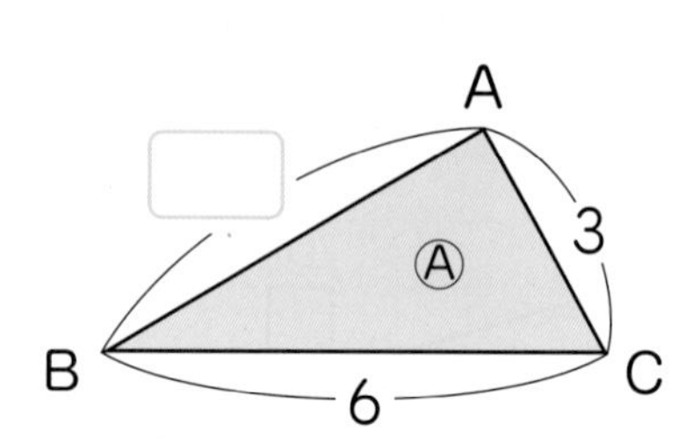

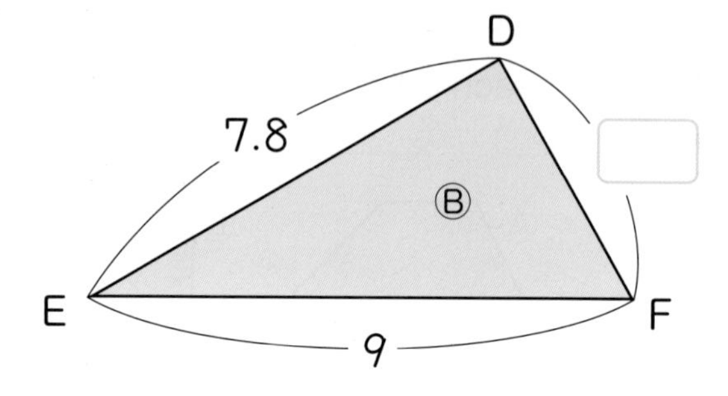

① What angle is corresponding to angle B?

② Let's find the ratio between the length of side BC and side EF.

③ How many times is triangle Ⓑ the enlargement of triangle Ⓐ?

④ Let's fill in the ▢ on triangle Ⓐ and triangle Ⓑ.

2 Let's draw in your notebook an enlargement of 2 times and a reduction of $\frac{1}{2}$ of the quadrilateral on the right.

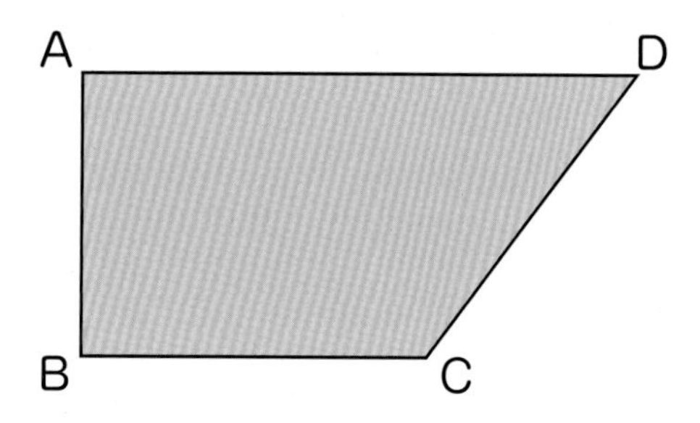

3 Let's find the original building height in the diagram on the right.

① Let's draw a reduction of $\frac{1}{500}$ the triangle ABC.

② Let's find the original building height.

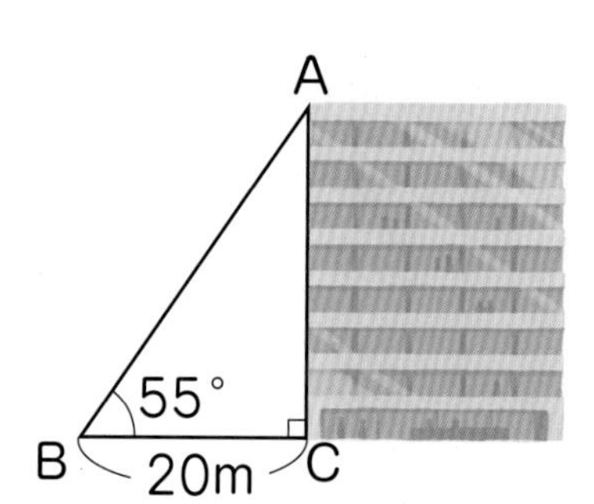

With the Way to See and Think Monsters...

Let's Reflect!

Let's reflect on which monster you used while learning "12 Enlargement and Reduction of Figures."

Why

By using the properties of enlargement and reduction of figures, we were able to explain why **it is in a relationship of enlargement and reduction.**

① Which of the following figures is always in the relationship of an enlargement and reduction of figures? Let's explain the reason.

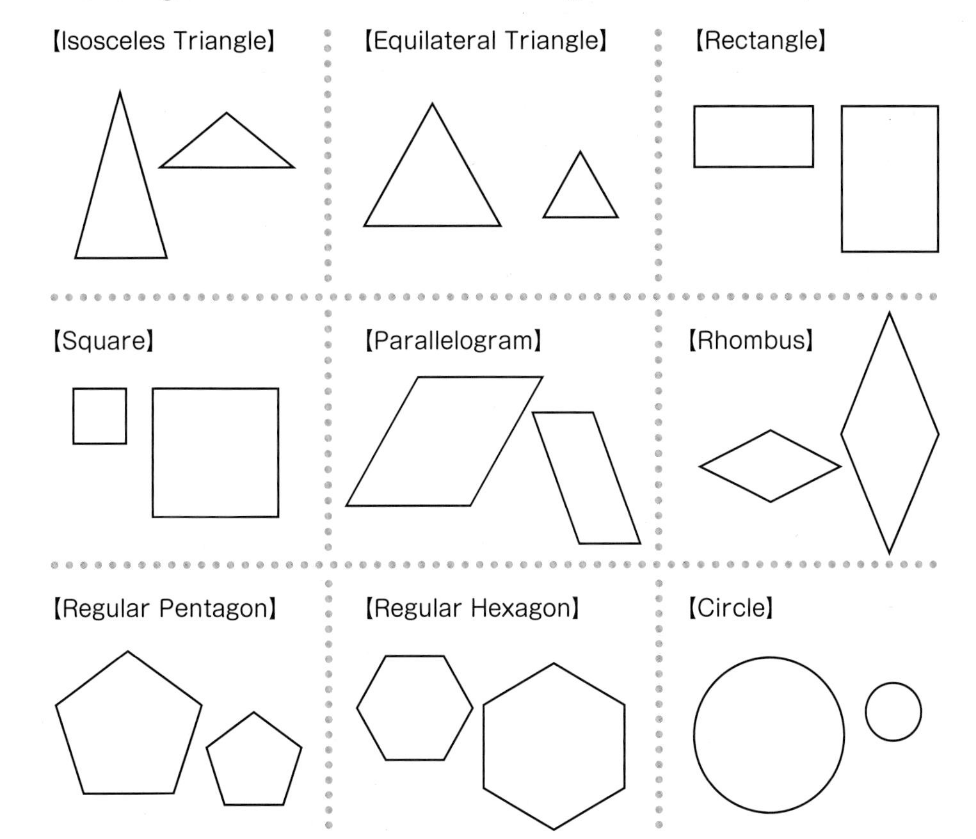

In the enlargement and the reduction, we can see that:
・The ratios of the corresponding sides are all equal.
・The corresponding angles are also equal.

For a rectangle, the corresponding angles are always equal in size at 90° ...

Let's deepen. → p.247

? Solve the ?

By examining the ratio of the lengths of the sides and the size of the corners, I could find out whether it is an enlarged or a reduced figure.

Akari

Want to Connect

To find out if there is a relationship between an enlarged or a reduced figure, do we have to find out the ratio of the lengths of all the sides or the size of all the angles?

Yu

Find the ? How do I find out the number of the sheets of paper?

We have decided to write our memories from 1st grade to 6th grade.
How many sheets of paper do we need?
Let's say we need 6 paper sheets per person...
1
There are 30 children in our class.
Then 180 would be nice.
2
We should take 180 sheets from this stack of paper....
It's so hard to count 1 by 1. I wonder if there is an easier way to do it...
3

How can we know how many sheets of paper are in a stack without counting each?

Proportion and Inverse Proportion

Let's explore the properties of 2 quantities changing and the characteristics of their correspondence.

1 Proportion

1 There is a stack of paper. Let's think about how to find the number of paper sheets without counting.

As the number of paper sheets increases, the weight is more...
It also gets thicker as the number of paper sheets increases...
If the weight and thickness are proportional to the number of paper sheets, we can find them without counting.
But since we don't know the weight or thickness of 1 paper sheet, how can we be sure that it is proportional?

Want to explore

Purpose How can we find out if the weight is proportional to the number of paper sheets?

Akari's idea

I investigate the relationship between the number of paper sheets and the weight.
(1) I measure the weight when the number of sheets is 10, 20, 30, and so on.
(2) Let's consider the relationship by summarizing on a table with the information of the number of paper sheets and the weight.

① Akari decided to investigate as shown on the right.
Let's discuss the reasons why.

② Let's write the weight of different numbers of paper sheets in the following table.

Number of paper sheets and weight

Number of paper sheet	10	20	30	40	50	
Weight (g)						

③ From exercise ② on the previous page, we got the weight of the paper shown on the following table. Let's discuss whether the weight is proportional to the number of paper sheets.

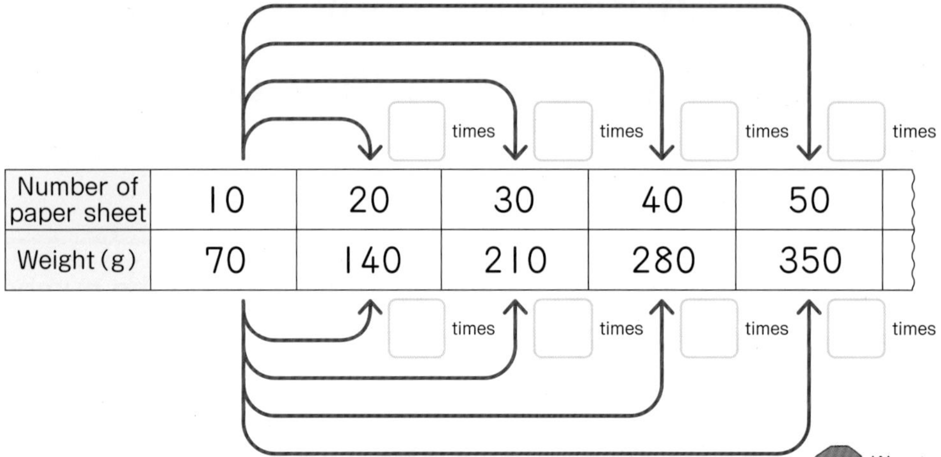

Number of paper sheet	10	20	30	40	50	
Weight (g)	70	140	210	280	350	

④ What is the weight of a paper sheet?

⑤ What would be the weight for the 180 paper sheets needed in g?

Way to see and think

We can find the number of paper sheets and their weights by using the idea of proportion.

! Summary

We can find out if they are proportional by using the weight of the total paper sheets.

? Is the thickness of the paper proportional to the weight as well?

2 **The following table shows the relationship between the number of paper sheets and their thickness. Let's answer the questions below.**

Number and thickness of paper sheets

Number of paper sheet	210	315	420	525	
Thickness (cm)	2	3	4	5	

① Can we say that the thickness of paper is proportional to the number of sheets?

If the number of paper sheets doubles, then the thickness...

② If the thickness of this stack of paper is 8 cm, how many sheets of paper can you say there are? Let's fill in the answer in the ☐ and let's explain the ideas of the 3 children.

\ Want to think /

? Purpose **Can we find the number of sheets from the thickness of the paper?**

Sara's idea

Since the thickness is 4 times as thick as 2 cm, the number of sheets is also 4 times as large.

☐ × 4 = ☐

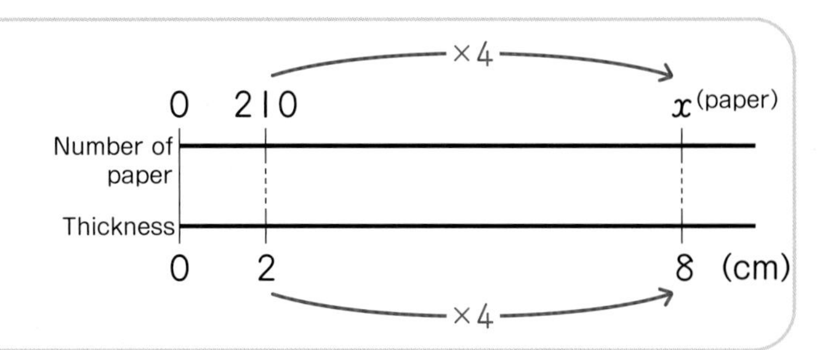

Haruto's idea

The number of sheets per cm is obtained by $210 \div 2 = 105$.
so, 8 times the number of pieces per centimeter.

☐ × 8 = ☐

×8

Number of paper sheet	105	x
Thickness (cm)	1	8

×8

Akari's idea

Let x be the number of paper sheets with 8 cm, and consider the ratio of the number of sheets to each other and the corresponding ratio of the thicknesses to each other.

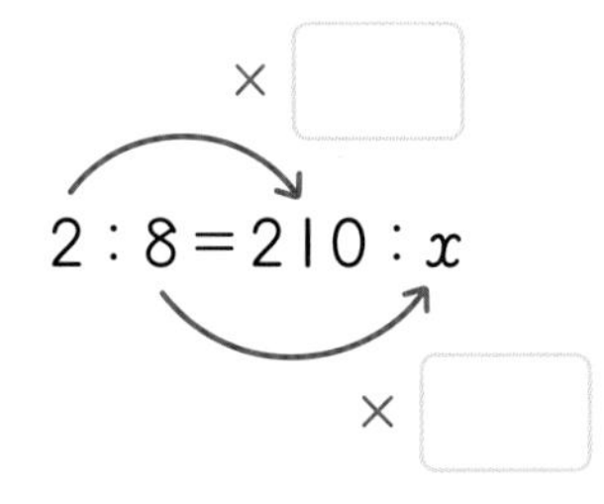

! Summary

We can find the number of paper sheets from the thickness by using the idea of times and the idea of ratio.

1 If the stack of paper on page 187 weighs 1400 g, how many sheets of this paper are there?

? Are there any other proportional relationships?

3

The following table was made after examining the relationship between the length and weight of a wire. Let's investigate the relationship between x and y, when the length of wire is x m and the weight is y g.

Length and weight of a wire

Length x (m)	2	3	4	5	6	9	18	
Weight y (g)	40	60	80	100	120	180	360	

① When the value of x changes by 2 times, 3 times, ..., how does the corresponding value of y change?

Akari
y is proportional to x.

What is the value of y when the value of x changes by 1.5 times or $\frac{1}{2}$ times?
Yu

Want to think

? Purpose How does the value of x and y change when y is proportional to x?

② When y is proportional to x and the value of x changes by 1.5 times and 2.5 times, how does the value of y change?

(1.5 times, 2.5 times, $\frac{1}{3}$ times, $\frac{1}{2}$ times)

Length x (m)	2	3	4	5	6	9	18	
Weight y (g)	40	60	80	100	120	180	360	

(□ times, □ times, □ times, □ times)

③ When y is proportional to x and the value of x changes by $\frac{1}{2}$ and $\frac{1}{3}$ times, how does the value of y change?

! Summary

When y is proportional to x, if the value of x changes by □ times, then the value of y also changes by □ times.

□ also applies to decimal numbers and fractions.

? Can we express the relationship between x and y in a math equation, as we did in 5th grade?

4 Water is poured into a tank. Let's investigate the relationship between the amount of water x L and the depth y cm of the accumulated water.

Amount and depth of water inside the tank

Water x (L)	1	2	3	4	5	6	7	8	9	
Depth y (cm)	2	4	6	8	10	12	14	16	18	

① Can we say that the depth y cm is proportional to the amount of water x L?

② Let's explore how much the value of y increases when the value of x increases by 1.

+1 　 +1 　 +1

x	1	2	3	4	5	6	7	
y	2	4	6	8	10	12	14	

+□ 　 +□ 　 +□

When the amount of water increases by 1 L, the depth increases by 2 cm.

Sara

＼ Want to think ／

? **Purpose** What is the relationship between x and y?

③ Let's calculate $y \div x$ using the corresponding values of x and y in the table above. What does the quotient of $y \div x$ represent?

④ Let's investigate the relationship between the amount and depth of water, using the fact that the depth of water is 2 cm per liter. Let's represent the relationship between x and y in a math equation.

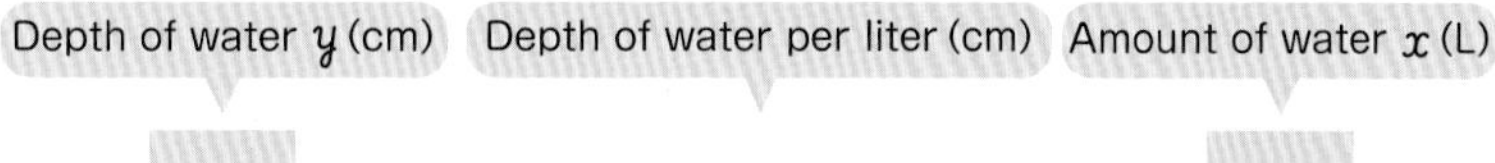

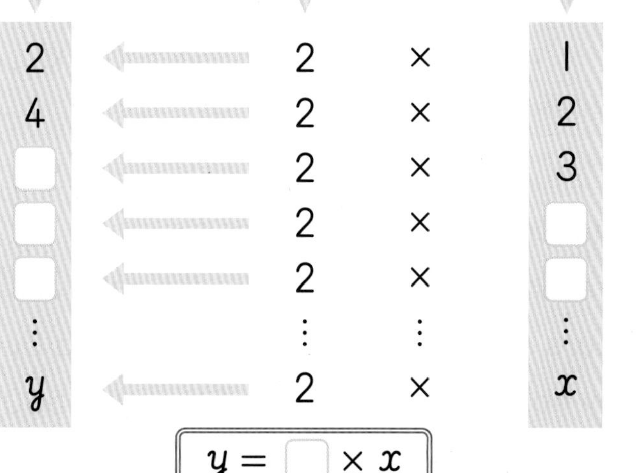

Depth of water y (cm)		Depth of water per liter (cm)		Amount of water x (L)
2	←	2	×	1
4	←	2	×	2
□	←	2	×	3
□	←	2	×	□
□	←	2	×	□
⋮		⋮	⋮	⋮
y	←	2	×	x

y = □ × x

! Summary

When there are 2 quantities x and y, and y is proportional to x, their relationship can be represented with the following math equation:

$$y = \text{constant} \times x$$

The constant in a proportional relationship represents:

① How much the value of x increases when the value of x increases by 1

② The quotient of $y \div x$

③ The value of y when the value of x is 1

5 Using the math equation in 4, let's find the depth of the water when 10 L and 20 L are poured into the tank.

1 Consider the wire on page 190. Let's represent the relationship between length x m and weight y g in a math equation.

Length and weight of a wire

Length x (m)	1	2	3	4	5	6	
Weight y (g)	20	40	60	80	100	120	

① Let's find the quotient of $y \div x$

② Let's represent the relationship between x and y in a math equation. $y = \square \times \square$

③ Let's find the weight of a wire that is 12 m long.

2 The relationship between time and distance when a car travels at 40 km per hour is shown in the table below. Let's look at this table and answer the following questions.

Time and distance with a speed of 40 km per hour

Time x (hours)	1	2	3	4	5	6	
Distance y (km)	40	80	120	160	200	240	

① Is the distance proportional to the time?

② Let's represent in a math equation the relationship between the time x hours and the distance y km traveled by a car.

③ How many hours will the car need to travel 560 km?

? What else can we learn by using the proportional equation?

5 Let's investigate the relationship between x and y, when the length of one side of an equilateral triangle is x cm and its perimeter is y cm.

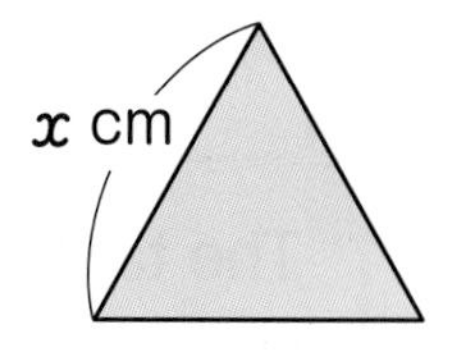

Length of side and perimeter of an equilateral triangle

Length of one side x(cm)	1	2	3	4	5	6	
Perimeter y (cm)	3	6					

1. Let's fill in the table.
2. Is y proportional to x?
3. Let's represent the relationship between x and y in a math equation. What does the constant represent?

\ Want to think /

Purpose Let's consider the relationship between the length of one side of a figure and the length of its circumference.

Haruto

When y is proportional to x, the relationship can also be represented with the following math equation: $y = x \times \text{constant}$

4. Let's find the perimeter when the length of one side is 20 cm and 35 cm.
5. Let's find the length of one side when the perimeter is 51 cm.

1 Consider the length of one side of a square as x cm and the perimeter as y cm. Let's represent the relationship between x and y in a math equation.

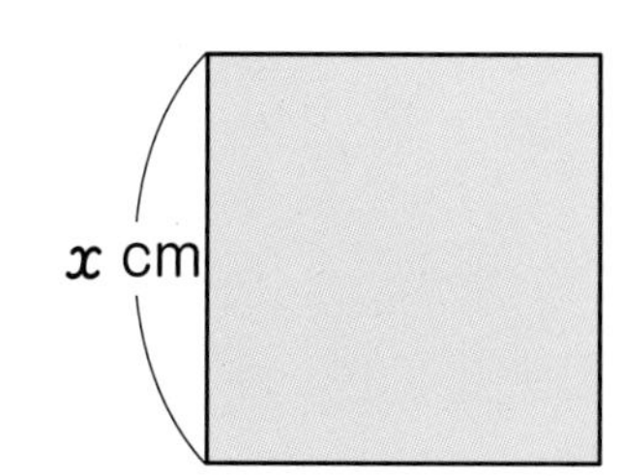

2 Write the following relationships between x and y in a table and represent them in a math equation. Also, what does the constant represent?

① Diameter of a circle as x cm and circumference as y cm.
② Number of 50 yen cards as x and total cost as y yen.
③ Length of one side x cm and perimeter y cm of a regular hexagon.

?

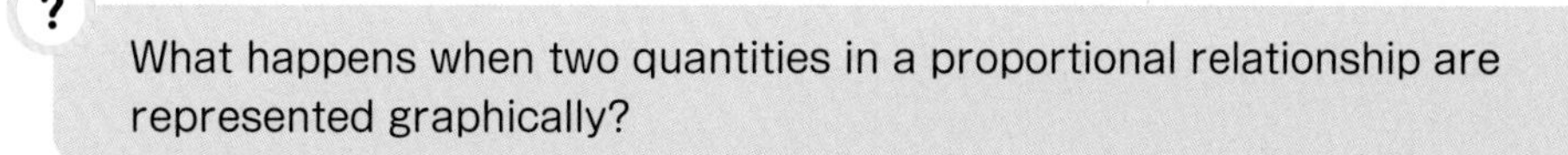
What happens when two quantities in a proportional relationship are represented graphically?

2 Proportional graph

1 **The table below shows the relationship between the amount of water x L and depth y cm inside a tank. Let's answer the following questions.**

Amount of water and depth inside the tank

water x (L)	1	2	3	4	5
Depth y (cm)	2	4	6	8	10

Way to see and think

The relationship between x and y shown in the table can be dotted with pairs of x and y values, just as in a line graph.

① Let's represent the relationship between x and y in a math equation.

② Let's draw in the diagram on the right, the points that represent the corresponding pair of x and y values from the table above .

A more detailed point....

Haruto

The point on the right indicates that the volume of water is 1 liter and the depth is 2 cm.

Amount of water and depth inside the tank

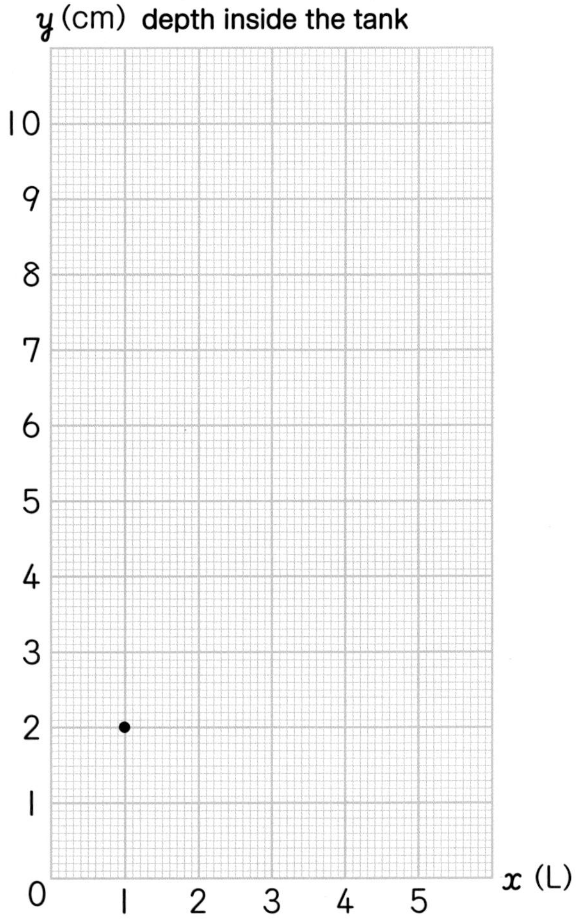

Want to think

Purpose **What kind of graph represents a proportional relationship?**

③ How are the points on the graph lining up?

④ Let's consider the corresponding pair of x and y on the previous page graph to complete the table below and draw the points.

Amount of water and depth inside the tank

Amount of water x (L)	0	0.1	0.2	0.5	1	2.4	3.9	
Depth y (cm)	0				2			

When the points that represent the corresponding pair of x and y values are connected, it becomes a straight line, as the one on the right.
This straight line is the graph of $y = 2 \times x$.

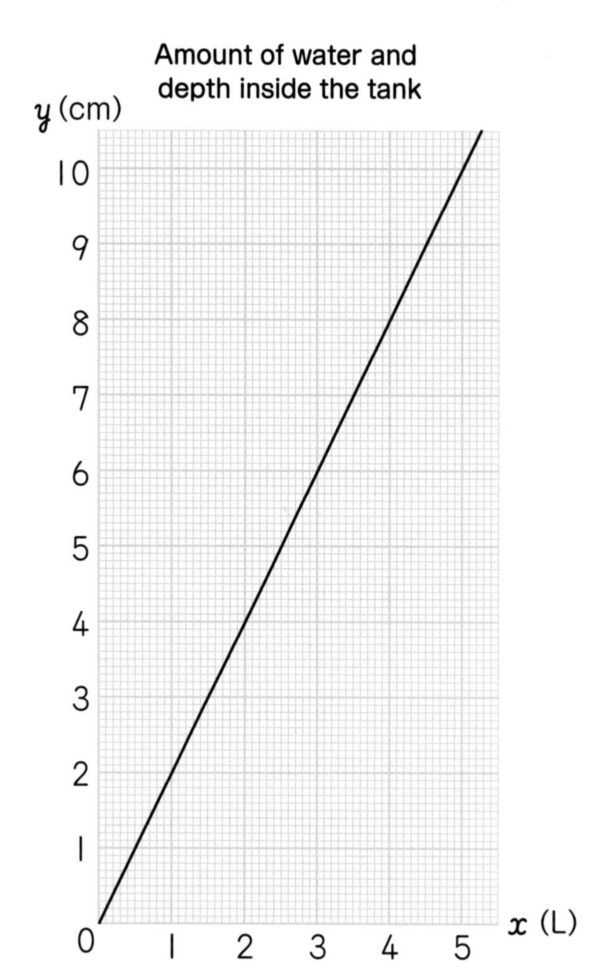

⑤ Using the graph, let's find the depth when the amount of water is 1.5 L and the amount of water when the depth is 5 cm.

A proportional relationship in a graph shows a straight line that goes through the intersection point of the vertical and horizontal axis at 0.

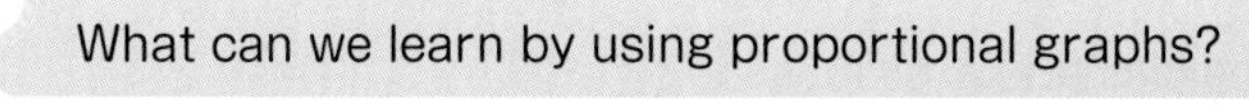

2

The graph below represents the relationships between the length x m and weight y g of two different wires Ⓐ and Ⓑ. Let's answer the following questions about these relationships.

① Can you say which wire weighs more? From which part of the graph can you say so?

\ Want to think /

Purpose What can we learn from a proportional graph?

② Let's read the following length or weight of each wire from the graph.

ⓐ weight of 2.4 m wire

ⓑ length of 48 g wire

③ What is the weight, in grams, is the weight per meter of each wire?

④ Let's represent the relationship between x and y from Ⓐ and Ⓑ in a math equation.

⑤ Choose the correct wire Ⓐ or Ⓑ given the following conditions:

ⓐ wire with a length of 3.8 m that weighs 114 g

ⓑ wire with a length of 4.2 m that weighs 168 g

⑥ Let's discuss whether the difference between the weight of Ⓐ and Ⓑ is proportional to the length.

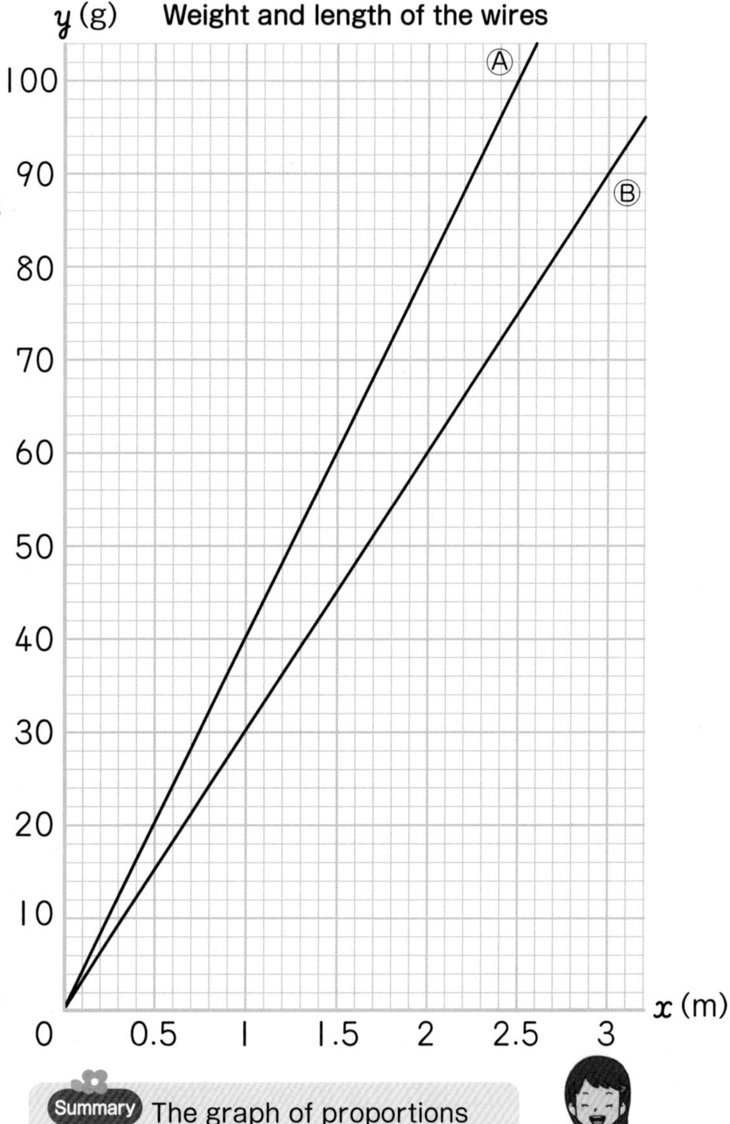

Summary The graph of proportions makes it easier to read the relationship between x and y.

Akari

? How else can we use the properties about proportion?

3 Use of proportionality properties

\ Want to try /

Purpose Let's think about various problems by using the property of proportionality.

Haruto

1 The table below represents the relationship between the amount of cola and amount of sugar in it. Let's answer the following questions about the relationship between the 2 amounts.

Amount of cola and sugar

Cola x (mL)	0	1	50	100	150	180	250	
Sugar y (g)	0		6	12	18			

① Is the amount of sugar y g proportional to the amount of cola x mL?

② What is the amount of sugar, in grams, inside 250 mL of cola?

Yu's idea

250 mL of cola is 5 times of 50 mL, therefore the amount of sugar is also 5 times more.

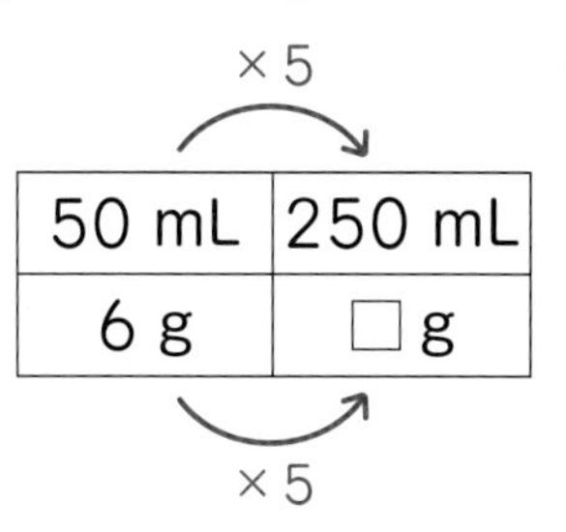

Sara's idea

The amount of sugar per mL of cola is constant. Therefore, I can make a math equation.

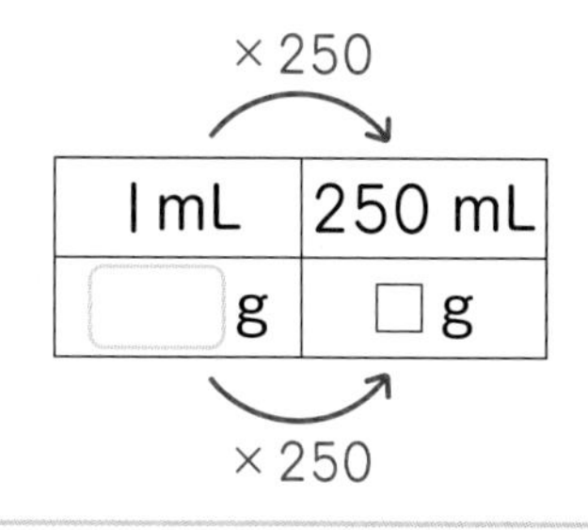

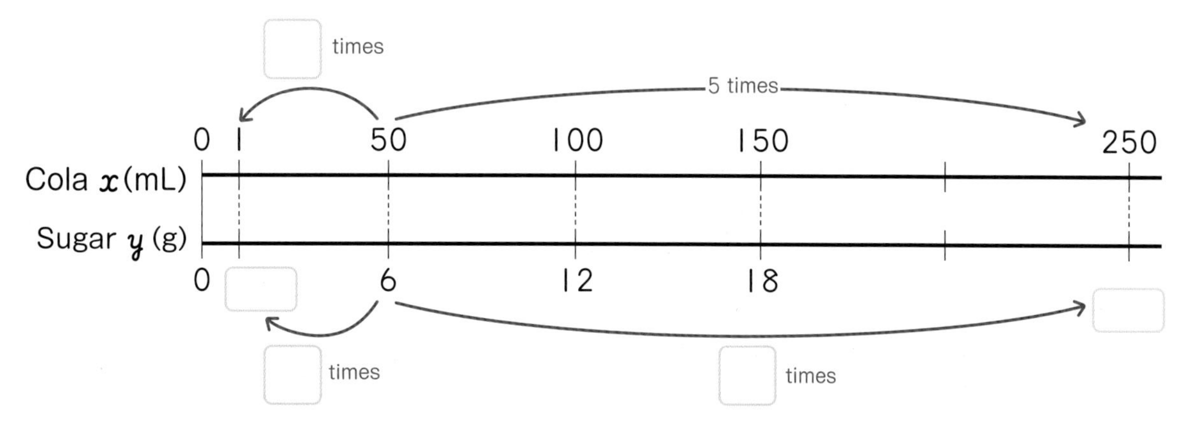

ⓐ Let's find the answer using Yu's idea.

ⓑ Using Sara's idea, let's represent the relationship between x and y in a math equation.

$y = \square \times x$

③ What is the amount of sugar, in grams, inside 180 mL of cola?

1 The table below represents the relationship between the number of nails x and its weight y g. Let's answer the following questions about the relationship between these two quantities.

Number and weight of nails

Number of nails x	0	1	50	100	150	200	250
Weight of nails y (g)	0	ⓐ	300	600	900	ⓑ	ⓒ

① Is y proportional to x?

② Let's find the appropriate answer for ⓐ, ⓑ, and ⓒ.

③ Let's represent the relationship between x and y in a math equation. Also, how many nails are there if the nail's weight is 240 g?

2 The graph below represents the relationship between the weight x g and elongation length y cm of a rubber band. Let's answer the following questions about the relationship between these two quantities.

Weight and elongation length of a rubber band

① How many cm will be the rubber band stretch if the weight is increased by 20 g?

② Represent the relationship between x and y in a math equation.

③ If a stone is attached to the rubber, the rubber stretched 13 cm. What is the weight of the stone?

? Is there any relationship other than proportionality between two quantities that change with each other?

Consider a rectangle with an area of 24 cm^2. Let's explore how the width and length change.

\ Want to explore /

Purpose What is the relationship between the width and the length of a rectangle with a fixed area?

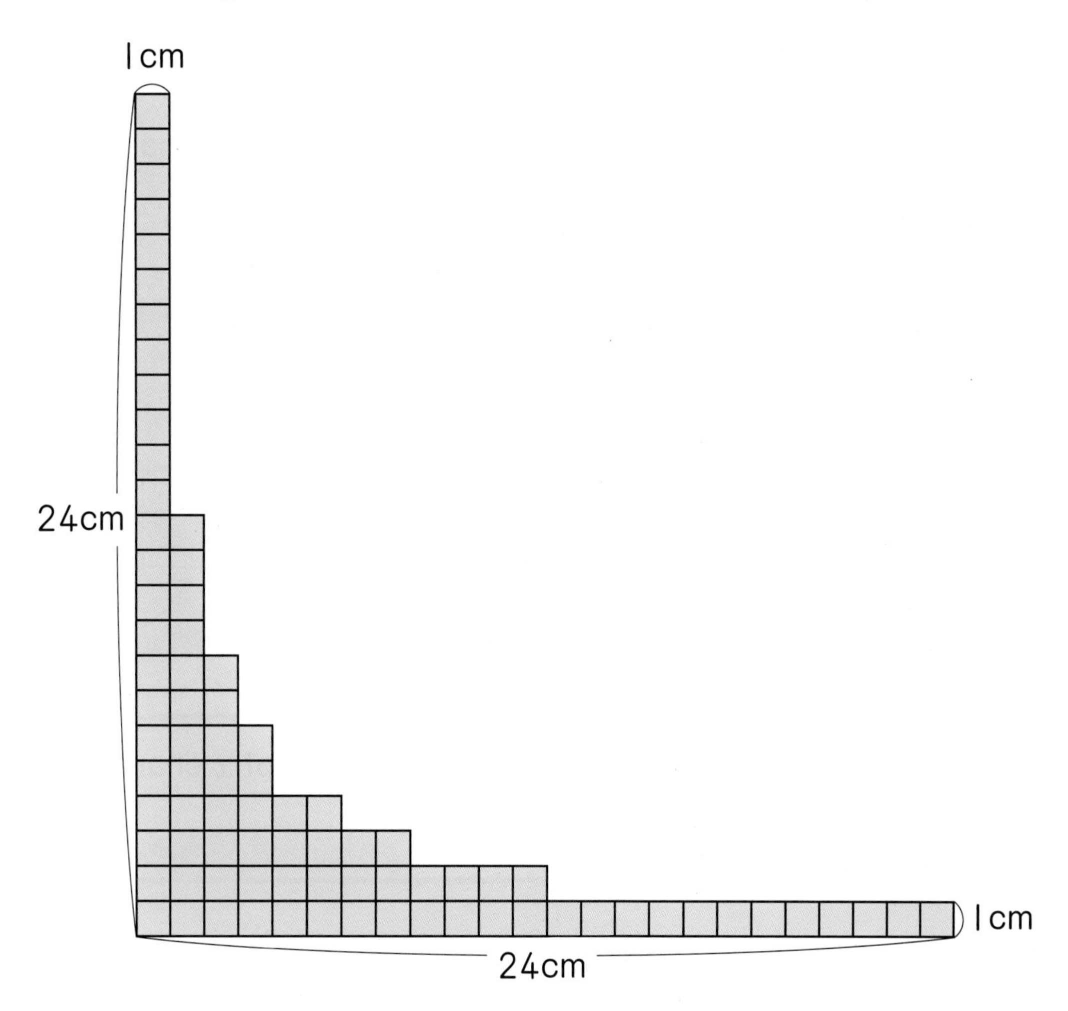

① Let's summarize in the following table about the width and length of various rectangles made by 24 squares of 1 cm^2.

Width and length of a rectangle with area of 24cm^2

Width x (cm)	1	2	3	4	6	8	12	24
Length y (cm)	24							

② If the value of x changes by 2 times, 3 times and so on, how does the corresponding value of y change?

2 times 3 times 2 times

Width x (cm)	1	2	3	4	6	8	12	24
Length y (cm)	24							

☐ times ☐ times ☐ times

We say that **y is inversely proportional to x**, when there are 2 quantities changing together, x and y, if the **value of x changes by 2, 3 times** and the corresponding **value of y changes by $\frac{1}{2}$, $\frac{1}{3}$ times** respectively.

In contrast to inverse proportion, proportion is sometimes called **direct proportion**.

③ If the value of x changes by $\frac{1}{2}$ times or $\frac{1}{3}$ times, how does the value of y change?

$\frac{1}{3}$ times $\frac{1}{2}$ times

Width x (cm)	2	3	6
Length y (cm)			

☐ times ☐ times

Summary

When y is inversely proportional to x, if the value of x changes by ☐ times, the value of y changes by $\frac{1}{☐}$ times.

1 If the perimeter of a rectangle is 24 cm, the width is x cm and the length is y cm. Can you say that y is inversely proportional to x?

Width and length of a rectangle with perimeter of 24 cm

Width x (cm)	1	2	3	4	5	6
Length y (cm)	11	10	9	8	7	6

? Can inverse proportion also be expressed in math equations and graphs?

2

Let's represent with a math equation and a graph regarding the relationship between x and y when the area of the rectangle is 24 cm^2, the width is x cm, and the length is y cm.

\ Want to think /

? Purpose What is the math equation and graph of an inverse proportion?

Width and length of a rectangle with area of 24 cm^2

Width x (cm)	1	2	3	4	6	8	12	24
Length y (cm)	24	12	8	6	4	3	2	1

1. What rule is between x and y?

2. Let's find the product of the corresponding x and y values. What does the product of x and y represent?

3. Let's represent the relationship between x and y in a math equation.

Width (cm)		Length (cm)		Area (cm^2)
1	×	24	=	24
2	×	12	=	24
3	×	8	=	☐
4	×	6	=	☐
x	×	y	=	☐

! 

Summary

The relationship of x and y changing together and if y is inversely proportional to x, then this relationship can be represented with the math equation: $x \times y =$ constant

4. Let' s find the value of y when the value of x is 5.

$$5 \times y = 24$$
$$y = 24 \div 5$$
$$= \square$$

When y is inversely proportional to x, it can also be represented with the following math equation: $y =$ constant $\div x$

⑤ In the following diagram, let's draw the points that represent the corresponding pair of x and y values from the table in the previous page.

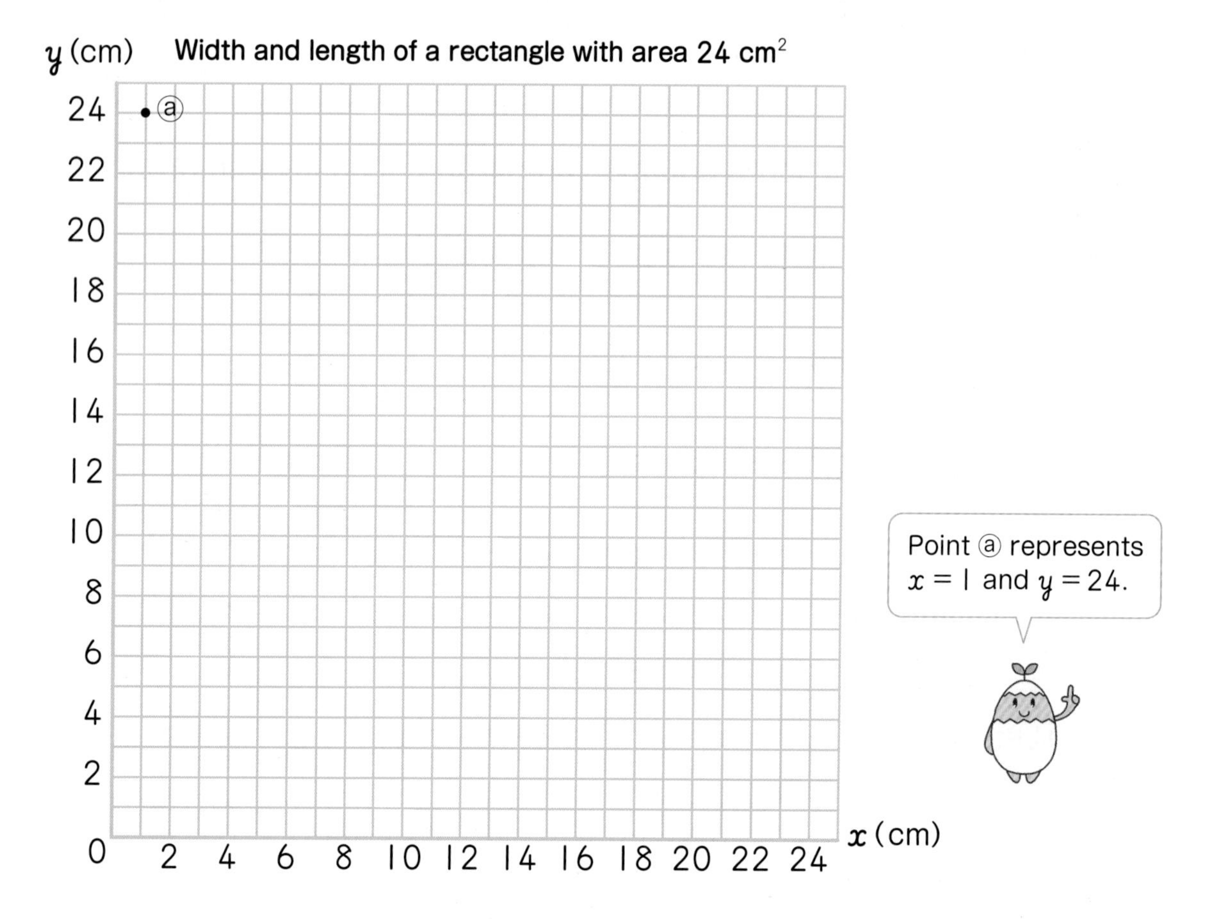

⑥ Let's complete the table below when x is 1.5, 2.5, 7.5, and 12.5, and draw the points that represent the corresponding pair of x and y values on the diagram.

Width and length of a rectangle with area 24 cm^2

Width x (cm)	1.5	2.5	7.5	12.5
Length y (cm)				

⑦ Let's compare with the proportional graph on page 195.

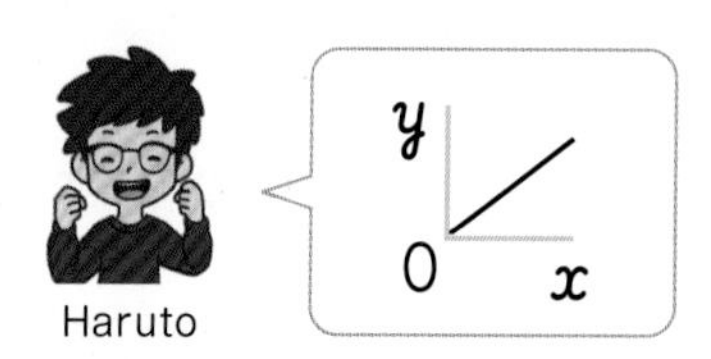

When the points from the previous page are drawn finely, it becomes as shown on the right.

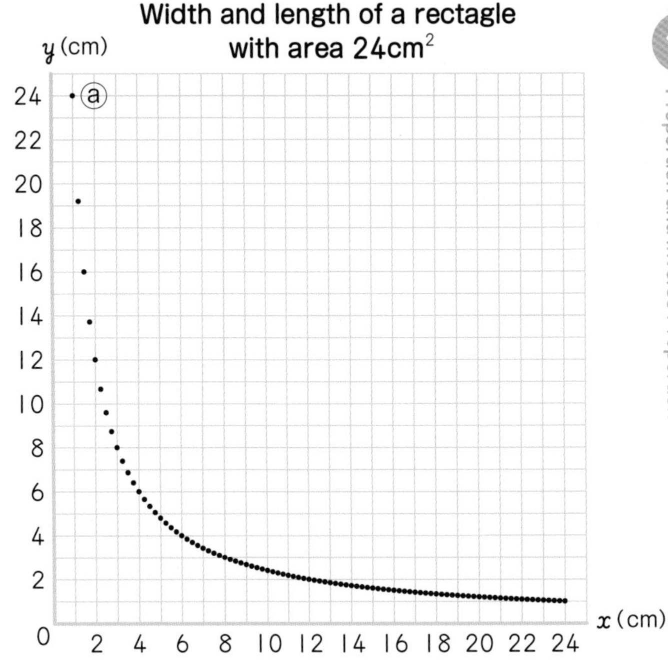

? Are there two quantities around us that are inversely proportional to each other?

3 There is a project that takes 60 days to be completed by 1 person that makes the same amount of work per day. If the number of people working is x and the number of days is y, let's think about the relationship between x and y.

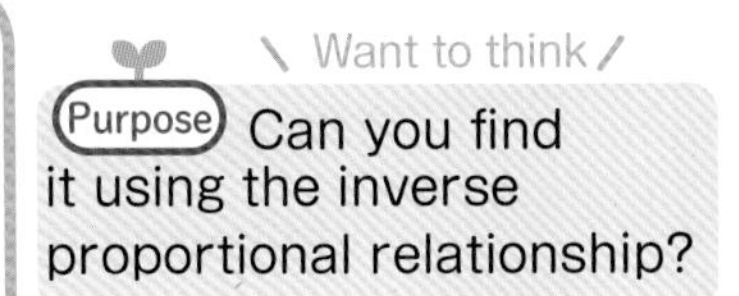

Want to think
Purpose Can you find it using the inverse proportional relationship?

Yu

1. Let's represent the relationship between x and y in a math equation.
2. How many days does it take to complete the work when 5 people are working?
3. How many people are needed to complete the work in 10 days?

1 Let's investigate the relationship between speed and time when the distance is 36 km.

① Let's summarize the relationship between speed and time in the following table.

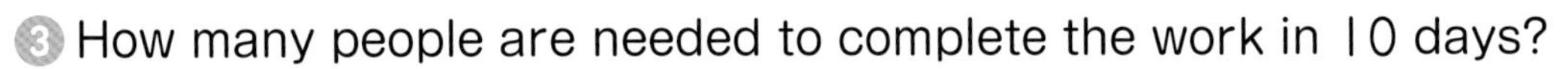

Speed and time

Distance per hour (km)	1	2	3	4	6	9	12	18	36
Time (hours)									

② Is time inversely proportional to the speed?

③ Let's consider speed as x km per hour and time as y hour, and represent the relationship between x and y in a math equation.

CAN What can you do?

☐ We understand the meaning of proportion. → pp.187 ~ 190

1 Let's fill in the blanks in the following tables with the appropriate answer.

① **Number of pencils and total cost**

Number of pencils x	0	1	2	3	4	5	
Cost y (yen)	0	50	100				

② **Walking time and distance**

Time x (hours)	0	1	2	3	4	5	
Distance y (km)	0	4	8				

☐ We can represent a proportional relationship with a math equation and a graph. → pp.191 ~ 195

2 There is a ribbon that costs 80 yen per meter.

① Let's represent the relationship between the length x m and cost y yen of the ribbon in the following table.

Length and cost of the ribbon

Length x (m)	0	1	2	3	4	5	
Cost y (yen)	0	80					

② Let's represent the relationship between x and y in a math equation.

③ Let's represent the relationship between the corresponding x and y values in a graph.

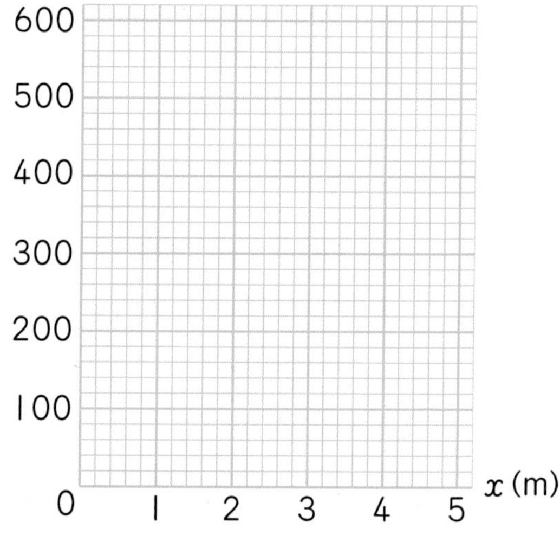

☐ We can solve problems using proportions. → pp.197 ~ 198

3 Let's answer the following questions about the table below.

Length and weight of a wire

Length x (cm)	0	1	2	3	4	5	6	
Weight y (g)	0	9	18	27	36	45	54	

① Let's represent the relationship between x and y in a math equation.

② What is the weight of a wire, in grams, that is 8 cm long?

③ What is the length of a wire, in centimeters, that weighs 117 g?

☐ We understand the meaning of inverse proportion. → pp.199 ~ 200

4 Let's fill in the blanks in the following tables with the number that applies.

① Base and height of a parallelogram with area 16 cm^2

Base x (cm)	1		4	5	8	
Height y (cm)		8			2	1

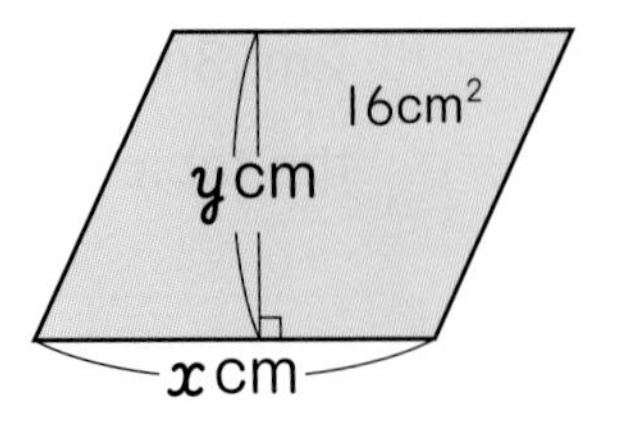

② Number of people and days to finish a work by a person in 45 days

People x	1	3	5	9	15	45
Days y						

☐ We can represent an inversely proportional relationship with a math equation and graph. → pp.201 ~ 203

5 Let's investigate the relationship between speed and time when the distance is 24 km.

① Let's summarize the relationship between speed x and time y in the following table.

Speed and time

Distance per hour x (km/h)	1	2	3	4	6	8	12	24
Time y (h)								

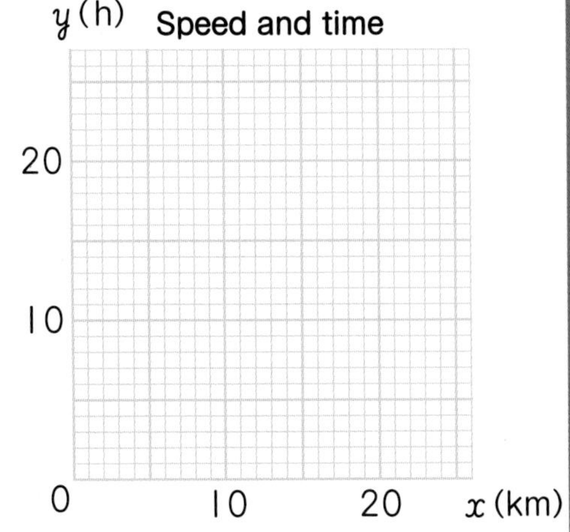

② Let's represent the relationship between x and y in a math equation.

③ Let's represent the relationship between the corresponding x and y values in the graph.

④ How many hours does it take with a speed of 10 km/h?

Supplementary Problems → p.243

Which "Way to See and Think Monsters" did you find in "13 Proportion and Inverse Proportion."

When looking at the proportionality relationship, I found "Other Way."

Haruto

When considering inversely proportional relationships....

Akari

Usefulness and Efficiency of Learning

1 There is a ribbon that costs 150 yen per meter. Let's answer the following questions.

① Consider the length of the ribbon as x m. When the length of the ribbon is 1m, 2m, 3m, and so on. Let's find the corresponding cost y yen and summarize it in the table.

Length and cost of the ribbon

Length x (m)	0	1	2	3	4	5	6	
Cost y (yen)	0							

② What is the cost of the ribbon, y yen, proportional to?

③ Let's represent the relationship between x and y in a math equation.

④ Let's represent the relationship between x and y with a graph.

⑤ How much is the cost when the ribbon's length is 2.5 m?

⑥ How many meters is the ribbon's length when the cost is 1800 yen?

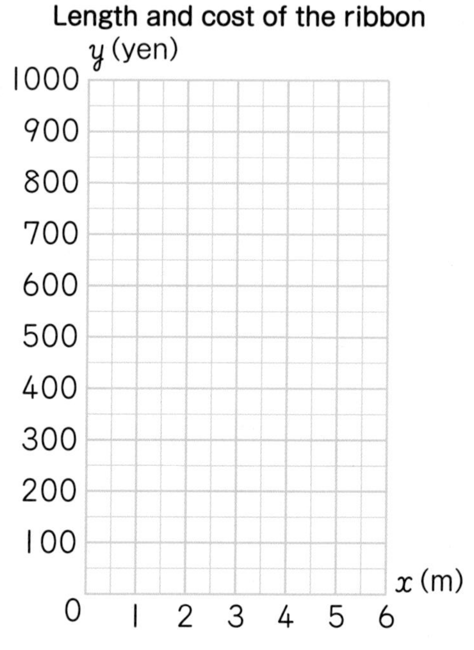

2 The graph on the right shows the relationship between the length x m and weight y g of wire Ⓐ and wire Ⓑ. Let's answer by looking at the graph.

① Considering both wires have the same length, which wire is heavier, Ⓐ or Ⓑ?

② What is the weight, in grams, is the weight per meter of wire Ⓐ and wire Ⓑ?

③ Let's represent, for both wires, the relationship between x m and y g in a math equation.

④ Which wire, Ⓐ or Ⓑ, weighs 21 g at a length of 3.5 m?

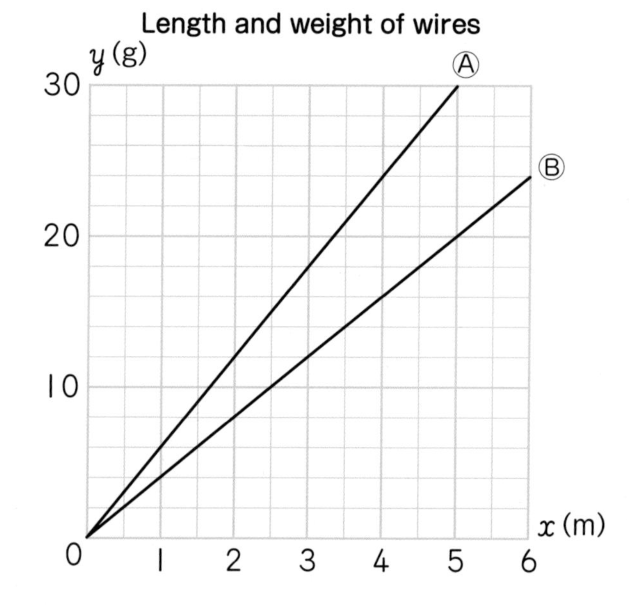

3 The figure on the right is a parallelogram with an area of 12 cm^2, base x cm and height y cm. Let's answer the following questions.

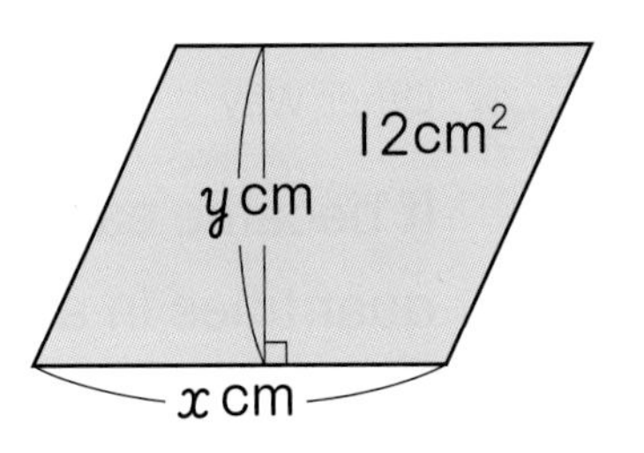

① The table below shows the relationship between x and y. Let's fill in the blanks with the appropriate answer.

Base and height of a parallelogram with area 12 cm^2

Base x (cm)	1		3	4	6	
Height y (cm)		6			2	1

② Let's represent the relationship between x and y in a math equation.

③ Let's draw the points that represent the corresponding pair of x and y values from the above table.

④ What is the height, in centimeters, if the base is 8 cm?

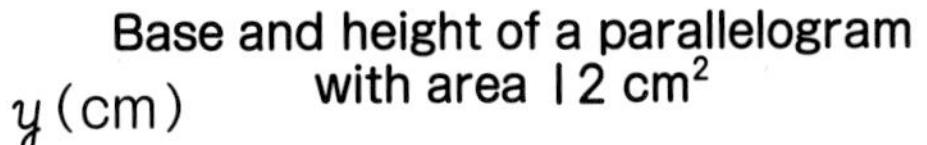

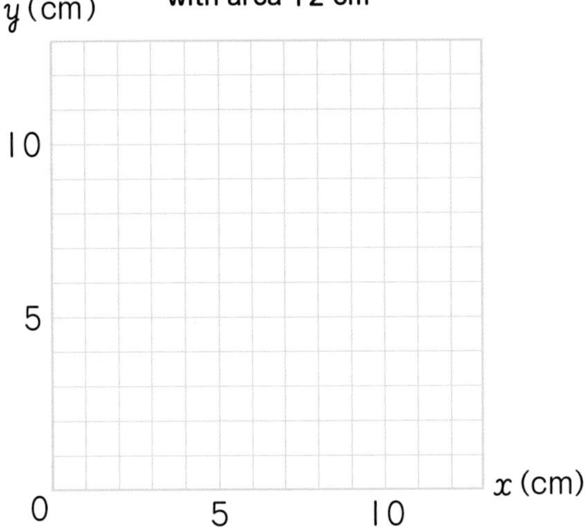

4 You take the highway to leave Tokyo and go to Shizuoka. It is about 160 km. Let's answer the following questions.

① Let's represent the relationship between x and y in a math equation when the speed is x km per hour and time is y hours.

② You arrived at Shizuoka in 1 hour and 36 minutes. How many km per hour was your speed?

(Shizuoka city, Shizuoka Prefecture)

With the Way to See and Think Monsters...

Let's Reflect!

Let's reflect on which monster you used while learning "13 Proportion and Inverse Proportion."

Other Way

It became easier to see what kind of relationship exists between two quantities in a proportional relationship by expressing them in a table, equation, or graph.

① The relationship between the amount of water x L in a container and the depth y cm is shown in the following table, math equation, and graph, respectively. What can you conclude from each?

【Table】

Amount of water and depth of water

Amount of water x (L)	1	2	3	4	5	
Depth y (cm)	2	4	6	8	10	

In the table, you can see that as x doubles, triples, ..., y also doubles, triples, ..., and that y is always twice as large as x.

Haruto

【Math equation】

$$y = 2 \times x$$

When represented in a math equation, it is easy to find the value of y if the value of x is known.

Yu

By using a graph, the values of x and y can be read without calculation.

Sara

【Graph】

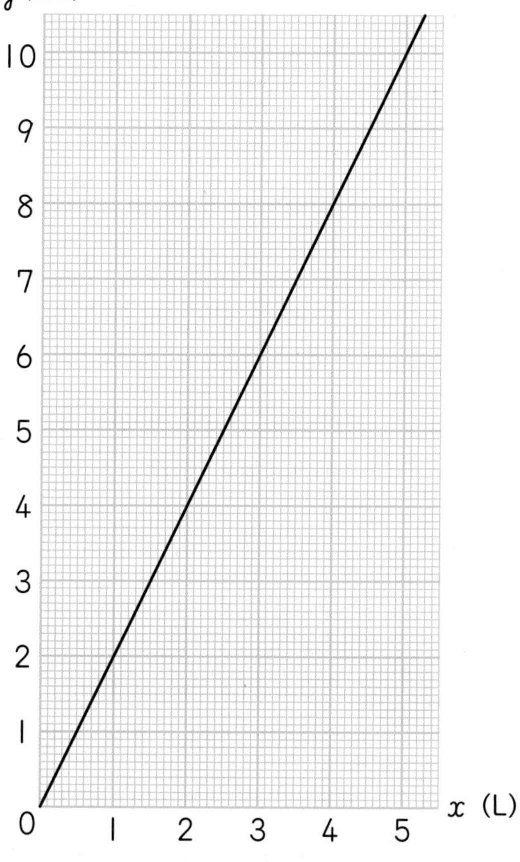

Solve the ?

Using the idea of proportionality, if one of the two quantities is known, the other could also be found out.

Akari

Want to Connect

What does the graph of inverse proportion look like?

Haruto

Proportion and inverse proportion are words that describe the relationship between two quantities that change together.
Make sure you understand the meaning of each of these terms.

① Choose one from the following Ⓐ to Ⓔ in which y is proportional to x and one in which y is inversely proportional to x.

Ⓐ the time y minutes it takes to travel 1500 m at a speed of x m per minute
Ⓑ the area y cm^2 of a square whose sides are x cm long
Ⓒ the number of pages y remaining in a 100-page book when x pages have been read
Ⓓ the price y of x notebooks that price 80 yen each
Ⓔ the length y m of a x m ribbon when three persons divide it into equal lengths

If we actually apply the numbers...

Sara

Yu

If we represent it in a math equation...

Frequently found mistake

- Consider proportionality to be a relationship in which "when one side increases, the other side also increases."
- Consider inverse proportion to be the relationship that "when one side increases, the other side decreases."

Be careful!

【Meaning of Proportionality】 When y is proportional to x, if the value of x changes by □ times, then the value of y also changes by □ times.
【Math Equation for Proportionality】
y = constant × x
【Meaning of Inverse Proportionality】When y is inversely proportional to x, if the value of x changes by □ times, the value of y changes by $\frac{1}{\square}$ times.
【Math Equation for Inverse Proportionality】
$x \times y$ = constant or y = constant ÷ x

The relationship between x and y for Ⓐ to Ⓔ, respectively, can be represented as follows:

Ⓐ (Time) = (Distance) ÷ (Speed), so $y = 1500 \div x$
Ⓑ (Area of a square) = (side) x (side), so $y = x \times x$
Ⓒ (Number of pages remaining)
= (Total number of pages) − (Number of pages read), so $y = 100 - x$
Ⓓ (Price) = (Price of one book) x (Number of books), so $y = 80 \times x$
Ⓔ (Length per person) = (Total length) ÷ (Number of people), so $y = x \div 3$

Let's actually apply a number in x and see how y changes.

Utilizing Math for SDGs

Let's think about a well-balanced diet!

We need energy to live our lives, and this energy comes from food.
Food contains a variety of nutrients, the main ones being "carbohydrates," "lipids," and "proteins". In addition, there are "inorganic nutrition" and "vitamins." These five are called the five macronutrients. It is very important to take these in good balance.
The traditional Japanese diet provided an ideal nutritional balance, but today's diet is said to be nutritionally unbalanced due to the spread of instant foods and fast foods, as well as meat-centered diets. In addition, vegetables and fruits are in short supply, especially among the younger generation.
Let's take the cooking card on page 261 as an example to think about dietary habits.

Way to explore the cooking card

The front side is the dish and the back side is list of the main ingredients needed to make the dish. (Seasonings, etc. are omitted.) The symbols in front of the ingredients are the main nutrients.

Ⓒ…Carbohydrate　L…Lipids　Ⓟ…Proteins
I…Inorganic nutrition (e.g. calcium)
Ⓥ…Vitamins・Inorganic nutrition

Front

Spaghetti with meat sauce

Back

Spaghetti with meat sauce
Ⓒ spaghetti
Ⓟ minced beef
Ⓥ onion

① How many combinations of main and side dishes are there?

② Haruto is thinking of having rice as his staple food and hamburgers as his main dish. What should he choose as his side dish if he is to eat all the five macronutrients? Let's think about it using the table.

		Foods that are primarily energy sources		Foods that are mainly a source of body building		Foods that are the main source of tonifying the body
		rice · bread · noodles, potatoes etc.	oil · butter · mayonese etc.	fish · meat · egg · bean products etc.	small fish · milk · dairy products · seaweed etc.	other vegetables · fruits · mushrooms etc.
Staple food	Rice	rice				
Soup						
Main dish	Hamburg steak			minced beef		onion
Side dish						
Nutrition contained		Carbohydrate	Lipids	Proteins	Inorganic nutrition (e.g. calcium)	Vitamins · Inorganic nutrition

③ Let's think of your own nutritionally balanced meal by combining the cooking cards on page 261. You can also look up the ingredients for foods that are not listed on the cards and think about them.

Think back on what you felt through this activity, and put a circle.

Let's reflect on yourself!

	Strongly agree	Agree	Don't agree
① I could think about a well-balanced menu.			
② I could think of various combinations and create a menu.			
③ I could think of combinations using diagrams and tables learned in mathematics.			

	Strongly agree
④ I am proud of myself because I did my best.	

Let's praise yourself with some positive words for trying hard to learn!

14

Utilization of Data

Let's solve various problems.

1 Let's find problems from our surroundings and use tables or graphs that have been learned so far to solve them.

I hear a lot of talk about global warming these days.
What exactly is global warming?
The temperature seems to be getting higher and higher.
How high are they?

Problem found by Yu's group

【Big problem】
I want to know if the earth is really warming.

【Problem】
Examine the temperatures of a certain year and the years before and consider whether the temperatures have increased.

First, we find "big problems" in our surroundings.
We are trying to think about it by changing it into a problem that we can think about using what we have learned in mathematics.

P Finding a problem

Where should we check the temperature at?
I am sure that they check the temperatures every day.I want to compare the temperatures in 2020 and 2000.
For example, we can choose a region and compare the temperatures there.
We can use the Internet to find out.

They are discussing what kind of data they need and how to collect the data. Depending on what we want to research, we can use the Internet or collect questionnaires.

P Making a plan

Using the Internet, we looked up the daily maximum temperatures in January 2000 and 2020 for Kagoshima City, Kagoshima Prefecture, and found the results in the following table.

Daily maximum temperatures in January 2000

Date	Temperature (℃)	Date	Temperature (℃)
1	17.4	17	16.7
2	17.1	18	16.0
3	16.2	19	12.4
4	17.8	20	8.5
5	17.0	21	8.6
6	21.2	22	14.0
7	14.2	23	19.3
8	12.5	24	15.3
9	14.3	25	11.5
10	16.7	26	6.0
11	15.3	27	8.1
12	15.8	28	10.5
13	16.8	29	14.1
14	13.5	30	12.1
15	13.7	31	9.4
16	15.1		

Daily maximum temperatures in January 2020

Date	Temperature (℃)	Date	Temperature (℃)
1	13.9	17	11.6
2	15.6	18	13.8
3	14.0	19	13.4
4	16.2	20	13.4
5	14.9	21	15.2
6	17.0	22	15.4
7	20.7	23	20.5
8	20.5	24	18.1
9	17.2	25	15.9
10	16.5	26	14.8
11	17.2	27	21.7
12	11.7	28	17.7
13	14.7	29	17.7
14	9.6	30	14.4
15	13.3	31	14.4
16	12.2		

Data was collected using the Internet. When collecting data via the Internet, be careful to ensure that the data is reliable.
Also, be careful to manage the data collected through questionnaires.
It is also important to tell them that you will not use the data for any purpose other than to solve the problem.

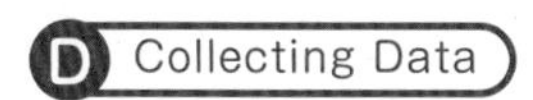

We looked up representative values and summarized them into a table.

Yu

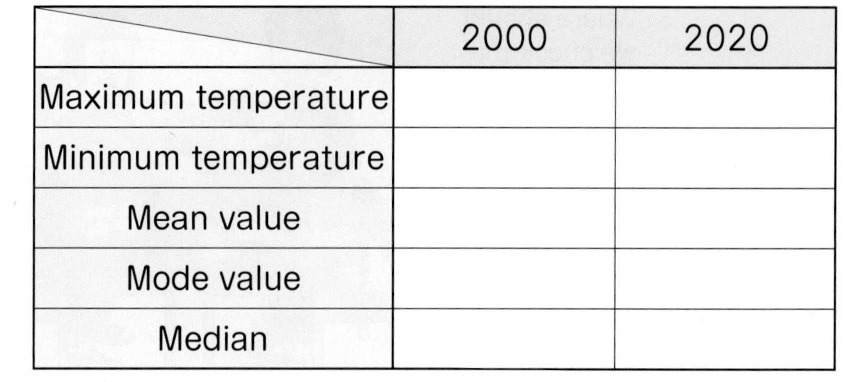

	2000	2020
Maximum temperature		
Minimum temperature		
Mean value		
Mode value		
Median		

Akari

I want to represent the distribution of data in a histogram.

If there are several values that appear most frequently, all of them become mode values.

They are going to analyze data in tables and graphs.
Let's not only compare the data with representative values and histograms, but also consider whether we can use the tables and graphs we have studied so far in our analysis.

A Analyzing data

● Based on what you have studied so far, explain how you would think about it.

There are fewer days with lower temperatures in 2020.

I checked the daily maximum temperatures, but I would like to check the minimum temperatures as well.

Since we only studied Kagoshima City, we cannot draw any conclusions. It would be better to look at a wider range of regions.

I wonder how it is in other countries.

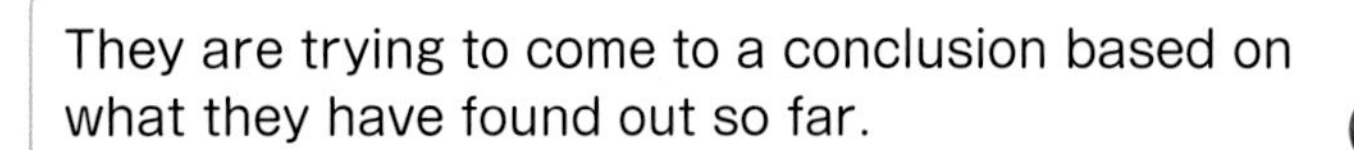

They are trying to come to a conclusion based on what they have found out so far.

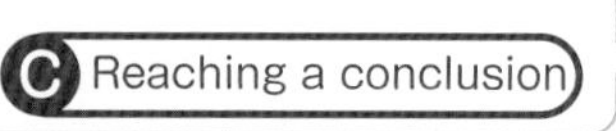

C Reaching a conclusion

On the other hand, they could not make a decision only based on what they have studied. They came up with new problems such as studying other cities than Kagoshima City, minimum temperatures, and other countries.

P Finding new problems

When there is a problem that you want to solve, one solving method is called the **PPDAC cycle**. This cycle has the following 5 ordered steps, each represented by the first letter of a word.

(1) Problem … finding a **problem**
(2) Plan … making a **plan**
(3) Data … collecting **data**
(4) Analysis … **Analyzing** data
(5) Conclusion … reaching a **conclusion**

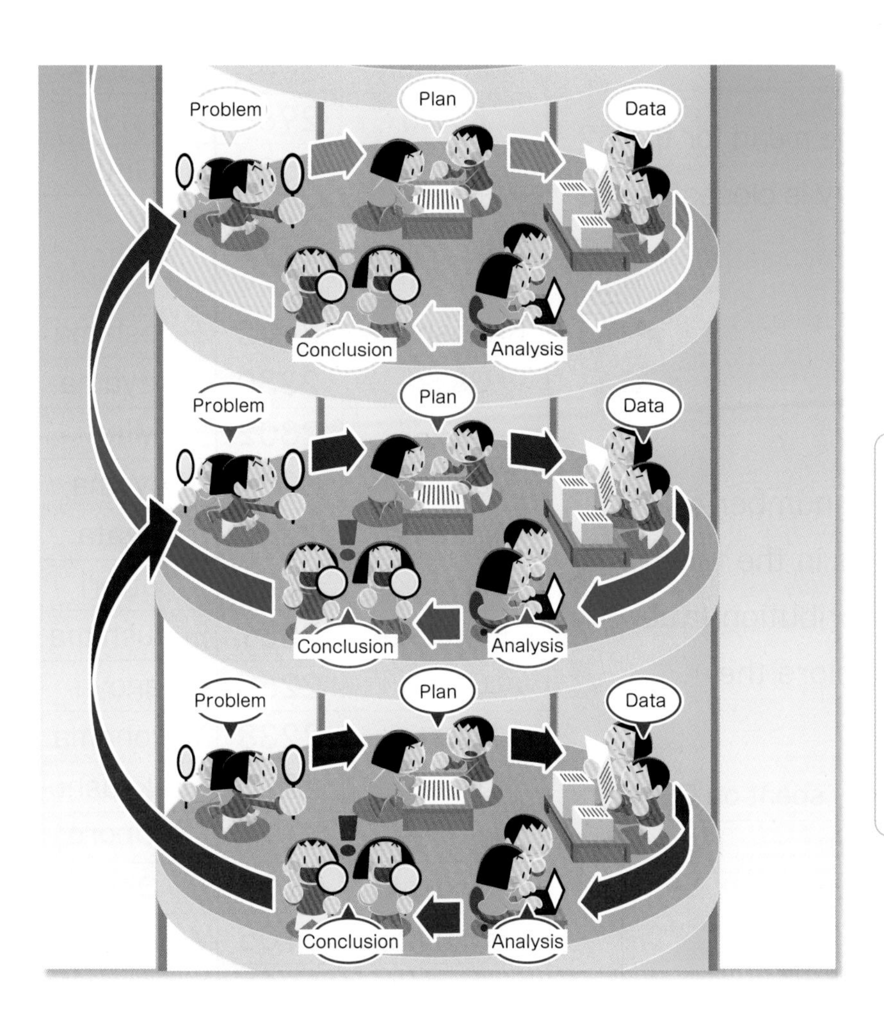

The illustration on the left represents the PPDAC cycle.

The PPDAC cycle does not always work this way. After you have collected the data, you may need to review and revise your plan. While Analyzing, you may want to re-collect the data.

Doing your own research →

2 **Let's find problems from our surroundings and examine using PPDAC cycle.**

3 The table on the right shows the results from a survey conducted in 52 cities about the amount of money spent on Gyoza per household per year (mean from 2018 to 2020). Let's answer the following questions.

Amount of money spent on Gyoza (yen)

City	Amount of money	City	Amount of money
Utsunomiya	3764	Takamatsu	1936
Hamamatsu	3590	Kanazawa	1933
Miyazaki	3053	Nagasaki	1928
Kyoto	2969	Morioka	1921
Otsu	2501	Gifu	1912
Tokyo Metropolitan Area	2471	Naha	1909
Kagoshima	2442	Hiroshima	1908
Osaka	2330	Okayama	1897
Sakai	2302	Mito	1896
Maebashi	2284	Toyama	1887
Saitama	2284	Niigata	1882
Shizuoka	2281	Tottori	1878
Fukuoka	2281	Fukushima	1878
Chiba	2267	Nagoya	1857
Matsue	2238	Yokohama	1856
Kobe	2210	Kitakyushu	1853
Fukui	2131	Sapporo	1847
Kawasaki	2105	Oita	1818
Sagamihara	2105	Matsuyama	1815
Kofu	2095	Kumamoto	1814
Saga	2055	Yamaguchi	1805
Nagano	2050	Tsu	1775
Sendai	2030	Aomori	1669
Tokushima	2027	Yamagata	1582
Nara	2003	Kochi	1557
Wakayama	1971	Akita	1527

① Let's find out the mean for the 52 cities. Which city is closest to the mean value?

② Let's write the number of cities per class in the following frequency distribution table in order to explore the distribution.

Amount of money spent on Gyoza

Class (yen)	City
greater than or equal to 1000 ～ less than 1500	
1500 ～ 2000	
2000 ～ 2500	
2500 ～ 3000	
3000 ～ 3500	
3500 ～ 4000	
Total	

③ Based on the table made in ②, let's draw a histogram.

④ Which class has the largest number of cities?

⑤ In which class is the median value included?

⑥ In which class is the mean value found in ① included?

⑦ Based on the investigation done so far, let's discuss which representative value described this data the best.

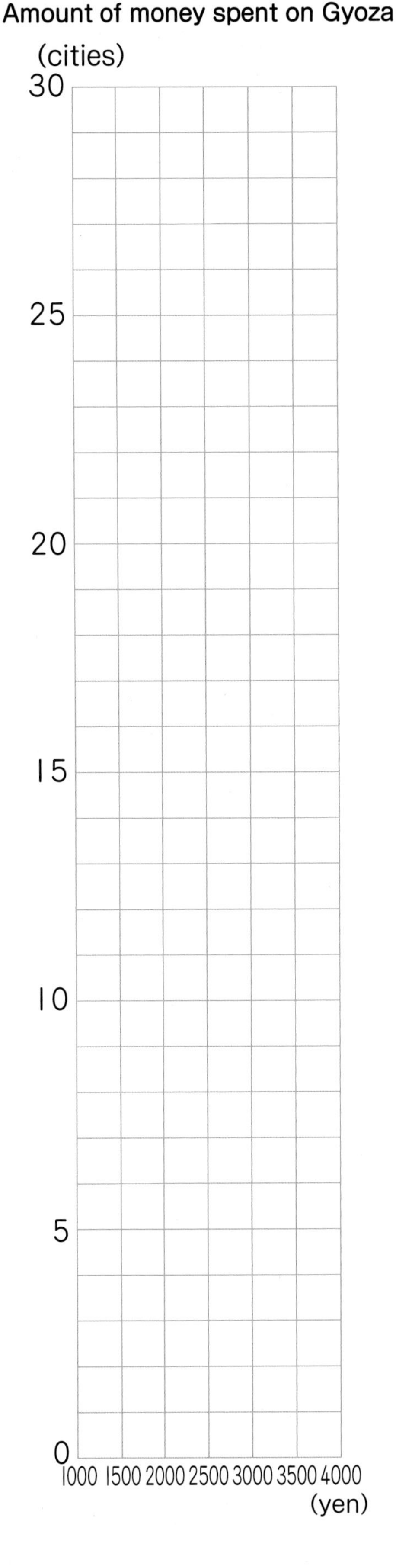

1 In the following cases ①~③, which is the representative value that best describes the data?

① At a store that sells indoor shoes, decide which size of indoor shoes to prepare this year based on the sizes of indoor shoes sold last year.

② Based on the records of the new physical fitness test of all your classmates, you want to find out if your score is higher than the middle scores.

③ As for a relay competition between Group 1 and Group 2, you want to predict the results based on the records of each student per group.

15

Summary

Let's review what you have learned in mathematics over the past 6 years.

Numbers and calculations, math equations

1 Let's summarize about whole numbers and decimal numbers. → 4th grade

① Let's write the numbers that are 10 times, 100 times, and 1000 times of 1.07.

② Let's write the number that is $\frac{1}{10}$ and $\frac{1}{100}$ of 521.

③ As for the following numbers, how many units are gathered using the number inside [ ] as the reference?

ⓐ 23000 [100] ⓑ 23000 [1000]

ⓒ 2.3 [0.1] ⓓ 2.3 [0.01]

2 Let's summarize about fractions. → 3rd grade 4th grade 5th grade

① Let's write the equality or inequality sign inside the following □.

ⓐ $\frac{2}{5}$ □ $\frac{3}{5}$ ⓑ $\frac{2}{5}$ □ $\frac{2}{7}$ ⓒ $\frac{2}{5}$ □ $\frac{8}{20}$

② Let' s write the number that applies inside the following □.

ⓐ $\frac{3}{5}$ has □ sets of $\frac{1}{5}$. ⓑ $\frac{9}{7}$ has 9 sets of □.

③ Let's change the following mixed fractions into improper fractions and the improper fractions into mixed fractions.

ⓐ $1\frac{2}{3}$ ⓑ $4\frac{3}{5}$ ⓒ $\frac{7}{4}$ ⓓ $\frac{8}{3}$

3 Let's summarize the properties of whole numbers. → 5th grade

① From the whole numbers up to 50, let's find the numbers that have three divisors.

② Let's find the least common multiple and the greatest common divisor from the following pair of numbers.

ⓐ (12, 18) ⓑ (8, 16)

4 Let's summarize the relationship between whole numbers, decimal numbers, and fractions.

→ 5th grade

① Let's change the following whole number and decimal numbers into fractions and the fractions into decimal numbers.

ⓐ 4 ⓑ 0.7 ⓒ 3.08 ⓓ $\frac{13}{25}$ ⓔ $1\frac{3}{4}$

② Let's arrange the following five numbers in ascending order.

$\frac{2}{5}$ $\frac{1}{3}$ $\frac{7}{15}$ 0.3 0.41

5 Let's summarize how to calculate.

→ 3rd grade 4th grade 5th grade 6th grade

① Let's calculate the following.

ⓐ $4+2\times6-3$ $(4+2)\times6-3$ $4+2\times(6-3)$

ⓑ $4.2+1.5$ $4.2-1.5$ 4.2×1.5 $4.2\div1.5$

ⓒ $64.8+1.8$ $64.8-1.8$ 64.8×1.8 $64.8\div1.8$

ⓓ $\frac{2}{5}+\frac{1}{3}$ $\frac{2}{5}-\frac{1}{3}$ $\frac{2}{5}\times\frac{1}{3}$ $\frac{2}{5}\div\frac{1}{3}$

ⓔ $6\div\frac{2}{7}\div\frac{3}{4}$ $1.2\div1.5\times\frac{3}{4}$

② Let's find out the number that applies for x.

ⓐ $8+x=15$ ⓑ $x\times7=56$

6 Let's write a math equation, using x, for the area of the following triangle and trapezoid. Let's find out the number that applies for x.

→ 5th grade 6th grade

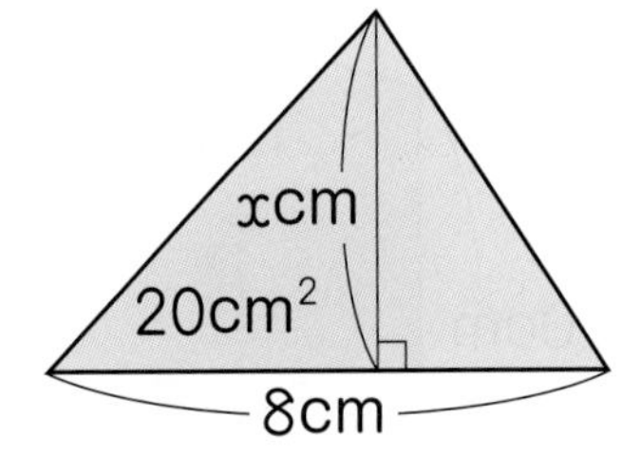

ⓑ

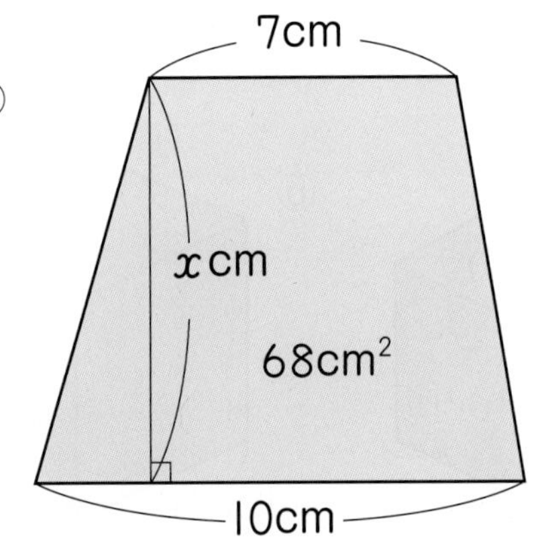

1 Let's summarize how to find out the area.

→ 4th grade
5th grade
6th grade

① Let's write the area formula for the following figures.

Area of a rectangle = ☐ × ☐

Area of a square = ☐ × ☐

Area of a parallelogram = ☐ × ☐

Area of a triangle = ☐ × ☐ ÷ ☐

Area of a circle = ☐ × ☐ × ☐

② Let's draw two figures with an area of 20 cm^2.

③ Let's find out the area of the following colored sections.

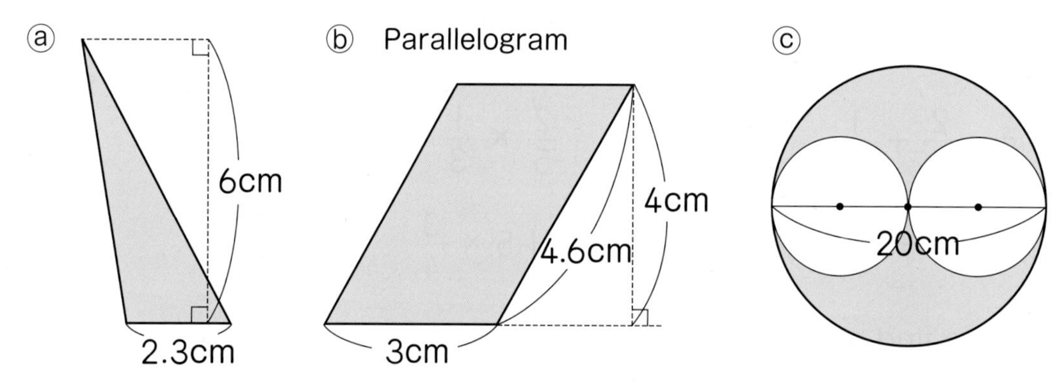

④ There is a flower bed with a length of 1 m and a width of 3 m. What is the area of this flower bed in square meters (m^2)? How many square centimeters (cm^2) is this?

2 Let's summarize how to find out the volume.

→ 5th grade
6th grade

① Let's write the formula to find out the volume of a cuboid and a cube.

② Let's find out the volume of the following solids.

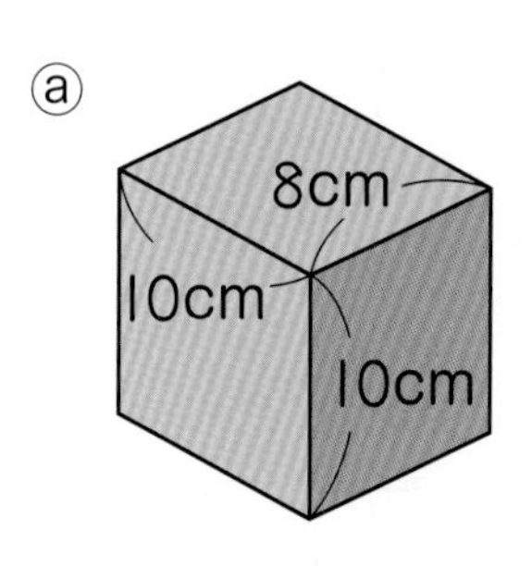

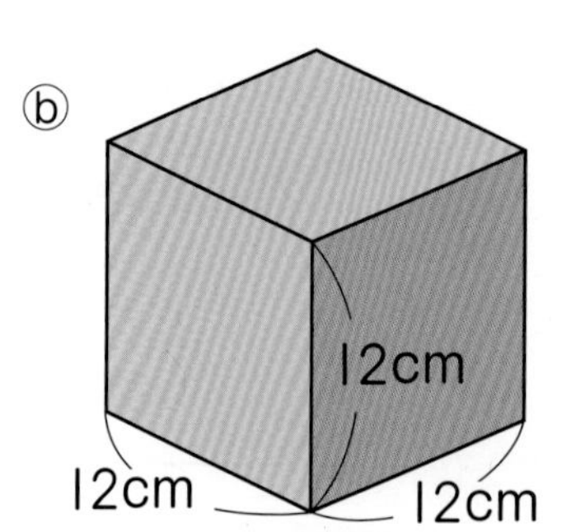

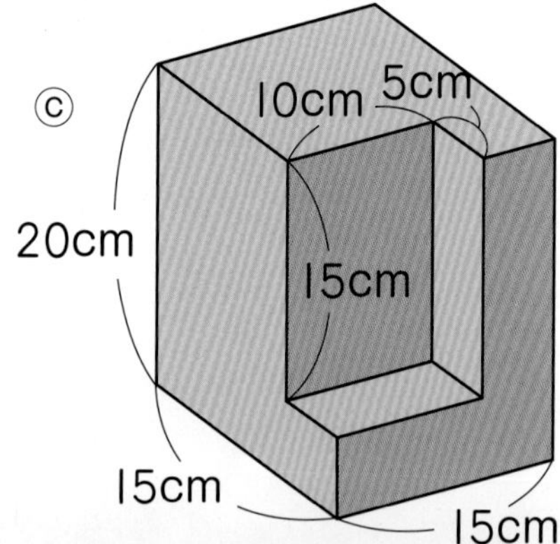

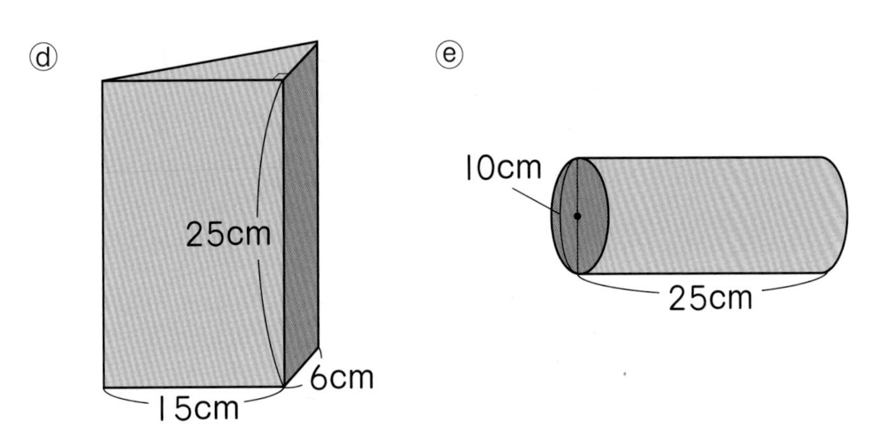

Let's summarize the properties of figures.

→ 2nd grade
4th grade
5th grade

① Let's write ○ for the properties that apply and × for the properties that do not apply about parallelograms, rhombuses, rectangles, and squares.

	Parallelogram	Rhombus	Rectangle	Square
2 pair of sides are parallel.				
All 4 angles are right angles.				
All 4 sides have the same length.				
2 diagonals intersect perpendicularly.				
The sum of adjacent angles is 180°.				

② Let's write the number that applies inside each ☐.

ⓐ

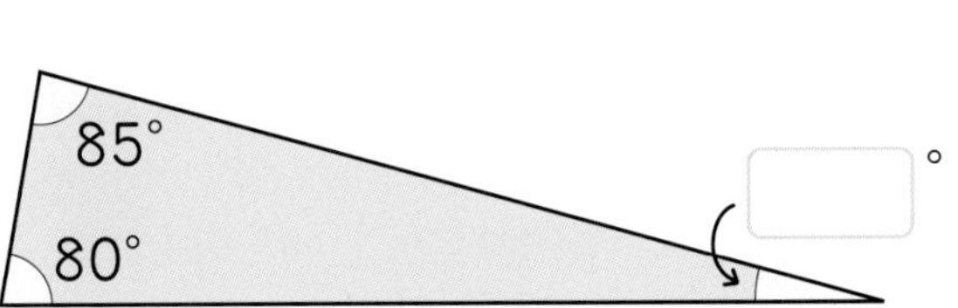

ⓑ

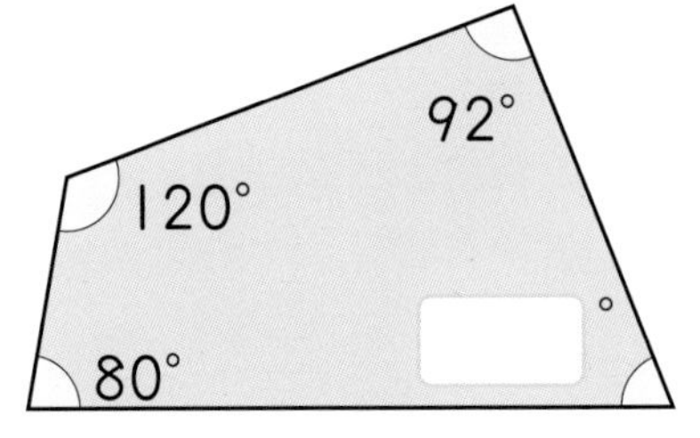

ⓒ Parallelogram

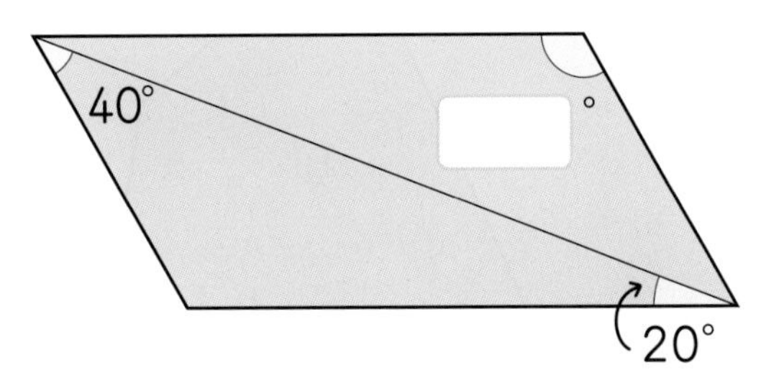

ⓓ Regular hexagon

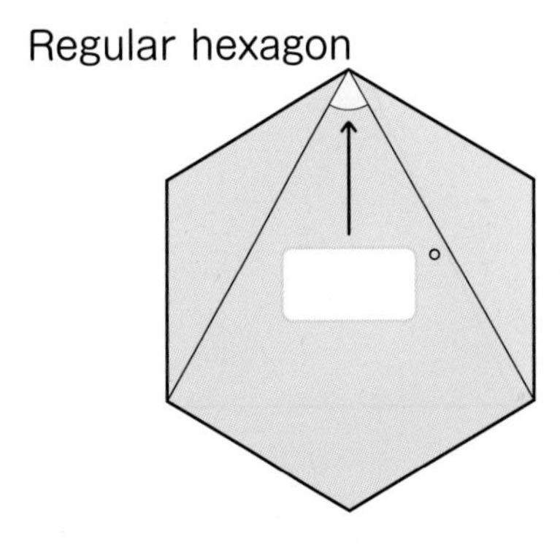

③ Let's explore about the cuboid shown on the right.

ⓐ Which face is parallel to face ABCD?

ⓑ Which side is parallel to side AB?

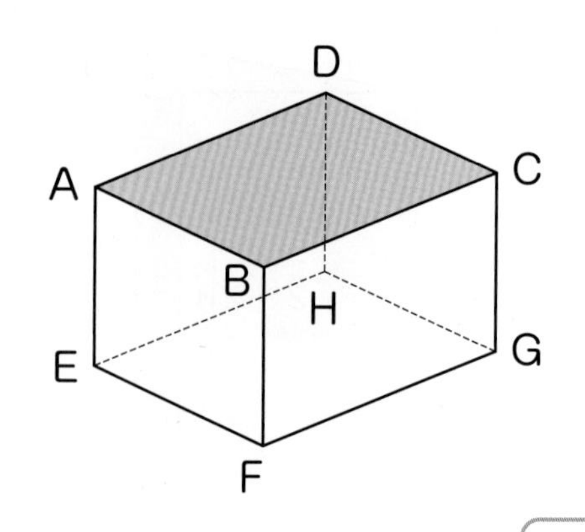

4 Let's draw the following figures.

→ 6th grade

① a line symmetric figures and straight line XY as the line symmetry

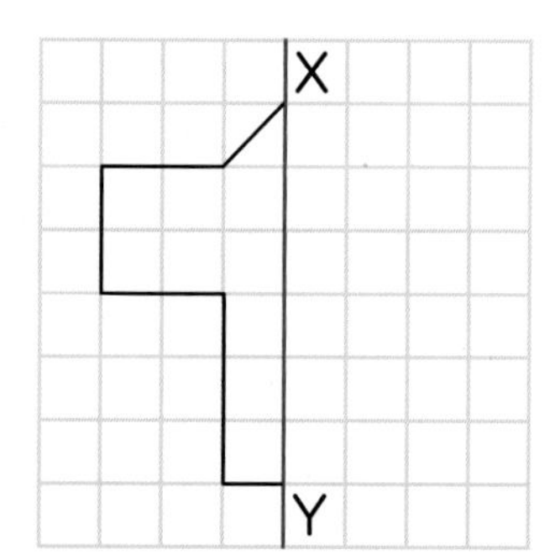

② a point symmetric figure and point O as the point of symmetry

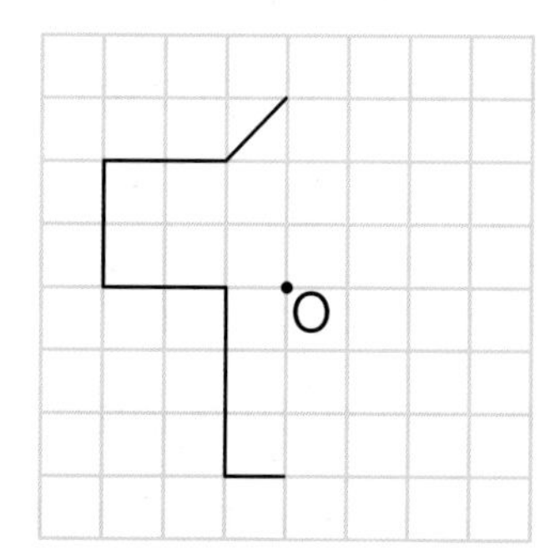

③ an enlargement of 2 times Ⓐ

④ a reduction of $\frac{1}{2}$ Ⓑ

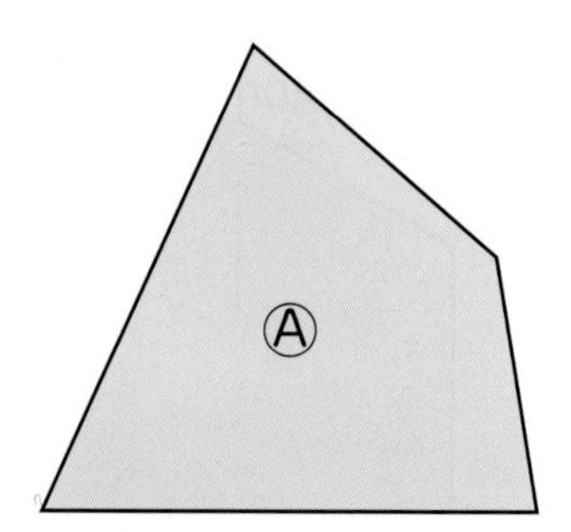

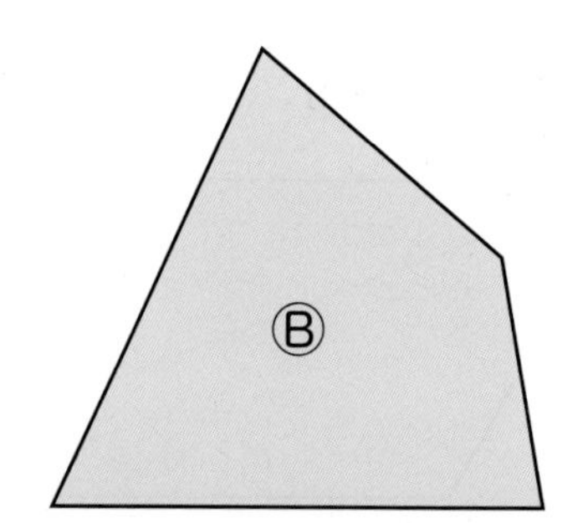

1 Let's summarize about the unit of quantities in your surroundings.

→ 2nd grade 3rd grade 4th grade

① Let's write the unit that applies inside the following ☐.

ⓐ The cover area of the mathematics textbook is about 540 ☐.

ⓑ The amount of milk inside a milk pack is 200 ☐.

ⓒ The weight of one egg is about 50 ☐.

ⓓ Shinano River is Japan's longest river with a length of about 367 ☐.

② Let's answer the following questions.

ⓐ Koharu walked 1.6 km. How many meters remain to complete walking 2 km?

ⓑ There are 4 water bottles with a capacity of 500 mL. In total, how many liters can be poured in? How many deciliters is equivalent to this?

2 Let's summarize about the size per unit quantity.

→ 5th grade

① Haruka's city has a population of 39,000 and an area of 50 km^2. The neighboring city has a population of about 45,000 and an area of about 65 km^2. Which city has a higher population density?

② There is a car that uses 15 liters of gasoline to run 360 km. How many kilometers can this car run with 27 liters of gasoline?

3 Let's summarize about speed.

→ 5th grade

① How many meters per second is 240 m per minute? How many kilometers per hour is this?

② I started walking from the station towards the library that is 1.5 km away. After 15 minutes, I arrived at the park that is 900 m far from the station. If I continue walking with the same speed, how long will it take me to arrive at the library from the park?

4 Let's summarize how to represent the relationship between quantities.

→ 3rd grade
4th grade
5th grade
6th grade

① What kind of graph should be used to represent the following?

ⓐ ratio of import value for each type of imported food

ⓑ changes in export values

ⓒ rice production in each country

② The table shown below summarizes the number of books and magazines published in one year.

Number of Books and Magazines

(hundred million books)

	1990	2020
Books	9.1	5.3
Weekly magazines	15.6	2.4
Monthly magazines	20.2	7.1
Total	44.9	14.8

ⓐ As for the ratio of monthly magazines, what is the percentage to all publications in each year? Round off to the nearest whole number.

ⓑ Let's represent the ratio of books and magazines published each year using a strip graph. Let's discuss what you noticed.

Number of Books and Magazines

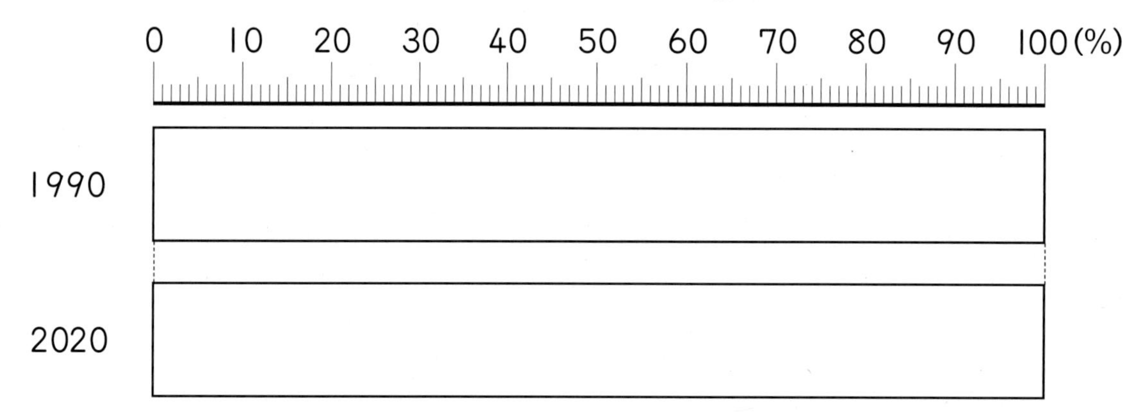

③ To make sweet soybean flour, 35 g of soybean flour and 14 g of sugar were mixed.

ⓐ If the value of the sugar is set to 2, what is the value of the soybean flour?

ⓑ Sweet soybean flour will be made. If there are 140 g of soybean flour, how many grams of sugar is needed?

5 Let's explore the relationship between x and y summarized in tables Ⓐ and Ⓑ. → 6th grade

Ⓐ

Number of people by which a string is divided x (people)	2	3	4	6	8	
Length of the string per person y (m)	12	8	6	4	3	

Ⓑ

Length of the string x (m)	0	1	2	3	4	5	
Weight of the string y (g)	0	8	16	24	32	40	

① In which table is y proportional to x? In which table is y inversely proportional to x?

② As for tables Ⓐ and Ⓑ, let's represent the relationship between x and y in a math equation.

③ Let's draw a graph for the proportional relationship.

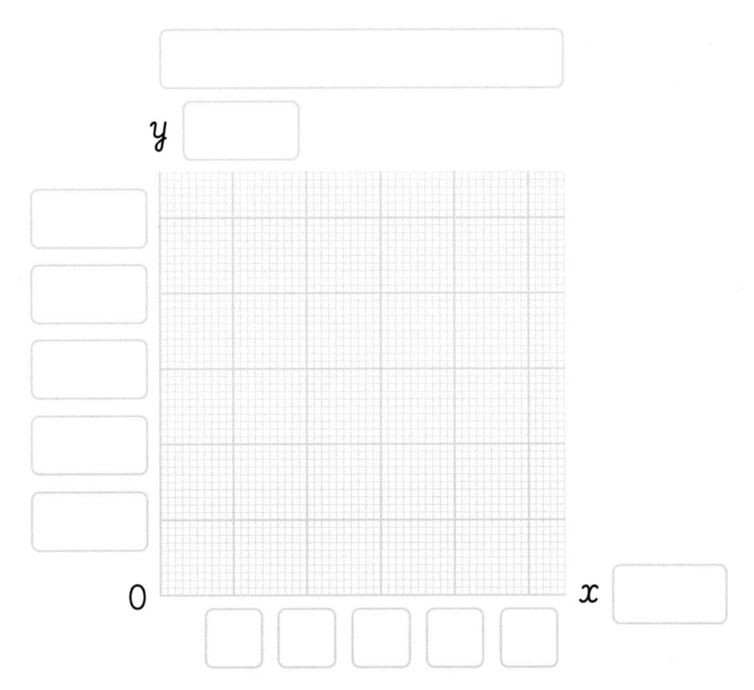

Introduction to Computational Thinking

Let's figure out how to transfer the "Tower of Hanoi" disk.

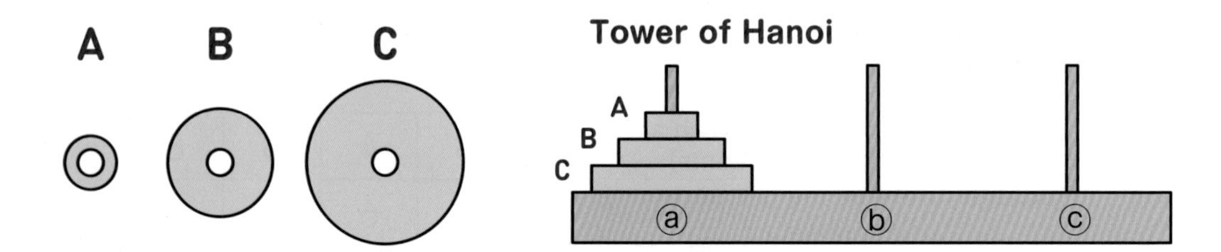

Rules

- Move the disks A, B, C, that are stuck on rods ⓐ onto ⓑ.
- Only one disk can be transferred at a time, and a larger disk cannot be placed on top of a smaller disk.

① From the starting position, Sara moved the disks three times until she got the following position. Let's explain how she moved the disks.

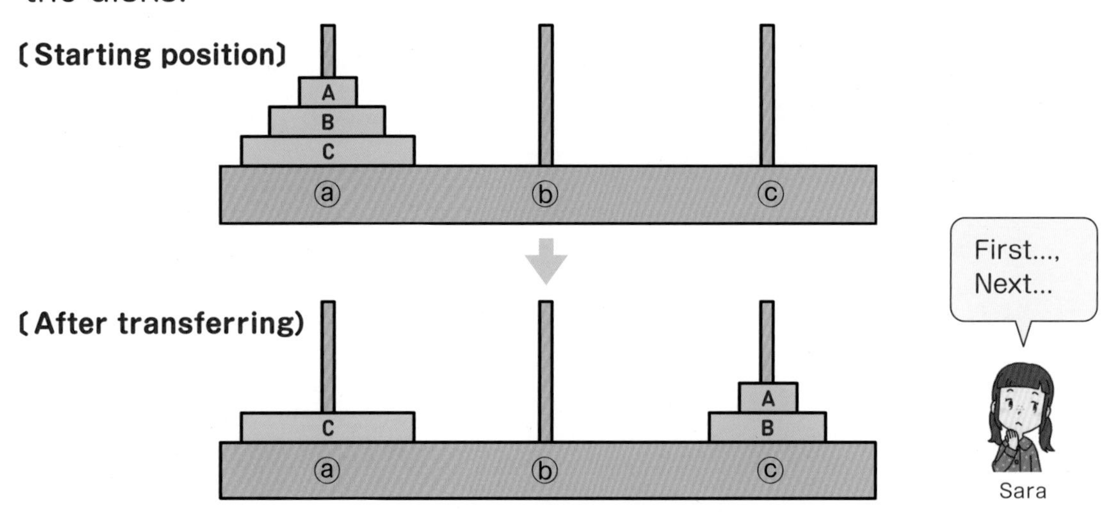

② From the state after the transfer of ①, continue transferring disks, and think about how to transfer all the disks to ⓑ.

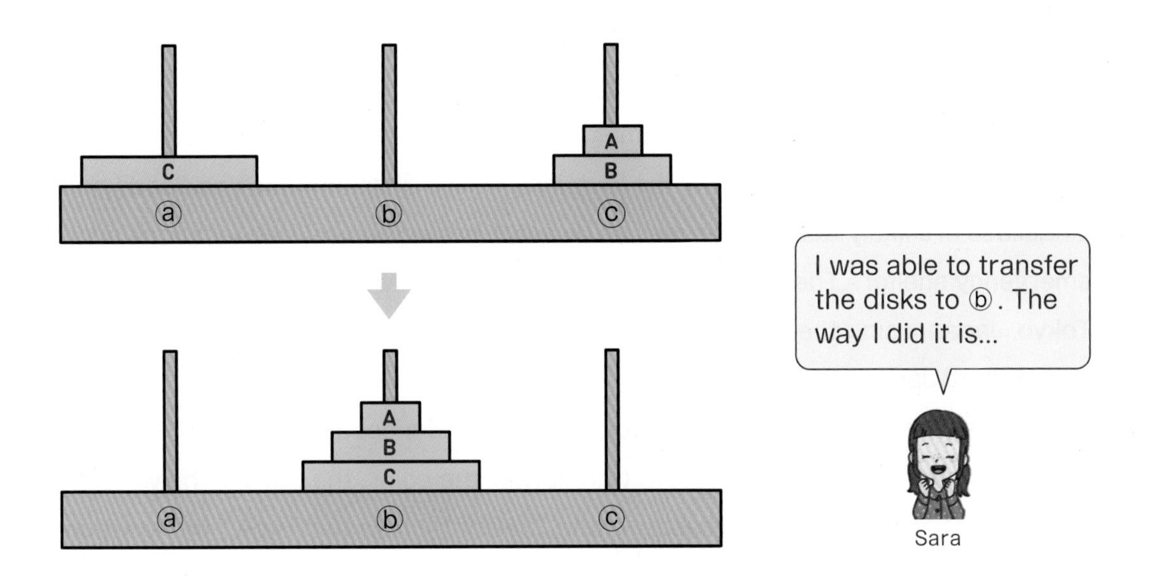

③ Think of a way to transfer the disks from ⓑ to ⓒ, and explain the order in which the disks were transferred.

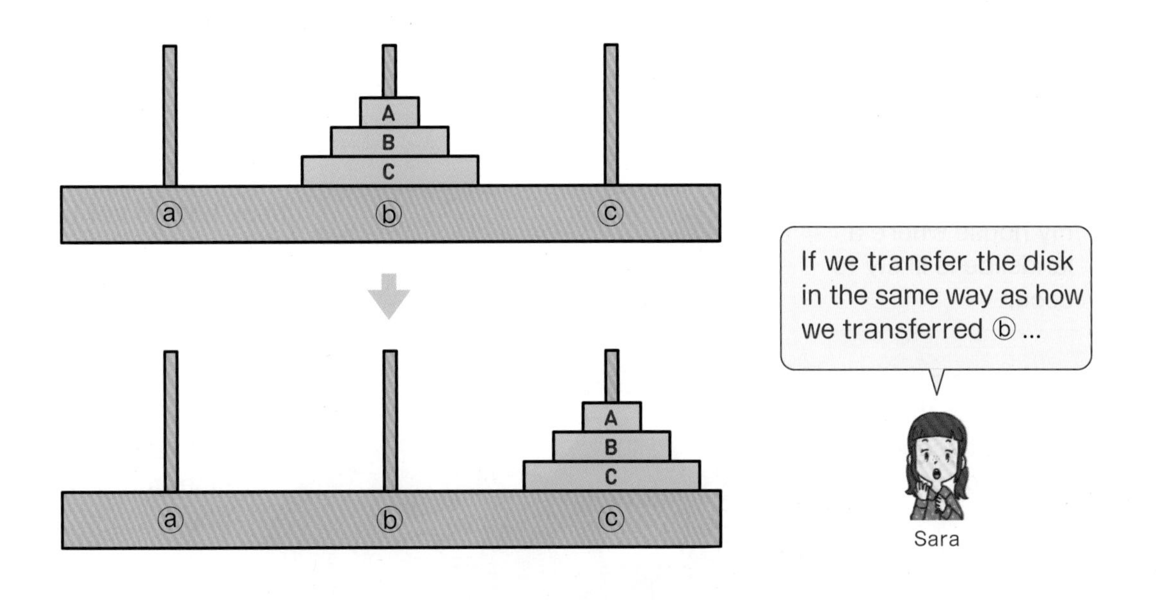

A
B
C
ⓐ
ⓑ
ⓒ
A
B
C
ⓐ
ⓑ
ⓒ
If we transfer the disk in the same way as how we transferred ⓑ ...
Sara

Utilizing Math for SDGs

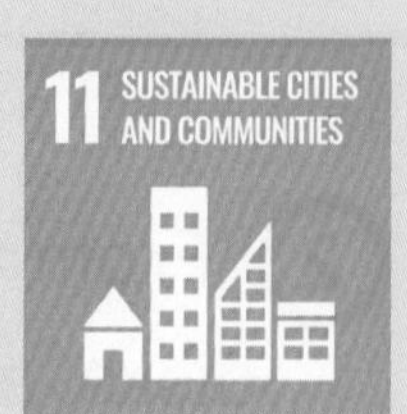

Let's take a look at a hazard map.

Do you know what a hazard map is?
Hazard maps are used to reduce damage caused by natural disasters and for disaster prevention measures. They show the locations where disasters such as floods and landslides are likely to occur, as well as disaster prevention-related locations such as emergency shelters.The figure on the right shows the hazard damage caused by flood in Tokyo, Japan, using the "Hazard Map Portal Site" published by the Geospatial Information Authority of Japan.
The map shows the locations that would be flooded if a flood occurs, and the maximum depth of water that can be expected (the darker the color, the deeper the water).
If you select another type of disaster in the "Select by Disaster Type," the locations where the other type of disaster will occur will be colored.
By looking at the hazard map, you can learn about disasters that may occur in your area and learn how to reduce the damage. Let's think about what you can do to reduce the damage.

① What is the scale of the hazard map on the right?
Find out by measuring the length of the 5 km in the diagram in the lower right.
② Look at the hazard map of your area and find out the route, distance, and time required to reach the evacuation site from your house.

There is a place near my house where a landslide is likely to occur.
There are places that get flooded by floods.
It would be safer to check how to get to the evacuation site in advance.
We may need to check our emergency carry-on bag.

Screen of the hazard map portal site

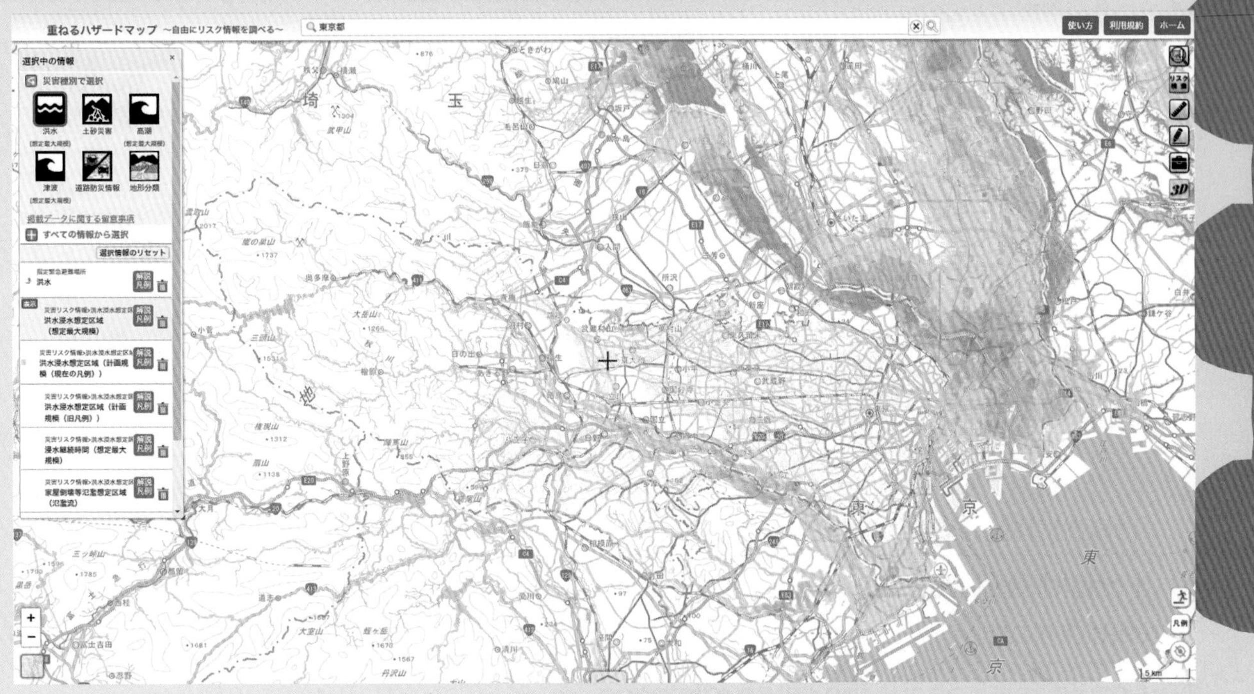

source: Geospatial Information Authority of Japan

Haruto

I chose "landslide" at the same location. The colored parts change.

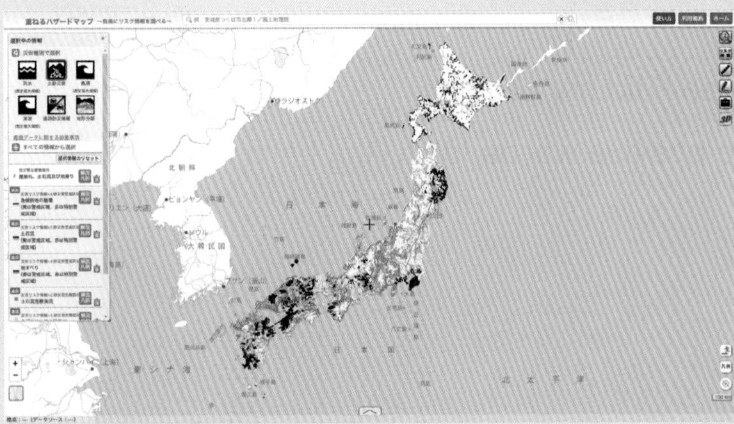

I tried to show the whole Japan. Different places are prone to different disasters.

Akari

Think back on what you felt through this activity, and put a circle.

Let's reflect on yourself!

❶ Competency towards learning	Strongly agree	Agree	Don't agree
① I could check the hazard map of the area where I live.			
② I could find out about disasters using hazard maps.			
③ I could learn about other regions with interest.			

❷ Competency to think, decide, and represent	Strongly agree	Agree	Don't agree
① I could find out the scale of the hazard map.			
② I could think about the route, distance, and the time required to reach the evacuation site.			

❸ What I know and what I can do	Strongly agree	Agree	Don't agree
① I know the evacuation site.			
② I could find out the distance and the time required to reach the evacuation site.			
③ I learned about possible disasters.			

❹ Encouragement for myself	Strongly agree
① I am proud of myself because I did my best.	

Let's praise yourself with some positive words for trying hard to learn!

More Math!

[Supplementary Problems]

[Let's deepen.]

[Answers]

1 Symmetry

→ pp.12～29

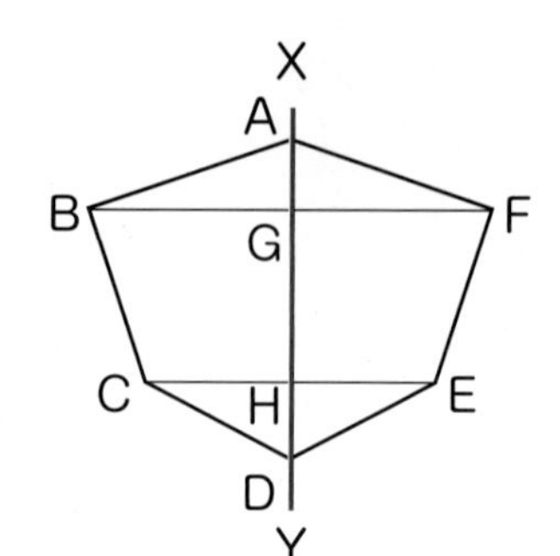

1 Let's answer about the line symmetric figure shown on the right.

① Which is the corresponding side to side BC?

② How is the intersection between the straight line BF and the line of symmetry?

③ Which other straight line has the same length as straight line EH?

2 In the following diagrams, let's complete the line symmetric figures considering straight line XY as the line of symmetry.

①

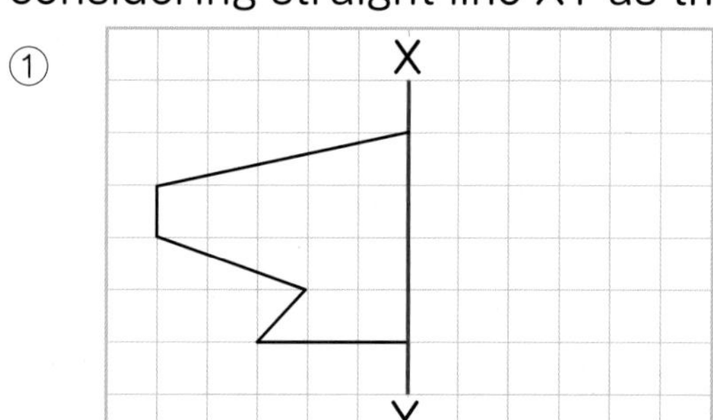

②

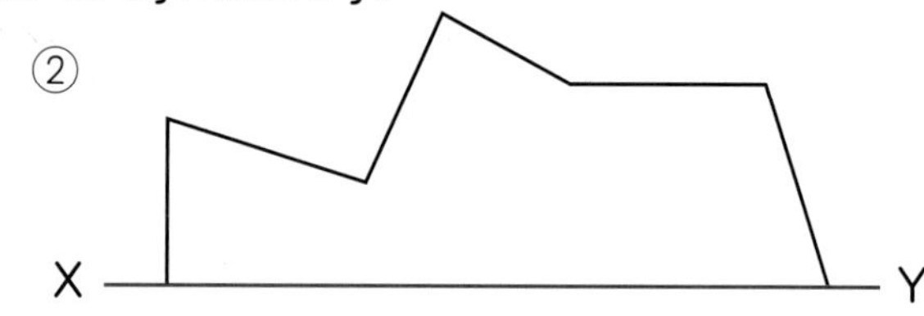

3 Let's answer about the point symmetric figure shown on the right.

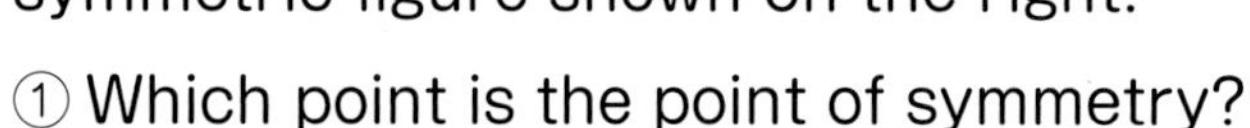

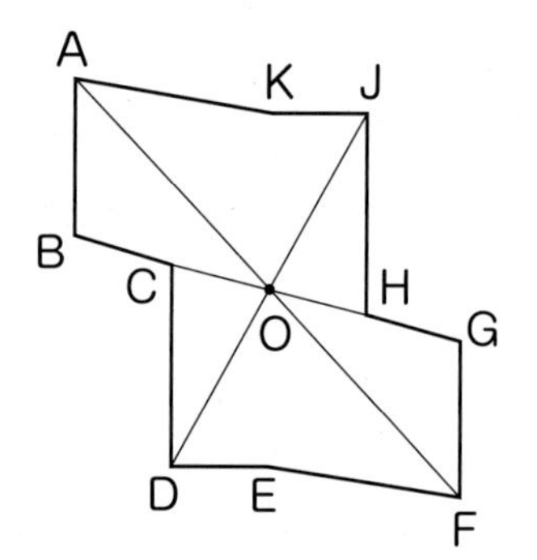

① Which point is the point of symmetry?

② Which is the corresponding point to point D?

③ Which is the corresponding side to side BC?

④ Which other straight line has the same length as straight line AO?

4 In the following diagrams, let's complete the point symmetric figures considering point O as the point of symmetry.

①

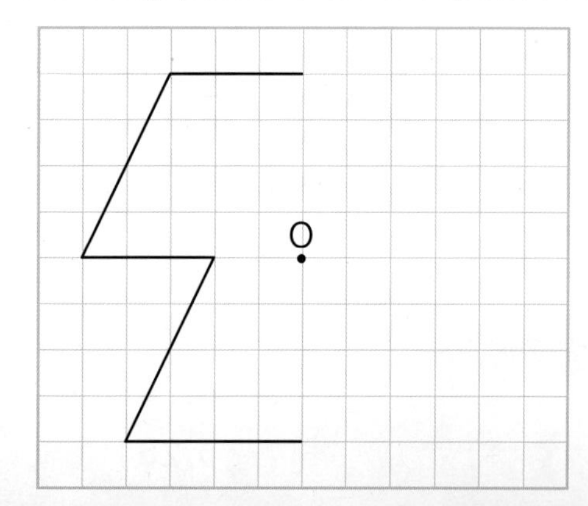

②

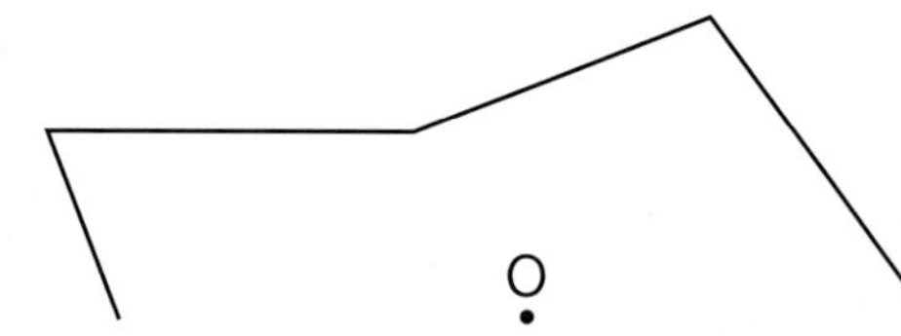

2 Mathematical Letter and Equation

→ pp.30 ~ 43

1 Some deciliters of juice were drunk from a total amount of 14 dL. The remaining amount became 5 dL. Let's answer the following questions.

① Let's write a math equation to find out the original amount of juice, considering that x dL of juice were drunk.

② How many deciliters of juice were drunk?

2 There is a square with a surrounding length of 26 cm. Let's answer the following questions.

① Let's write a math equation to find out the surrounding length, considering that the length of one side of the square is x cm.

② How many centimeters is the length of one side of the square?

3 Let's find the number that applies for x.

① $x+8=44$ ② $x+19=27$ ③ $27+x=50$ ④ $46+x=61$

⑤ $x-7=15$ ⑥ $x-18=22$ ⑦ $x-43=19$ ⑧ $x-25=68$

⑨ $9\times x=54$ ⑩ $3\times x=21$ ⑪ $x\times 8=32$ ⑫ $x\times 6=9$

4 There are 3 boxes and 5 cookies. Let' s answer the following questions.

① Let' s write a math expression that represents the total number, considering that 1 box has x cookies.

② There were 59 cookies in total. How many cookies did each box have? Let' s explore the number that applies for x by placing 16, 17, 18, ... as x.

5 The following math expressions ①~③ represent the area of the flower bed shown on the right.

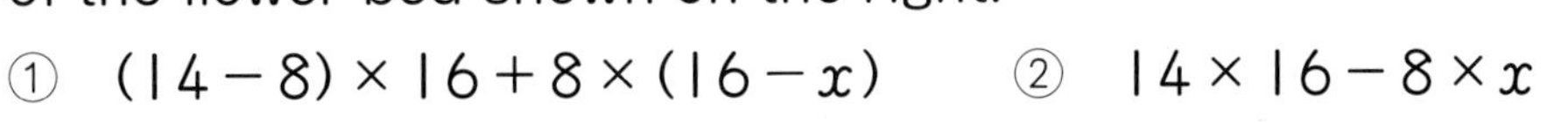

① $(14-8)\times 16+8\times(16-x)$ ② $14\times 16-8\times x$

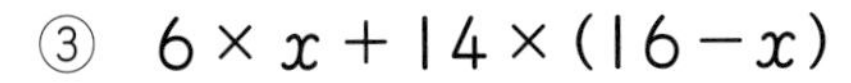

③ $6\times x+14\times(16-x)$

Which of the following diagrams Ⓐ~Ⓒ represent each of the above math expression?

Ⓐ

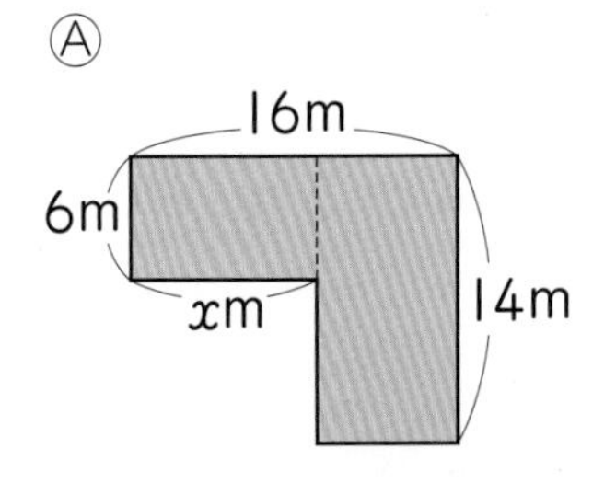

Ⓑ

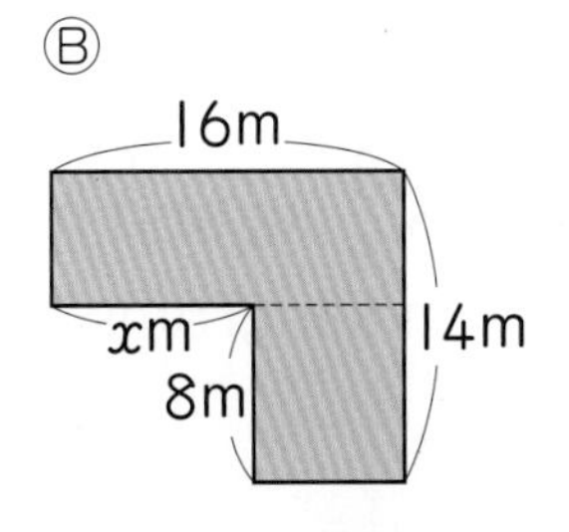

Ⓒ

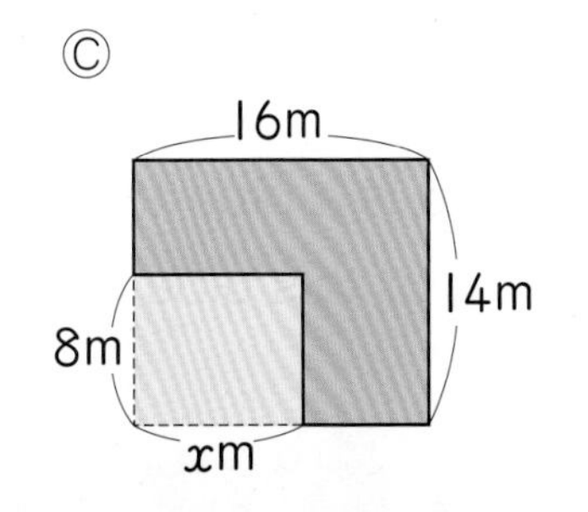

3 Multiplication and Division of Fractions and Whole Numbers

→ pp.44 ~ 57

1 Let's calculate the following.

① $\frac{2}{7} \times 2$ ② $\frac{3}{10} \times 3$ ③ $\frac{1}{5} \times 3$ ④ $\frac{3}{4} \times 5$

⑤ $\frac{4}{9} \times 4$ ⑥ $\frac{5}{8} \times 3$ ⑦ $\frac{5}{12} \times 2$ ⑧ $\frac{4}{15} \times 5$

⑨ $\frac{1}{6} \times 4$ ⑩ $\frac{2}{9} \times 6$ ⑪ $\frac{9}{8} \times 12$ ⑫ $\frac{6}{5} \times 10$

⑬ $1\frac{2}{9} \times 4$ ⑭ $1\frac{5}{11} \times 2$ ⑮ $1\frac{1}{10} \times 4$ ⑯ $2\frac{4}{9} \times 3$

⑰ $2\frac{1}{4} \times 2$ ⑱ $2\frac{4}{15} \times 5$

2 4 pieces of tape will be made, each with a length of $\frac{5}{8}$ m. What is the length of the tape, in meters, that will be needed in total?

3 What is the area of a rectangular field, in square meters, that has a length of $3\frac{1}{2}$ m and a width of 4 m?

4 Let's calculate the following.

① $\frac{1}{3} \div 3$ ② $\frac{5}{7} \div 4$ ③ $\frac{3}{5} \div 2$ ④ $\frac{3}{8} \div 5$

⑤ $\frac{4}{9} \div 3$ ⑥ $\frac{3}{4} \div 7$ ⑦ $\frac{4}{7} \div 2$ ⑧ $\frac{5}{16} \div 5$

⑨ $\frac{8}{9} \div 6$ ⑩ $\frac{8}{15} \div 4$ ⑪ $\frac{4}{3} \div 2$ ⑫ $\frac{20}{7} \div 16$

⑬ $2\frac{4}{9} \div 3$ ⑭ $1\frac{7}{10} \div 2$ ⑮ $2\frac{1}{2} \div 5$ ⑯ $3\frac{1}{5} \div 6$

⑰ $1\frac{5}{7} \div 4$ ⑱ $2\frac{8}{11} \div 4$

5 $\frac{9}{8}$ L of juice will be equally poured into 3 bottles. What is the amount of juice, in liters, that each bottle contain?

6 There is a pipe that weighs $3\frac{3}{4}$ kg and has a length of 5 m. What is the weight of this pipe per meter in kilograms?

Fraction × Fraction

→ pp.60 ~ 73

1 Let's calculate the following.

① $\frac{5}{8} \times \frac{1}{9}$　② $\frac{7}{9} \times \frac{1}{2}$　③ $\frac{2}{3} \times \frac{2}{5}$　④ $\frac{7}{6} \times \frac{5}{2}$

⑤ $\frac{9}{8} \times \frac{5}{4}$　⑥ $\frac{10}{9} \times \frac{2}{3}$　⑦ $\frac{5}{6} \times \frac{2}{9}$　⑧ $\frac{7}{12} \times \frac{2}{3}$

⑨ $\frac{3}{10} \times \frac{5}{6}$　⑩ $\frac{8}{15} \times \frac{9}{10}$　⑪ $\frac{5}{3} \times \frac{9}{11}$　⑫ $\frac{3}{8} \times \frac{12}{5}$

⑬ $4 \times \frac{3}{8}$　⑭ $5 \times \frac{7}{10}$　⑮ $6 \times \frac{2}{3}$　⑯ $1\frac{1}{2} \times 1\frac{2}{5}$

⑰ $5\frac{5}{6} \times 2\frac{1}{7}$　⑱ $2\frac{2}{3} \times 2\frac{1}{4}$　⑲ $4\frac{2}{9} \times 6\frac{3}{4}$　⑳ $3\frac{1}{5} \times \frac{5}{8}$

㉑ $1\frac{5}{6} \times \frac{3}{10}$　㉒ $\frac{5}{12} \times 1\frac{3}{5}$　㉓ $\frac{2}{3} \times 2\frac{1}{4}$

2 $\frac{7}{10}$ m^2 of a wall can be painted per liter of paint. What is the area, in square meters, that can be painted with $\frac{5}{6}$ dL of this paint?

3 There is an iron bar that weighs $1\frac{1}{8}$ kg per meter. What is the weight, in kilograms, for $2\frac{2}{3}$ m of this iron bar?

4 Let's write the equality or inequality sign that applies in each ☐.

① $4 \times \frac{9}{10}$ ☐ 4　② $\frac{4}{5} \times \frac{4}{3}$ ☐ $\frac{4}{5}$　③ $\frac{5}{9} \times 1\frac{1}{8}$ ☐ $\frac{5}{9} \times \frac{9}{8}$

5 Let's fill in the ☐ with numbers.

① $\frac{4}{5} \times \frac{3}{8} = \square \times \frac{4}{5}$　② $\left(\frac{5}{7} \times \frac{8}{9}\right) \times \frac{3}{4} = \frac{5}{7} \times \left(\square \times \frac{3}{4}\right)$

③ $\left(\frac{2}{3} + \frac{2}{5}\right) \times \square = \frac{2}{3} \times \frac{3}{10} + \frac{2}{5} \times \frac{3}{10}$

④ $\left(\frac{7}{8} - \frac{2}{3}\right) \times 24 = \frac{7}{8} \times 24 - \frac{2}{3} \times \square$

6 Let's find out the reciprocal of the following numbers.

① $\frac{3}{8}$　② $1\frac{1}{2}$　③ $2\frac{2}{5}$　④ $\frac{1}{6}$　⑤ 0.8　⑥ 1.3　⑦ 4

5 Fraction ÷ Fraction

→ pp.74 ~ 85

1 Let's calculate the following.

① $\frac{2}{9} \div \frac{3}{5}$ ② $\frac{5}{6} \div \frac{6}{7}$ ③ $\frac{8}{3} \div \frac{1}{2}$ ④ $\frac{3}{4} \div \frac{8}{7}$

⑤ $\frac{7}{8} \div \frac{1}{2}$ ⑥ $\frac{3}{10} \div \frac{5}{8}$ ⑦ $\frac{12}{5} \div \frac{8}{7}$ ⑧ $\frac{7}{4} \div \frac{14}{3}$

⑨ $\frac{5}{3} \div \frac{5}{6}$ ⑩ $\frac{5}{9} \div \frac{10}{3}$ ⑪ $5 \div \frac{4}{7}$ ⑫ $3 \div \frac{2}{9}$

⑬ $6 \div \frac{3}{8}$ ⑭ $15 \div \frac{3}{5}$ ⑮ $\frac{4}{7} \div 1\frac{4}{5}$ ⑯ $\frac{5}{8} \div 2\frac{6}{7}$

⑰ $8 \div 1\frac{7}{10}$ ⑱ $11 \div 2\frac{3}{4}$ ⑲ $2\frac{1}{4} \div \frac{3}{7}$ ⑳ $3\frac{3}{5} \div \frac{9}{10}$

㉑ $1\frac{7}{8} \div 1\frac{2}{3}$ ㉒ $5\frac{5}{6} \div 3\frac{1}{2}$

2 There is a metal bar that is $\frac{4}{5}$ m long and weighs $\frac{8}{9}$ kg. How many kilograms is the weight per meter of this metal bar?

3 Let's write the inequality sign that applies in each □.

① $9 \div \frac{2}{3}$ □ 9 ② $\frac{5}{6} \div \frac{6}{5}$ □ $\frac{5}{6}$ ③ $\frac{7}{10} \div \frac{7}{8}$ □ $\frac{7}{10}$

4 $2\frac{1}{4}$ dL of paint were used to paint each square meter of a wall. What is the area, in square meters, that can be painted with 18 dL of this paint?

5 There is a rectangular flower bed with a length of $1\frac{5}{7}$ m and an area of $3\frac{3}{5}$ m^2. What is the width of this flower bed in meters?

6 Nozomi cut and used a piece of tape with a length of $1\frac{1}{3}$ m. This piece represented $\frac{3}{5}$ of the initial tape. What was the length of the initial tape in meters?

7 Keita drank $\frac{2}{9}$ of the total amount of juice. The amount of juice that Keita drank was 180 mL. What was the initial amount of juice in milliliters?

6 Data Arrangement

→ pp.86 ~ 101

1 Look at the records for throwing a softball and answer the following questions.

Records at throwing a softball

Number	Distance (m)	Number	Distance (m)	Number	Distance (m)
1	28	8	29	15	31
2	33	9	43	16	23
3	20	10	34	17	37
4	26	11	46	18	19
5	16	12	27	19	28
6	35	13	26	20	44
7	24	14	21		

① Let's summarize the records in the following frequency distribution table.

Records at throwing a softball

Distance (m)	Number of children
greater than or equal to 15 ~ less than 20	
20 ~ 25	
25 ~ 30	
30 ~ 35	
35 ~ 40	
40 ~ 45	
45 ~ 50	
Total	

Records at throwing a softball

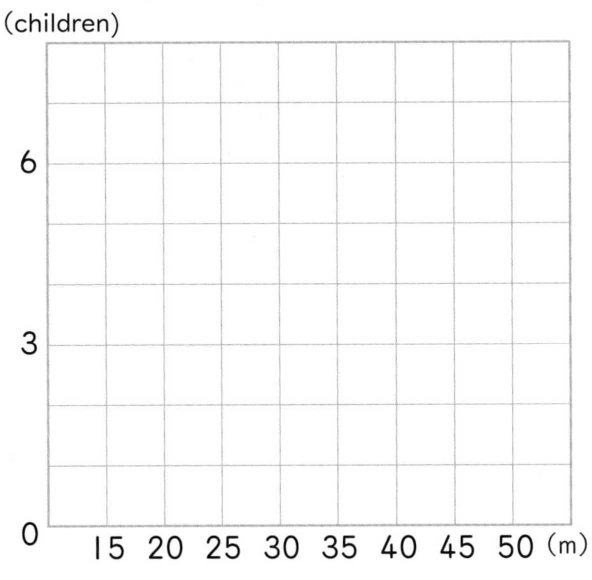

② Let's represent the records with a histogram.

③ Which class has the most number of children?

④ Based on the child with the longest distance, which class does the 5th place child belong to?

2 The 15 scores on the right represent the mathematics test score of children based on a 10 point score. Let's answer the following questions.

(points)

5	7	4	9	3	6	5	10
4	6	8	2	5	9	7	

① How many points is the mean value?

② How many points is the median value?

③ How many points is the mode value?

Ways of Ordering and Combining

→ pp.106 ~ 116

1 Four children, Tsukushi, Takuma, Yota, and Maimi will sit on the four chairs shown on the right. Let's answer the following questions.

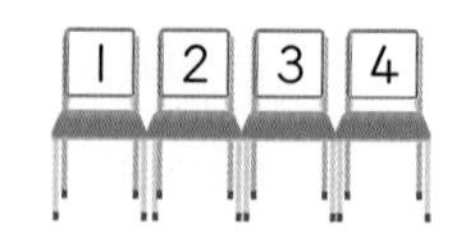

① How many ways of seating are there in the case that Tsukushi sits on seat 1?

② How many ways of seating are there, respectively, in the cases that Takuma, Yota, and Maimi sit on seat 1?

③ How many ways of seating are there in total?

2 There is one card for each of the following numbers: [2], [4], [6], [8]. From these 4 cards, use 3 cards to create 3-digit whole numbers. How many ways are there in total?

3 There is one card for each of the following numbers: [0], [3], [6], [9]. From these 4 cards, use 2cards to create 2-digit whole numbers. Let's answer the following questions.

① What number card can be used in the tens place other than [3]?

② How many 2-digit whole numbers can be made in total?

4 Three coins, A, B, and C, are thrown. How many ways are there in total as for how the front and back of coins come out?

5 Five teams, A, B, C, D, and E, will play soccer games. When each team competes with the other team only once, how many games are there in total?

6 There are 4 sheets of colored paper, red, blue, yellow, and green. When you choose 3 sheets from these, how many ways of choosing are there in total?

Calculations with Decimal Numbers and Fractions

→ pp.117 ~ 123

1 Which of the following calculations Ⓐ~Ⓓ cannot be calculated exactly if it is changed to decimal numbers?

Ⓐ $0.7 + \frac{2}{5}$ Ⓑ $1.3 + \frac{2}{3}$ Ⓒ $\frac{5}{6} - 0.2$ Ⓓ $0.92 - \frac{3}{4}$

2 Let's calculate the following.

① $0.8 + \frac{5}{6}$ ② $0.15 + \frac{3}{4}$ ③ $\frac{7}{9} + 0.5$ ④ $\frac{3}{5} + 0.24$

⑤ $\frac{3}{8} - 0.2$ ⑥ $\frac{3}{4} - 0.7$ ⑦ $\frac{2}{3} - 0.16$ ⑧ $0.45 - \frac{1}{6}$

3 Let's calculate the following using fractions.

① $\frac{5}{8} \times 0.6 \div \frac{1}{2}$ ② $0.25 \div \frac{7}{10} \times 0.9$

③ $0.9 \div 0.63 \times 1.75$ ④ $10 \div 8 \times 6 \div 15$

Area of a Circle

→ pp.128 ~ 142

1 Let's find out the circumference and the area of the following circles.

①
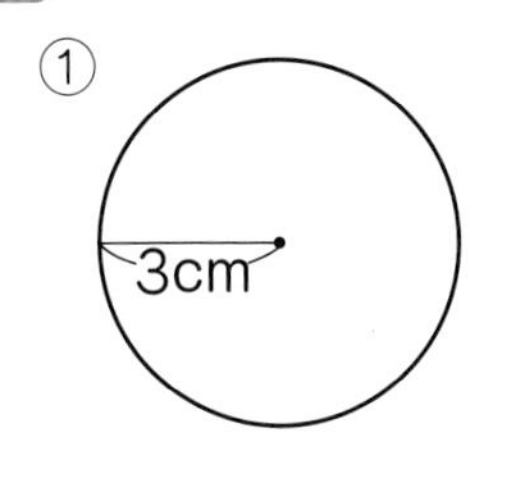

②
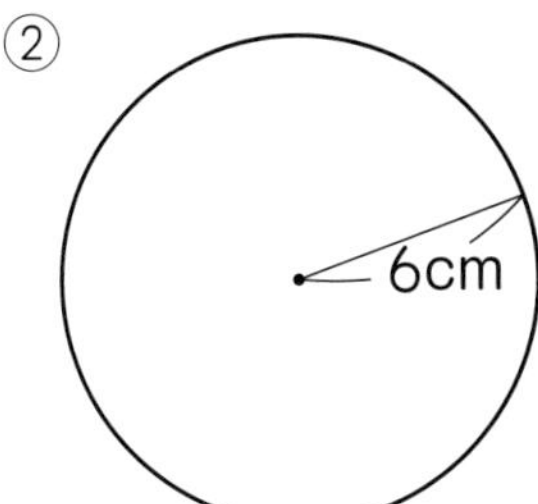

③
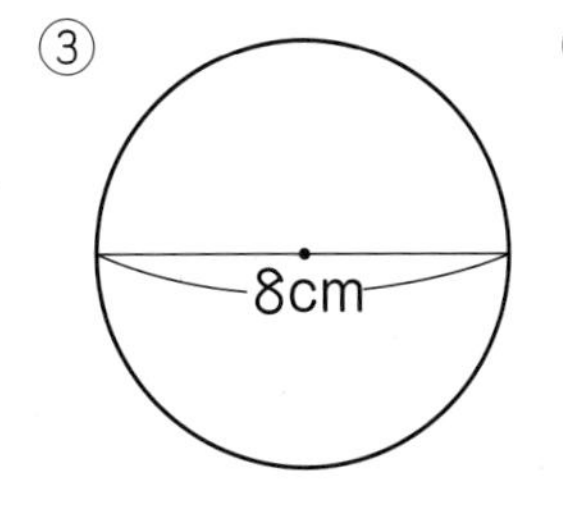

④
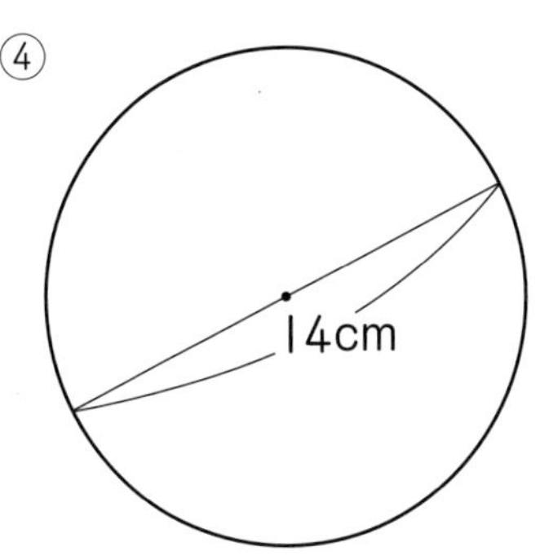

2 Let's find out the area of the colored parts of for the following diagrams.

①
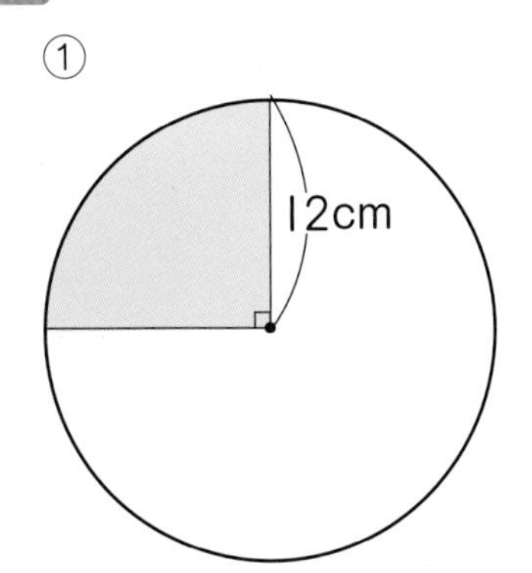

②
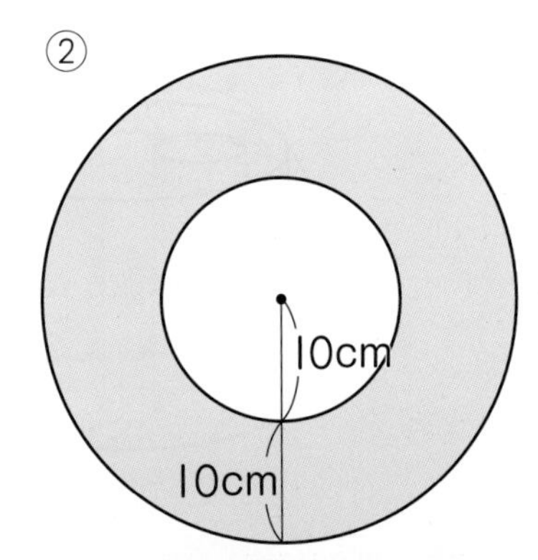

③
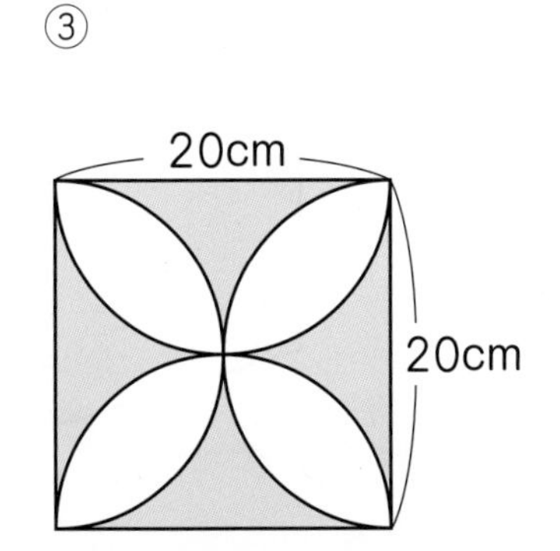

10 Volume of Solids

→ pp.143 ~ 153

1 There is a quadrangular prism like the one shown on the right. Let's answer the following questions, when the rectangular face ⓐ is the base.

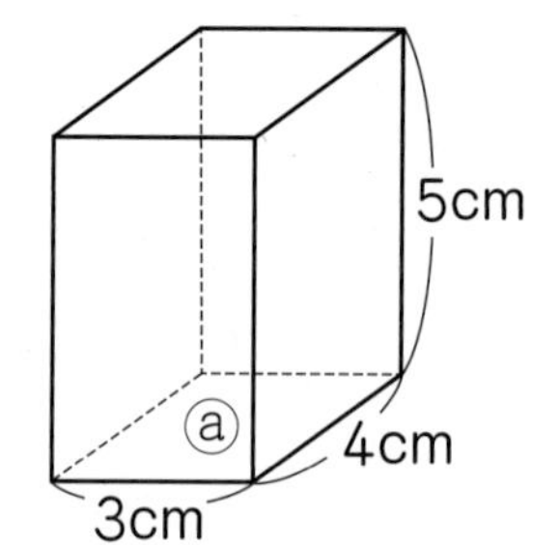

① What is the area of the base in square centimeters?

② What is the height in centimeters?

③ What is the volume in cubic centimeters?

2 Let's find out the volume of the following prisms.

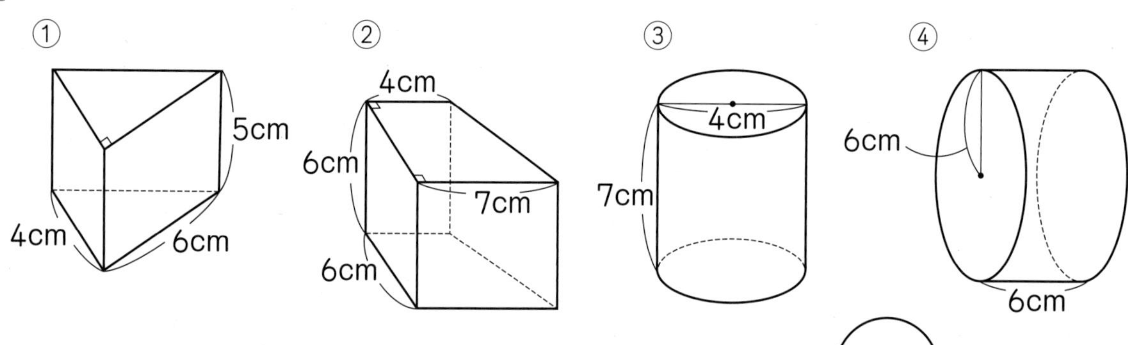

3 The diagram on the right is the net of a cylinder. Let's find out the volume of the cylinder that can be assembled.

6cm

31.4cm

4 Let's find out the volume of the following solids.

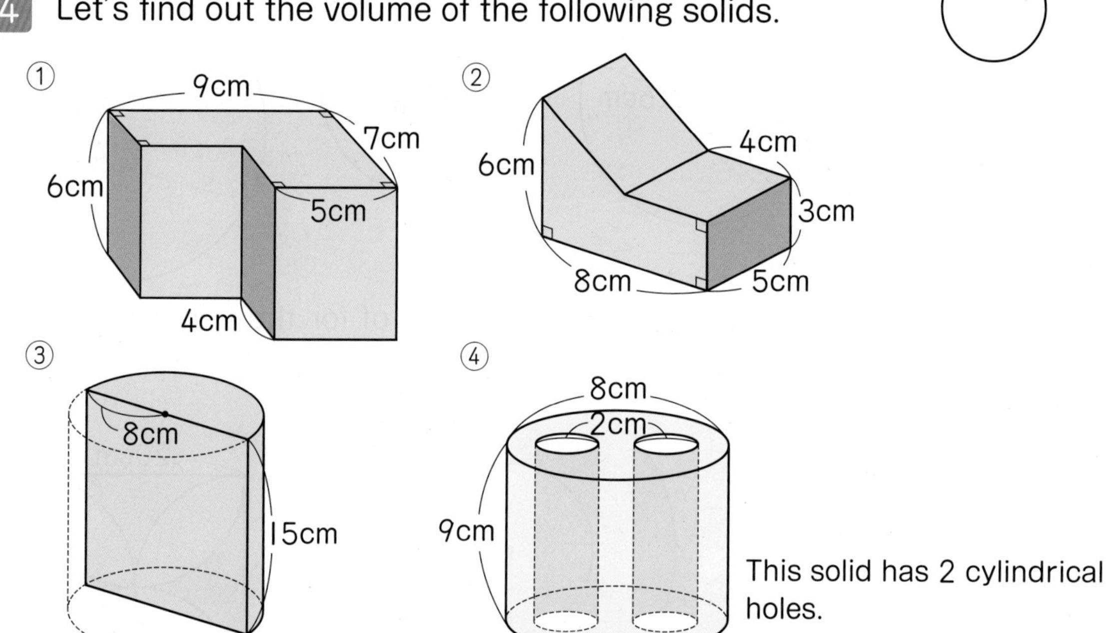

11 Ratio and its Applications

→ pp.158 ~ 169

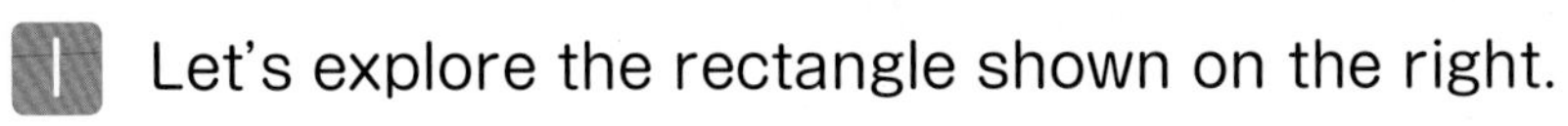

1 Let's explore the rectangle shown on the right.

① Let's write the ratio between the length and the width.

② Let's find out the value of the ratio in ①.

2 Let's find out the value of the following ratios.

① 4 : 9　② 3 : 2　③ 5 : 10　④ 9 : 12

3 Let's write three equal ratios to the following ratios.

① 4 : 1　② 2 : 7　③ 5 : 8　④ 6 : 8

4 Let's find out the number that applies for x.

① $8:5=x:25$　② $7:6=21:x$　③ $4:9=x:36$

④ $3:10=18:x$　⑤ $5:x=1:6$　⑥ $x:20=5:4$

⑦ $28:x=7:3$　⑧ $x:45=8:15$

5 Let's simplify the following ratios.

① 16 : 20　② 36 : 32　③ 150 : 200　④ 144 : 12

⑤ 2.4 : 3.2　⑥ 2 : 0.8　⑦ $\frac{1}{6}:\frac{3}{10}$　⑧ $\frac{2}{7}:\frac{4}{9}$

6 The sisters have some sheets of colored paper. The ratio between the elder sister's number of sheets and that of the younger sister is 5 : 4, and the elder sister has 40 sheets of colored paper. How many sheets does the younger sister have?

7 The shadow of a tree that was measured in the school yard was 8 m. A 0.9 m wooden stick was placed at the school yard and the length of the shadow was 1.2 m. What is the height of the tree in meters?

8 An elder brother and younger brother need to divide 2000 yen which they received from their mother, in a 3 : 2 ratio. How much will the elder brother receive?

12 Enlargement and Reduction of Figures

→ pp.170 ～ 185

1 In the diagram on the right, triangle DEF is an enlargement of triangle ABC. Let's answer the following questions.

① Let's find out the ratio between the length of side BC and side EF.

② What is the length, in centimeters, is the length of sides DE and DF?

③ How many times of triangle ABC is enlarged triangle DEF?

2 In the following on the right, point B is the center of enlarging. Let's draw an enlargement of 2 times triangle ABC.

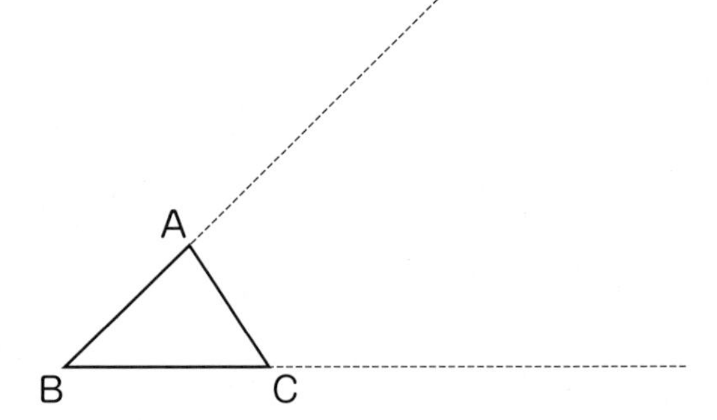

3 Let's draw a reduction of $\frac{1}{2}$ quadrilateral ABCD, using point E as the center point.

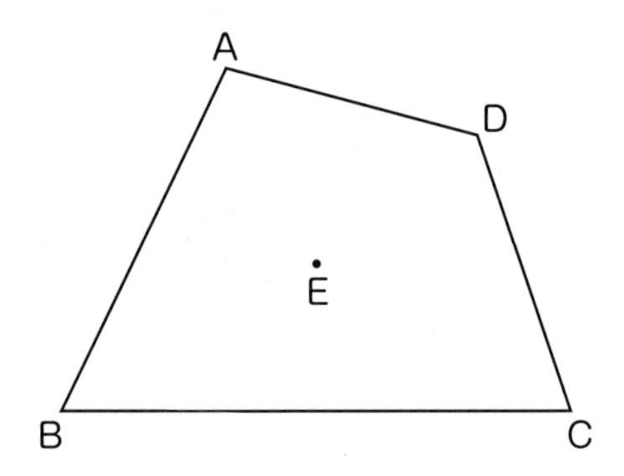

4 In the figure on the right, how many meters is the actual height of the tree? Let's find out the answer by drawing a reduced triangle in $\frac{1}{100}$ reduced scale.

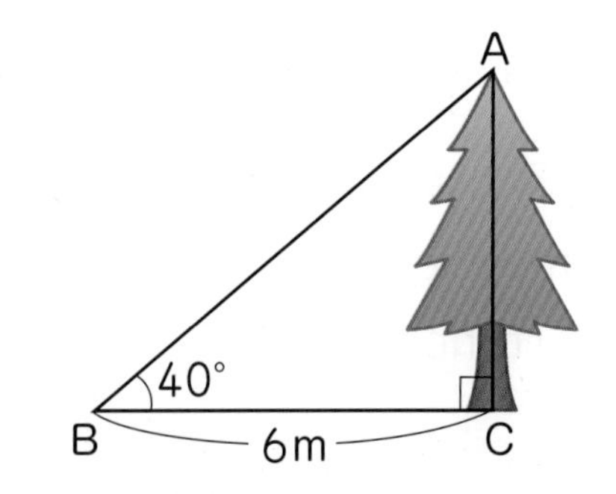

13 Proportion and Inverse Proportion

→ pp.186 ～ 209

1 The table on the right shows the relationship between the water depth y cm and time x minutes when water is poured into a cuboid. Let's answer the following questions.

Water Depth and Pouring Time

Time x (min)	1	2	3	4	5	6
Water depth y (cm)	3	6	9	12	15	18

① When the value of x changes by 2 times, 3 times, ... and so on, how does the value of y change?

When the value of x changes by $\frac{1}{2}$ times, $\frac{1}{3}$ times, ... and so on, how does the value of y change?

② Let's represent the relationship between x and y in a math equation.

③ How many centimeters is the water depth when the water is poured for 9 minutes?

④ Let's represent the relationship between x and y with a graph.

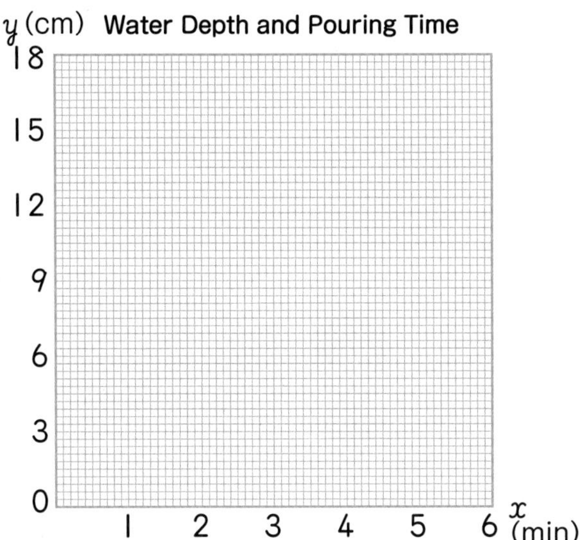

2 Consider the speed as x km per hour and the traveling time as y hours to move a distance of 60 km. Let's answer the following questions.

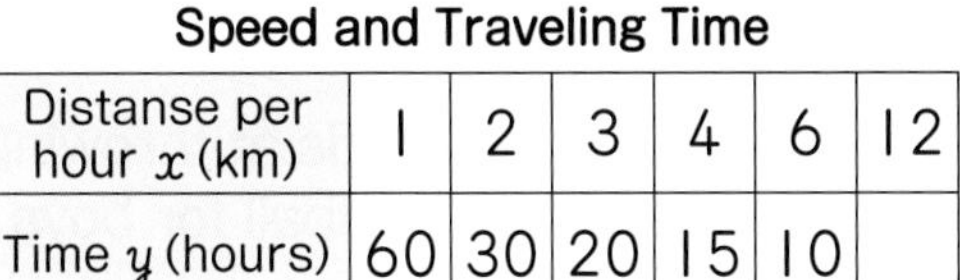

Speed and Traveling Time

Distanse per hour x (km)	1	2	3	4	6	12
Time y (hours)	60	30	20	15	10	

① Let's fill in the following table with the number that applies.

② When the value of x changes by 2 times, 3 times, ... and so on, how does the value of y change?

③ Is y inversely proportional to x?

④ When the value of x changes by $\frac{1}{2}$ times, $\frac{1}{3}$ times, ... and so on, how does the value of y change?

3 Choose from the following ⓐ～ⓒ in which y is inversely proportional to x.

ⓐ a rectangle with a fixed perimeter of 20 cm, length of x cm, and width of y cm

ⓑ a rectangle with a length of 4 cm, width x cm, and an area of y cm^2

ⓒ a rectangle with an area of 40 cm^2, length of x cm, and width of y cm

Let's separate and think.

Based on the histogram of throwing records of the current 6th graders on page 94, Yu considered the following.

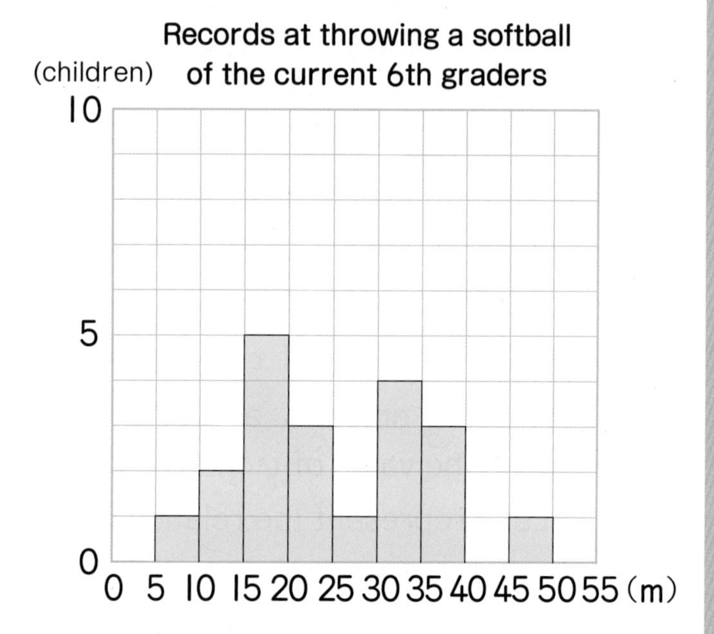

① Yu wanted to separate the results from boys and girls. He knew the results for boys, so the histogram was drawn as shown below. Using the results from this histogram, let's draw the histogram for girls.

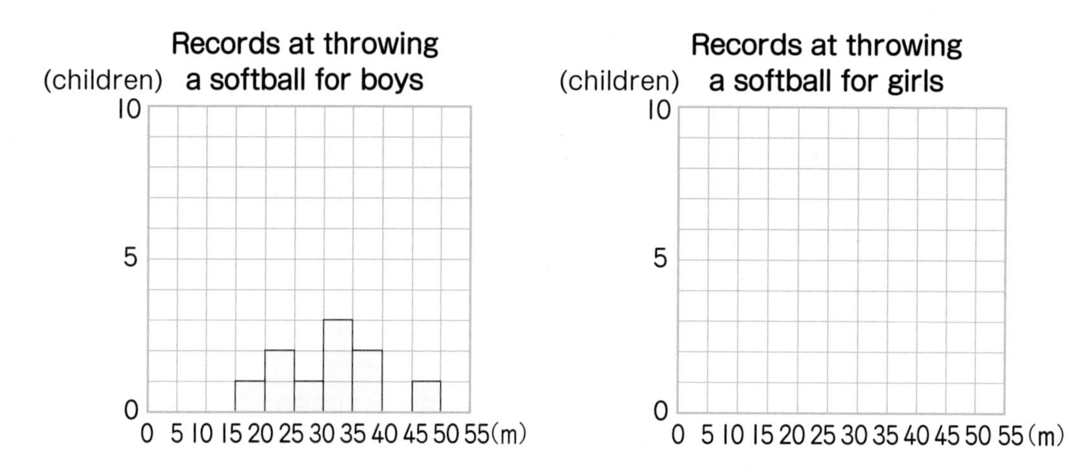

② Let's discuss what you noticed by looking at the graphs divided into boys and girls and the graph in which they are together.

Let's find out the area of a hand fan.

The area of a circle separated by two radii is called a **circular sector**. The angle formed by the two radii is called a **center angle**.

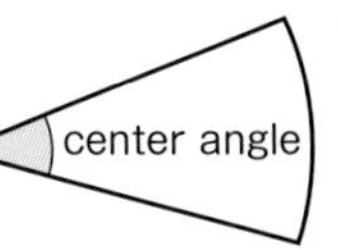

As shown in the figure below, the center angles are doubles, tripled, ..., without changing the radius of the circular sector.

ⓐ

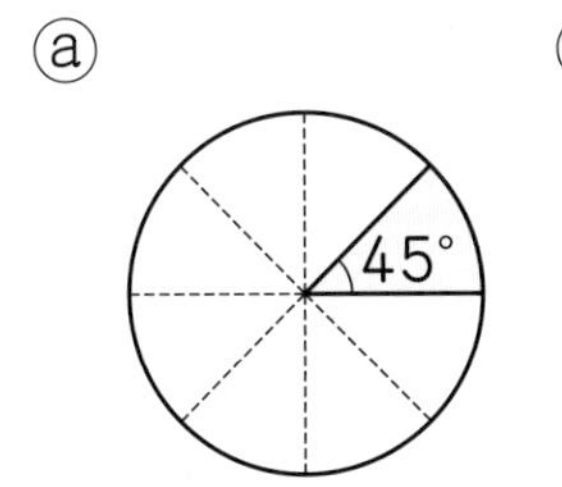

ⓑ

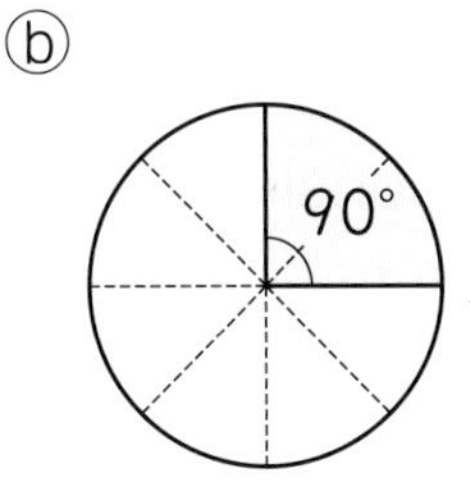

ⓒ

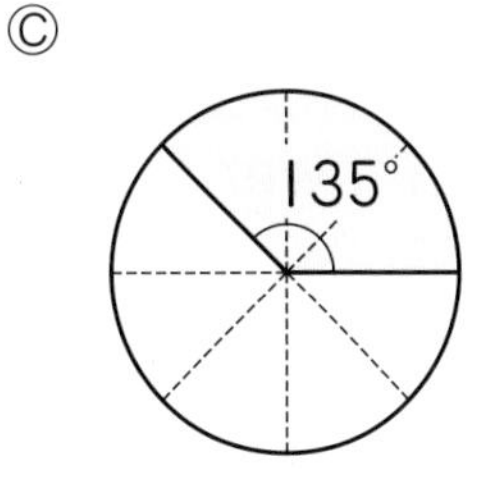

① Let's fill in the following table about the relationship between the center angle and the area when the area of a circular sector with center angle 45° is x cm^2.

Center angle (°)	45	90	135	180
Area (cm^2)	x	$x \times 2$		

The area of the circular sector is proportional to the center angle.

② Let's find out the area of a circular sector with center angle 45° and radius 4 cm.

Beautiful Ratios

There are a lot of things that are said to be beautiful in the world. The ratio 1 : 1.6 is commonly used in some of those beautiful things. This ratio is called the golden ratio. It is used in many structures and artworks as a ratio that is harmonic and beautiful. Let's think about this ratio.

Parthenon (Greece)

① Let's find out the ratio between the two lengths in tower Ⓐ. Let's consider A as 1, and find out the ratio A : B by rounding off the answer to the tenths place.

A : B = ☐ : 553 = 1 : ☐

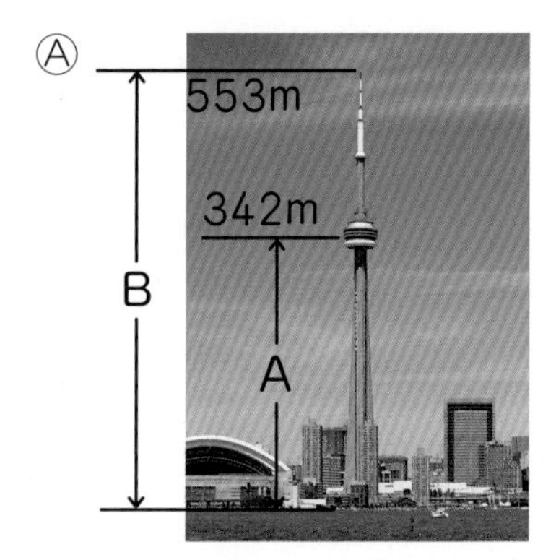

CN Tower (Canada)

There is another harmonic and beautiful ratio, which is 1 : 1.4. It is called the silver ratio. This ratio is used in many things in Japan.

② Let's find out the ratio between the two lengths in tower Ⓑ. Let's consider A as 1, and find out the ratio A : B by rounding off the answer to the tenths place.

A : B = ☐ : 634 = 1 : ☐

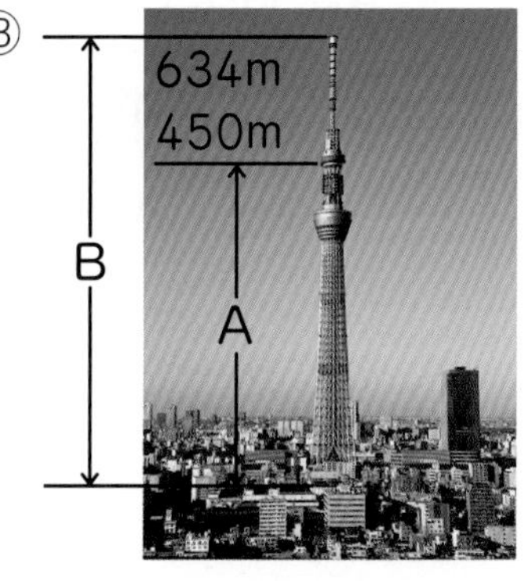

Tokyo Sky Tree (Sumida-ku, Tokyo)

Let's find out the actual distance on the map.

— Let's use a scale —

The diagram on the right is a map of the road through the sea in Okinawa. The map is in $\frac{1}{50000}$ reduced scale. Let's think about this map.

① How many centimeters represent the actual distance of 5 km on this map?

② What is the actual distance in km from point A to point B?

③ Let's find out the actual distance between the points C and D, E and F, and G and H on the map.

④ Yuta is walking from point B to point A at the speed of 4 km per hour. He left point B at 10 : 40 a.m. What time will he arrive at point A?

Kaichu-doro (Uruma City, Okinawa Prefecture)

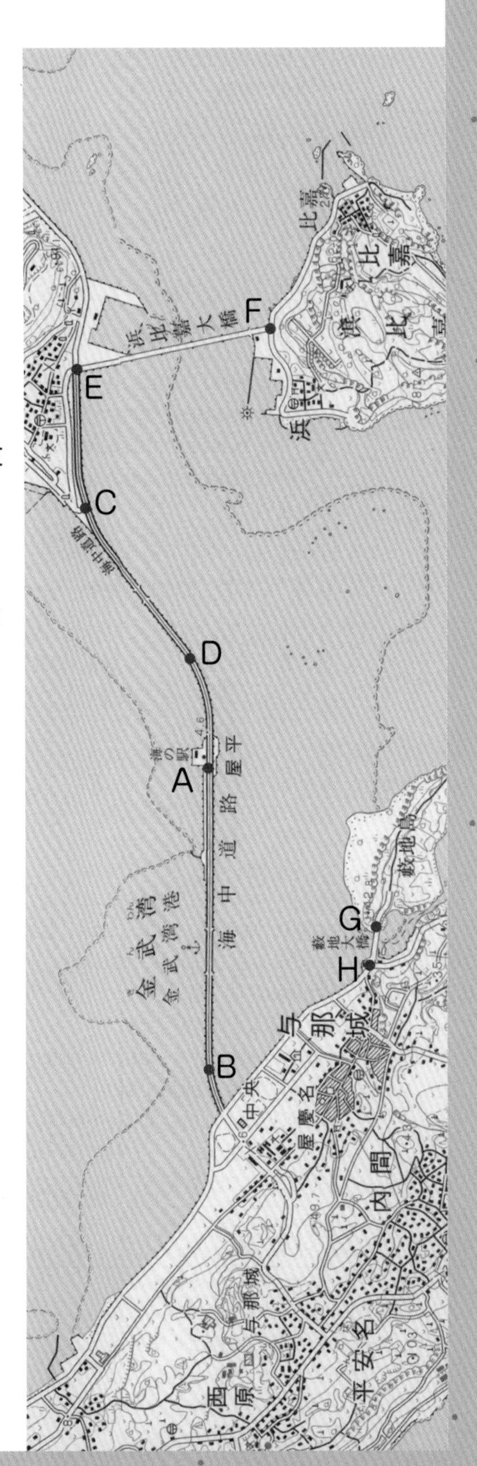

Answers

[Summary]

Numbers and calculations, math equations

→ pp.218 ~ 219

1 ① 10 times…10.7　100 times…107
1000 times…1070
② $\frac{1}{10}$…52.1　$\frac{1}{100}$…5.21
③ ⓐ 230 ⓑ 23 ⓒ 23 ⓓ 230

2 ① ⓐ $<$ ⓑ $>$ ⓒ $=$
② ⓐ 3 ⓑ $\frac{1}{7}$
③ ⓐ $\frac{5}{3}$ ⓑ $\frac{23}{5}$ ⓒ $1\frac{3}{4}$ ⓓ $2\frac{2}{3}$

3 ① 4, 9, 25, 49
② ⓐ the least common multiple … 36
the greatest common divisor… 6
ⓑ the least common multiple … 16
the greatest common divisor… 8

4 ① ⓐ $\frac{4}{1}$ ⓑ $\frac{7}{10}$ ⓒ $\frac{77}{25}$ $\left(3\frac{2}{25}\right)$
ⓓ 0.52 ⓔ 1.75
② 0.3, $\frac{1}{3}$, $\frac{2}{5}$, 0.41, $\frac{7}{15}$

5 ① ⓐ 13, 33, 10 ⓑ 5.7, 2.7, 6.3, 2.8
ⓒ 66.6, 63, 116.64, 36
ⓓ $\frac{11}{15}$, $\frac{1}{15}$, $\frac{2}{15}$, $\frac{6}{5}$ $\left(1\frac{1}{5}\right)$ ⓔ 28, $\frac{3}{5}$
② ⓐ 7 ⓑ 8

6 ⓐ $8 \times x \div 2 = 20$　x…5
ⓑ $(7 + 10) \times x \div 2 = 68$　x…8

Figures

→ pp.220 ~ 222

1 ① length, width, side, side, base, height, base, height, 2, radius, radius, 3.14
② (omitted) ③ ⓐ 6.9cm² ⓑ 12cm² ⓒ 157cm²
④ 3m², 30000cm²

2 ① cuboid…length × width × height
cube… side x side x side
② ⓐ 800cm³ ⓑ 1728cm³ ⓒ 3750cm³
ⓓ 1125cm³ ⓔ 1962.5cm³

3 ①

	Parallelogram	Rhombus	Rectangle	Square
2 pair of sides are parallel.	○	○	○	○
All 4 angles are right angles.	×	×	○	○
All 4 sides have the same length.	×	○	×	○
2 diagonals intersect perpendicularly.	×	○	×	○
The sum of adjacent angles is 180°.	○	○	○	○

② ⓐ 15° ⓑ 68° Ⓒ 120° ⓓ 60°
③ ⓐ face EFGH ⓑ sideDC, sideEF, sideHG

4 ① ② ③ ④

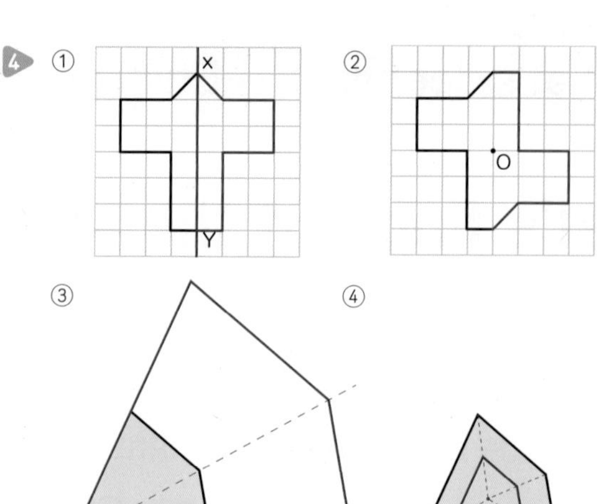

Measurement per unit quantity · Utilization of Data

→ pp.223 ~ 225

1 ① ⓐ cm² ⓑ mL ⓒ g ⓓ km
② ⓐ 400m ⓑ 2L, 20dL

2 ① Haruka's city ② 648 km

3 ① 4m per second, 14.4km per hour
② 10 minutes

4 ① ⓐ strip graph, pie chart
ⓑ line graph ⓒ bar graph
② ⓐ 1990…45%, 2020…48%
ⓑ

0 10 20 30 40 50 60 70 80 90 100 (%)

1990: Books | Weekly magazines | Monthly magazines
2020: Books | Weekly magazines | monthly Magazines

③ ⓐ 5 ⓑ 56g

5 ① proportional…Ⓑ　inverse proportional…Ⓐ
② Ⓐ $x \times y = 24$　Ⓑ $y = 8 \times x$
③

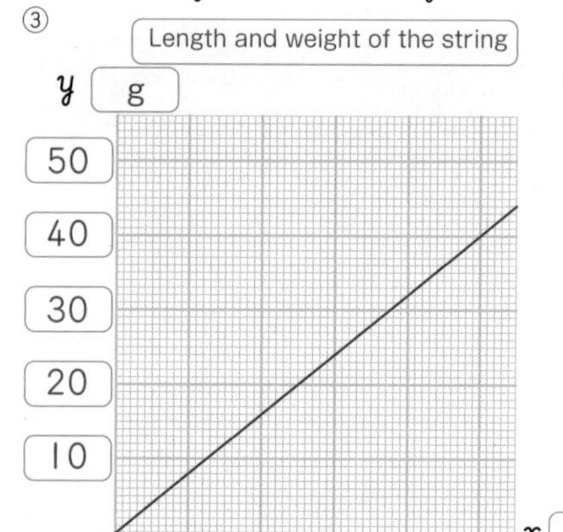

[Supplementary Problems]

1 Symmetry

→ p.232

1 ① side FE ② perpendicular ③ straight line CH

2 ①

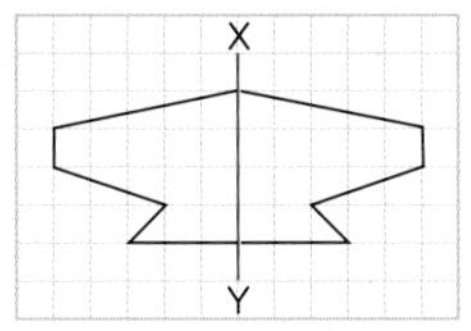

②

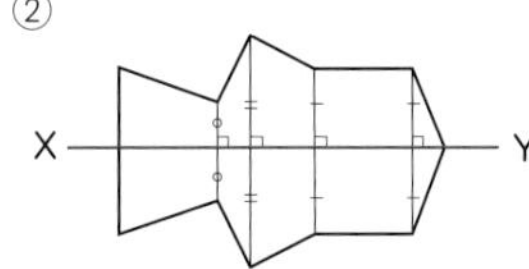

3 ① Point O ② Point J ③ side GH
④ straight line FO

4 ①

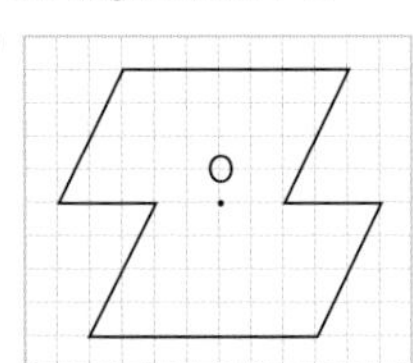

②

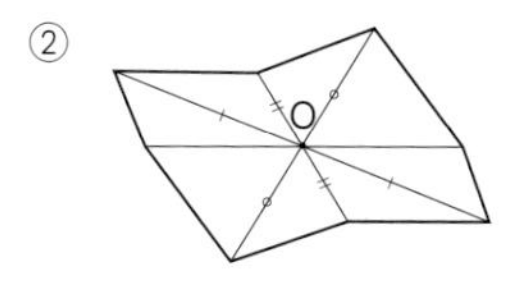

2 Mathematical Letter and Equation

→ p.233

1 ① $x + 5 = 14$ ② 9dL

2 ① $x \times 4 = 26$ ② 6.5cm

3 ① 36 ② 8 ③ 23 ④ 15 ⑤ 22 ⑥ 40
⑦ 62 ⑧ 93 ⑨ 6 ⑩ 7 ⑪ 4 ⑫ 1.5

4 ① $x \times 3 + 5$ ② 18 cookies

5 ① Ⓑ ② Ⓒ ③ Ⓐ

3 Multiplication and Division of Fractions and Whole Numbers

→ p. 234

1 ① $\frac{4}{7}$ ② $\frac{9}{10}$ ③ $\frac{3}{5}$ ④ $3\frac{3}{4}$ $\left(\frac{15}{4}\right)$
⑤ $1\frac{7}{9}$ $\left(\frac{16}{9}\right)$ ⑥ $1\frac{7}{8}$ $\left(\frac{15}{8}\right)$ ⑦ $\frac{5}{6}$
⑧ $1\frac{1}{3}$ $\left(\frac{4}{3}\right)$ ⑨ $\frac{2}{3}$ ⑩ $1\frac{1}{3}$ $\left(\frac{4}{3}\right)$
⑪ $13\frac{1}{2}$ $\left(\frac{27}{2}\right)$ ⑫ 12 ⑬ $4\frac{8}{9}$ $\left(\frac{44}{9}\right)$
⑭ $2\frac{10}{11}$ $\left(\frac{32}{11}\right)$ ⑮ $4\frac{2}{5}$ $\left(\frac{22}{5}\right)$ ⑯ $7\frac{1}{3}$ $\left(\frac{22}{3}\right)$
⑰ $4\frac{1}{2}$ $\left(\frac{9}{2}\right)$ ⑱ $11\frac{1}{3}$ $\left(\frac{34}{3}\right)$

2 $2\frac{1}{2}$m $\left(\frac{5}{2}\text{m}\right)$

3 14m^2

4 ① $\frac{1}{9}$ ② $\frac{5}{28}$ ③ $\frac{3}{10}$ ④ $\frac{3}{40}$ ⑤ $\frac{4}{27}$ ⑥ $\frac{3}{28}$
⑦ $\frac{2}{7}$ ⑧ $\frac{1}{16}$ ⑨ $\frac{4}{27}$ ⑩ $\frac{2}{15}$ ⑪ $\frac{2}{3}$ ⑫ $\frac{5}{28}$
⑬ $\frac{22}{27}$ ⑭ $\frac{17}{20}$ ⑮ $\frac{1}{2}$ ⑯ $\frac{8}{15}$ ⑰ $\frac{3}{7}$ ⑱ $\frac{15}{22}$

5 $\frac{3}{8}$L

6 $\frac{3}{4}$kg

4 Fraction × Fraction

→ p.235

1 ① $\frac{5}{72}$ ② $\frac{7}{18}$ ③ $\frac{4}{15}$ ④ $2\frac{11}{12}$ $\left(\frac{35}{12}\right)$
⑤ $1\frac{13}{32}$ $\left(\frac{45}{32}\right)$ ⑥ $\frac{20}{27}$ ⑦ $\frac{5}{27}$ ⑧ $\frac{7}{18}$ ⑨ $\frac{1}{4}$
⑩ $\frac{12}{25}$ ⑪ $1\frac{4}{11}$ $\left(\frac{15}{11}\right)$ ⑫ $\frac{9}{10}$ ⑬ $1\frac{1}{2}$ $\left(\frac{3}{2}\right)$
⑭ $3\frac{1}{2}$ $\left(\frac{7}{2}\right)$ ⑮ 4 ⑯ $2\frac{1}{10}$ $\left(\frac{21}{10}\right)$
⑰ $12\frac{1}{2}$ $\left(\frac{25}{2}\right)$ ⑱ 6 ⑲ $28\frac{1}{2}$ $\left(\frac{57}{2}\right)$ ⑳ 2
㉑ $\frac{11}{20}$ ㉒ $\frac{2}{3}$ ㉓ $1\frac{1}{2}$ $\left(\frac{3}{2}\right)$

2 $\frac{7}{12}$m^2

3 3kg

4 ① $<$ ② $>$ ③ $=$

5 ① $\frac{3}{8}$ ② $\frac{8}{9}$ ③ $\frac{3}{10}$ ④ 24

6 ① $2\frac{2}{3}$ $\left(\frac{8}{3}\right)$ ② $\frac{2}{3}$ ③ $\frac{5}{12}$ ④ 6
⑤ $1\frac{1}{4}$ $\left(\frac{5}{4}\right)$ ⑥ $\frac{10}{13}$ ⑦ $\frac{1}{4}$

5 Fraction ÷ Fraction

→ p. 236

1 ① $\frac{10}{27}$ ② $\frac{35}{36}$ ③ $5\frac{1}{3}$ $\left(\frac{16}{3}\right)$ ④ $\frac{21}{32}$
⑤ $1\frac{3}{4}$ $\left(\frac{7}{4}\right)$ ⑥ $\frac{12}{25}$ ⑦ $2\frac{1}{10}$ $\left(\frac{21}{10}\right)$ ⑧ $\frac{3}{8}$
⑨ 2 ⑩ $\frac{1}{6}$ ⑪ $8\frac{3}{4}$ $\left(\frac{35}{4}\right)$ ⑫ $13\frac{1}{2}$ $\left(\frac{27}{2}\right)$
⑬ 16 ⑭ 25 ⑮ $\frac{20}{63}$ ⑯ $\frac{7}{32}$
⑰ $4\frac{12}{17}$ $\left(\frac{80}{17}\right)$ ⑱ 4 ⑲ $5\frac{1}{4}$ $\left(\frac{21}{4}\right)$ ⑳ 4
㉑ $1\frac{1}{8}$ $\left(\frac{9}{8}\right)$ ㉒ $1\frac{2}{3}$ $\left(\frac{5}{3}\right)$

2 $1\frac{1}{9}$kg $\left(\frac{10}{9}\text{kg}\right)$

3 ① $>$ ② $<$ ③ $>$

4 8m^2

5 $2\frac{1}{10}$m $\left(2\frac{21}{10}\text{m}\right)$

6 $2\frac{2}{9}$m $\left(\frac{20}{9}\text{m}\right)$

7 810mL

6 Data Arrangement

→ p.237

1 ① Records at throwing a softball

Distance (m)	Number of children
greater than or equal to 15 ～ less than 20	2
20 ～ 25	4
25 ～ 30	6
30 ～ 35	3
35 ～ 40	2
40 ～ 45	2
45 ～ 50	1
Total	20

② Records at throwing a softball

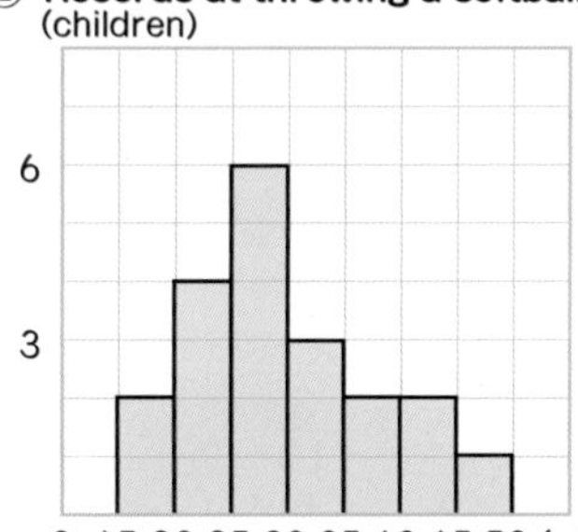

③ greater than or equal to 25m and less than 30m

④ greater than or equal to 35m and less than 40m

2 ① 6 points ② 6 points ③ 5 points

7 Ways of Ordering and Combining

→ p.238

1 ① 6 ways ② 6 ways each ③ 24 ways

2 24 ways

3 ① 6, 9 ② 9 ways

4 8 ways

5 10 games

6 4 ways

8 Calculations with Decimal Numbers

→ p.239

1 ⓑ, ⓒ

2 ① $1\frac{19}{30}\left(\frac{49}{30}\right)$ ② $\frac{9}{10}$ ③ $1\frac{5}{18}\left(\frac{23}{18}\right)$ ④ $\frac{21}{25}$

⑤ $\frac{7}{40}$ ⑥ $\frac{1}{20}$ ⑦ $\frac{38}{75}$ ⑧ $\frac{17}{60}$

3 ① $\frac{3}{4}$ ② $\frac{9}{28}$ ③ $2\frac{1}{2}\left(\frac{5}{2}\right)$ ④ $\frac{1}{2}$

9 Area of a Circle

→ p.239

1 ① circumference…18.84cm area…28.26cm^2

② circumference…37.68cm area…113.04cm^2

③ circumference…25.12cm area…50.24cm^2

④ circumference…43.96cm area…153.86cm^2

2 ① 113.04cm^2 ② 942cm^2 ③ 172cm^2

10 Volume of Solids

→ p.240

1 ① 12cm^2 ② 5cm ③ 60cm^3

2 ① 60cm^3 ② 198cm^3 ③ 87.92cm^3 ④ 678.24cm^3

3 471cm^3

4 ① 282cm^3 ② 150cm^3 ③ 1507.2cm^3

④ 395.64cm^3

11 Ratio and its Applications

→ p.241

1 ① 3 : 5 ② $\frac{3}{5}$

2 ① $\frac{4}{9}$ ② $\frac{3}{2}$ ③ $\frac{1}{2}$ ④ $\frac{3}{4}$

3 ① (examples) 8 : 2, 12 : 3, 16 : 4

② (examples) 4 : 14, 6 : 21, 8 : 28

③ (examples) 10 : 16, 15 : 24, 20 : 32

④ (examples) 3 : 4, 9 : 12, 12 : 16

4 ① 40 ② 18 ③ 16 ④ 60 ⑤ 30 ⑥ 25

⑦ 12 ⑧ 24

5 ① 4 : 5 ② 9 : 8 ③ 3 : 4 ④ 12 : 1

⑤ 3 : 4 ⑥ 5 : 2 ⑦ 5 : 9 ⑧ 9 : 14

6 32 sheets

7 6m

8 1200 yen

12 Enlargement and Reduction of Figures

→ p.242

1
① 1 : 2
② side DE…10cm side DF…8cm
③ 2 times

2

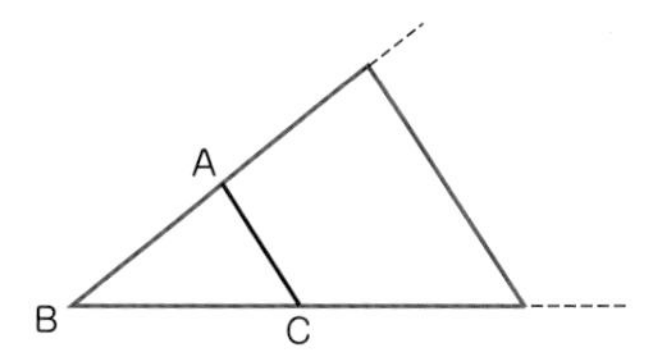

3

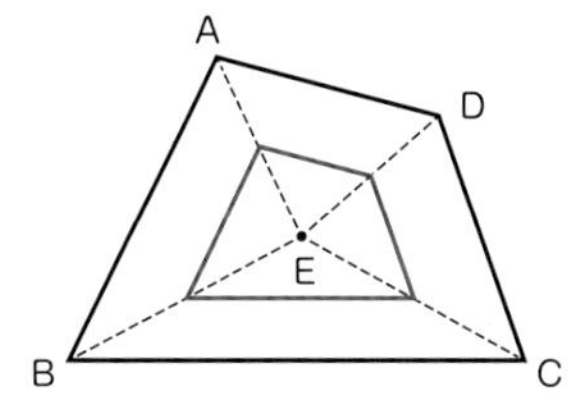

4 In the reduced drawing, the length of AC is about 5cm.
$5 \times 100 = 500$ (cm) About 5m

13 Proportion and Inverse Proportion

→ p.243

1
① It changes by 2 times, 3 times, and so on.
It changes by $\frac{1}{2}$ times, $\frac{1}{3}$ times, and so on.
② $y = 3 \times x$ ③ 27cm
④

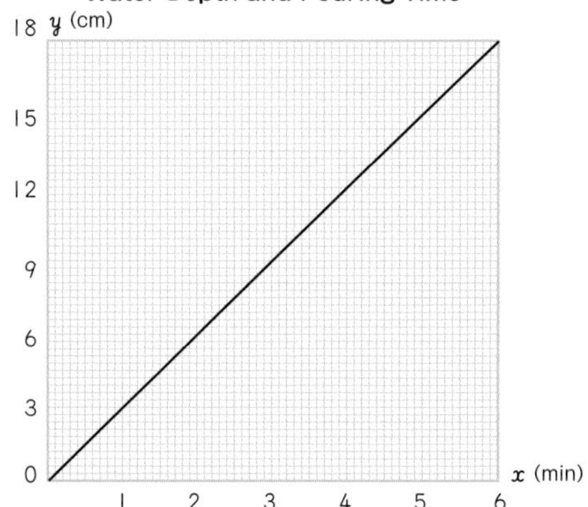

2
① 5
② It changes by $\frac{1}{2}$ times, $\frac{1}{3}$ times, and so on.
③ Yes
④ It changes by 2 times, 3 times, and so on.

3 ⓒ

[Let's deepen.]

Let's separate and think.

→ p.244

①

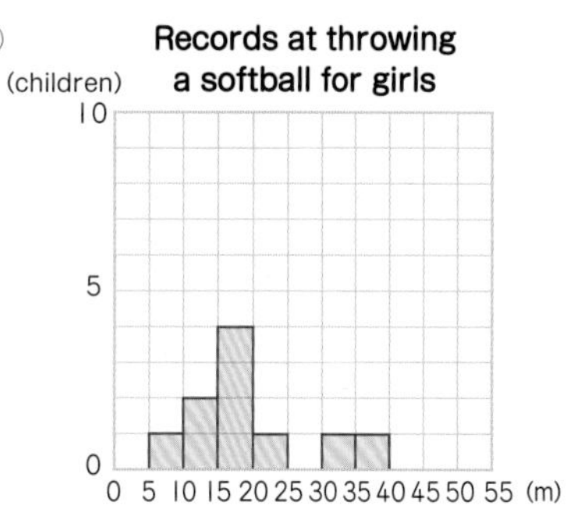

② (omitted)

Let's find out the area of a hand fan.

→ p.245

① (from the left) $x \times 3$, $x \times 4$
② 6.28cm^2

Beautiful Ratios

→ p.246

① 342, 1.6
② 450, 1.4

Let's find out the actual distance on the map.

→ p.247

① 10cm
② 2km
③ CD…1.25km EF…1.35km GH…250m
④ 11:10 a.m.

which we learned in this textbook

What kind of “Way to See and Think Monsters” have you found while learning mathematics so far?
Even if we focus on only one monster, we could find it in many different places.
Let’s summarize where you found each “Way to See and Think Monster.”

Think about where you found the monster. Let's write about the place where you found it.

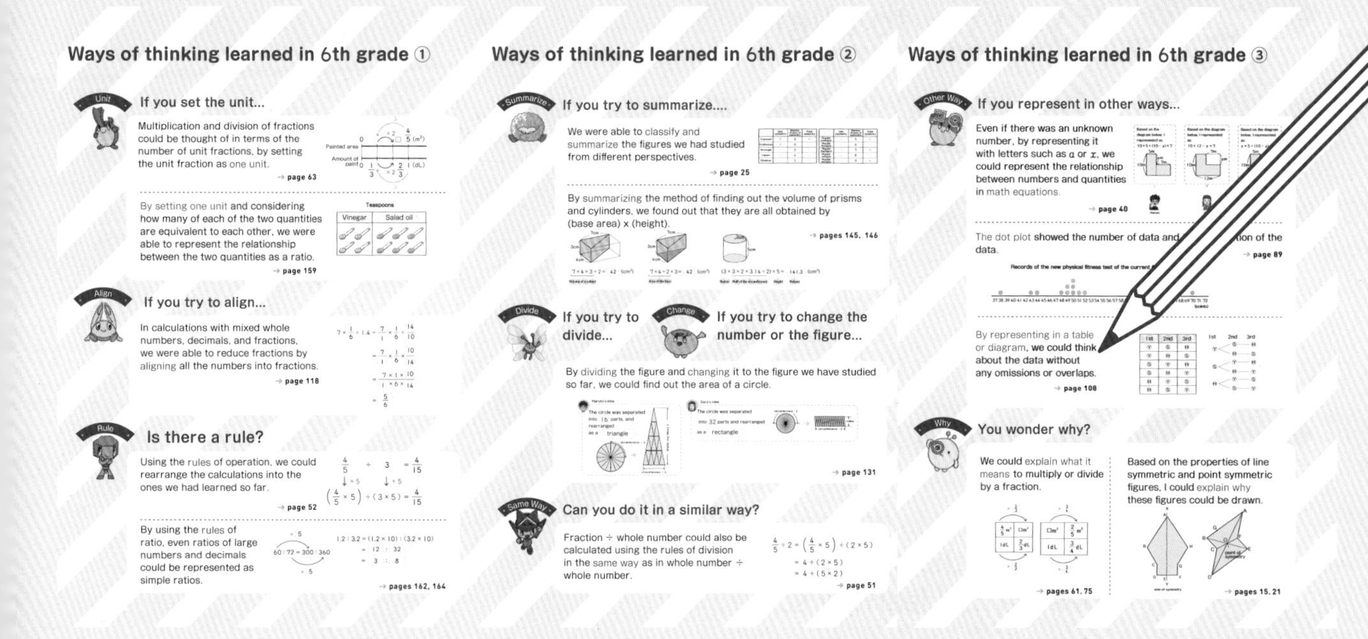

Ways of thinking learned in 6th grade ①

Unit — **If you set the unit...**

Multiplication and division of fractions could be thought of in terms of the number of unit fractions, by setting the unit fraction as one unit.
→ page 63

By setting one unit and considering how many of each of the two quantities are equivalent to each other, we were able to represent the relationship between the two quantities as a ratio.
→ page 159

Align — **If you try to align...**

In calculations with mixed whole numbers, decimals, and fractions, we were able to reduce fractions by aligning all the numbers into fractions.
→ page 118

Rule — **Is there a rule?**

Using the rules of operation, we could rearrange the calculations into the ones we had learned so far.
→ page 52

By using the rules of ratio, even ratios of large numbers and decimals could be represented as simple ratios.
→ pages 162, 164

Ways of thinking learned in 6th grade ②

Summarize — **If you try to summarize....**

We were able to classify and summarize the figures we had studied from different perspectives.
→ page 25

By summarizing the method of finding out the volume of prisms and cylinders, we found out that they are all obtained by (base area) x (height).
→ pages 145, 146

Divide — **If you try to divide...** Change — **If you try to change the number or the figure...**

By dividing the figure and changing it to the figure we have studied so far, we could find out the area of a circle.
→ page 131

Same Way — **Can you do it in a similar way?**

Fraction ÷ whole number could also be calculated using the rules of division in the same way as in whole number ÷ whole number.
→ page 51

Ways of thinking learned in 6th grade ③

Other Way — **If you represent in other ways...**

Even if there was an unknown number, by representing it with letters such as a or x, we could represent the relationship between numbers and quantities in math equations.
→ page 40

The dot plot showed the number of data and ... of the data.
→ page 89

By representing in a table or diagram, we could think about the data without any omissions or overlaps.
→ page 108

Why — **You wonder why?**

We could explain what it means to multiply or divide by a fraction.
→ pages 61, 75

Based on the properties of line symmetric and point symmetric figures, I could explain why these figures could be drawn.
→ pages 15, 21

There was also a place where we found two monsters at the same time.

Sara

If there is a newly found monster, write down what kind of learning experience you found it in.

Akari

Ways of thinking learned in 6th grade ①

If you set the unit...

Multiplication and division of fractions could be thought of in terms of the number of unit fractions, by setting the unit fraction as one unit.

→ page 63

Painted area: 0, $\square$, $\frac{4}{5}$ (m²)

Amount of paint: 0, $\frac{1}{3}$, $\frac{2}{3}$, 1 (dL)

$\times 2$, $\div 3$

By setting one unit and considering how many of each of the two quantities are equivalent to each other, we were able to represent the relationship between the two quantities as a ratio.

→ page 159

Teaspoons

Vinegar	Salad oil
4 teaspoons	6 teaspoons

If you try to align...

In calculations with mixed whole numbers, decimals, and fractions, we were able to reduce fractions by aligning all the numbers into fractions.

→ page 118

$$7 \times \frac{1}{6} \div 1.4 = \frac{7}{\boxed{1}} \times \frac{1}{6} \div \frac{\boxed{14}}{10}$$

$$= \frac{7}{\boxed{1}} \times \frac{1}{6} \times \frac{\boxed{10}}{\boxed{14}}$$

$$= \frac{7 \times 1 \times \boxed{10}}{\boxed{1} \times 6 \times \boxed{14}}$$

$$= \boxed{\frac{5}{6}}$$

Is there a rule?

Using the rules of operation, we could rearrange the calculations into the ones we had learned so far.

→ page 52

$$\frac{4}{5} \div 3 = \frac{4}{15}$$

$\downarrow \times 5$ $\quad \downarrow \times 5$

$$\left(\frac{4}{5} \times 5\right) \div (3 \times 5) = \frac{4}{15}$$

By using the rules of ratio, even ratios of large numbers and decimals could be represented as simple ratios.

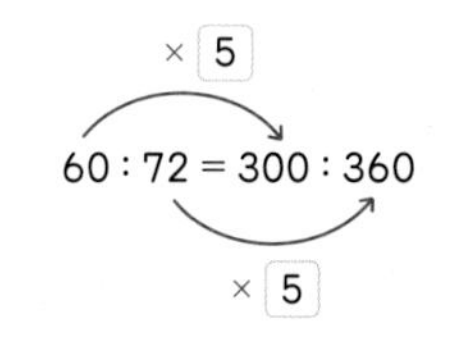

$$1.2 : 3.2 = (1.2 \times 10) : (3.2 \times 10)$$

$$= \boxed{12} : \boxed{32}$$

$$= \boxed{3} : \boxed{8}$$

→ pages 162, 164

Ways of thinking learned in 6th grade ②

If you try to summarize....

We were able to classify and summarize the figures we had studied from different perspectives.

	Line symmetry	Number of axes of symmetry	Point symmetry
Trapezoid	×	0	×
Parallelogram	×	0	○
Rectangle	○	2	○
Square	○	4	○
Rhombus	○	2	○

	Line symmetry	Number of axes of symmetry	Point symmetry
Regular pentagon	○	5	×
Regular hexagon	○	6	○
Regular heptagon	○	7	×
Regular octagon	○	8	○
Regular nonagon	○	9	×

→ page 25

By summarizing the method of finding out the volume of prisms and cylinders, we found out that they are all obtained by (base area) x (height).

→ pages 145, 146

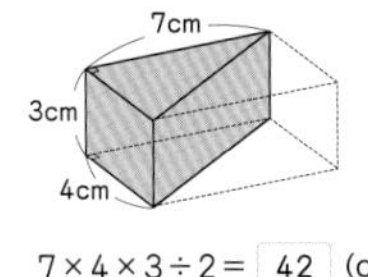

$7 \times 4 \times 3 \div 2 =$ 42 (cm³)

Volume of a cuboid

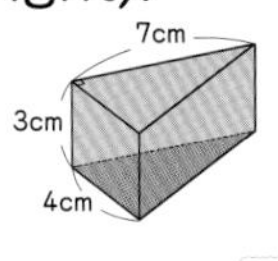

$7 \times 4 \div 2 \times 3 =$ 42 (cm³)

Area of the base

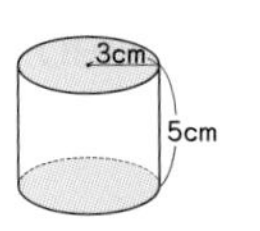

$(3 \times 3 \times 2 \times 3.14 \div 2) \times 5 =$ 141.3 (cm³)

Radius　Half of the circumference　Height　Volume

If you try to divide...

If you try to change the number or the figure...

By dividing the figure and changing it to the figure we have studied so far, we could find out the area of a circle.

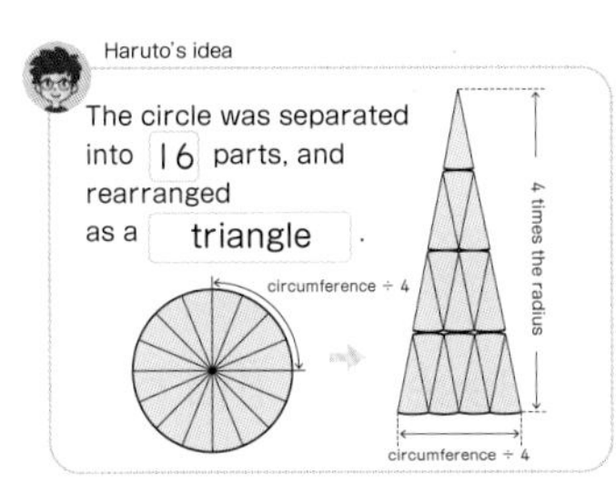

Haruto's idea
The circle was separated into 16 parts, and rearranged as a triangle.
circumference ÷ 4
4 times the radius
circumference ÷ 4

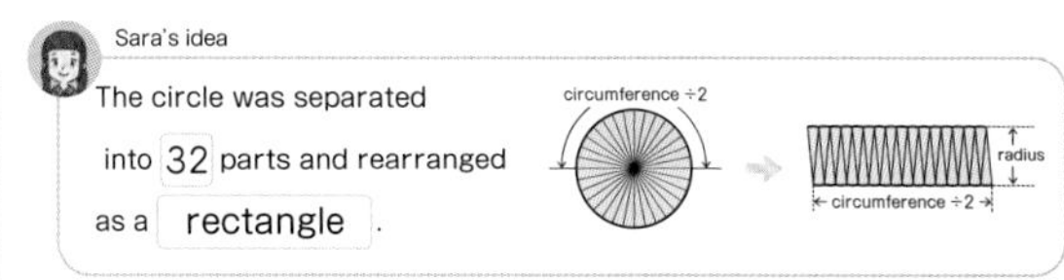

Sara's idea
The circle was separated into 32 parts and rearranged as a rectangle.
circumference ÷2
radius
circumference ÷2

→ page 131

Can you do it in a similar way?

Fraction ÷ whole number could also be calculated using the rules of division in the same way as in whole number ÷ whole number.

$$\frac{4}{5} \div 2 = \left(\frac{4}{5} \times 5\right) \div (2 \times 5)$$
$$= 4 \div (2 \times 5)$$
$$= 4 \div (5 \times 2)$$

→ page 51

Ways of thinking learned in 6th grade ③

If you represent in other ways...

Even if there was an unknown number, by representing it with letters such as a or x, we could represent the relationship between numbers and quantities in math equations.

→ page 40

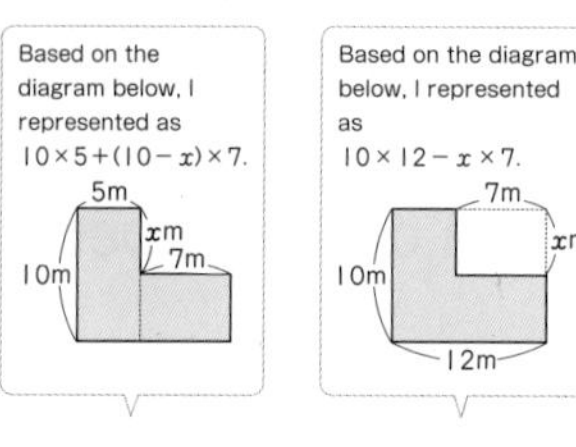

Based on the diagram below, I represented as $10\times5+(10-x)\times7$.
5m
xm
7m
10m
Based on the diagram below, I represented as $10\times12-x\times7$.
7m
xm
10m
12m

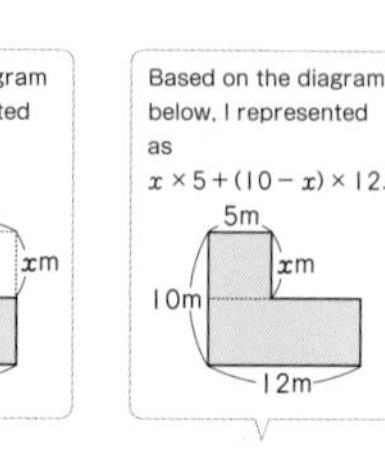

Based on the diagram below, I represented as $x\times5+(10-x)\times12$.
5m
xm
10m
12m

The dot plot showed the number of data and the distribution of the data.

→ page 89

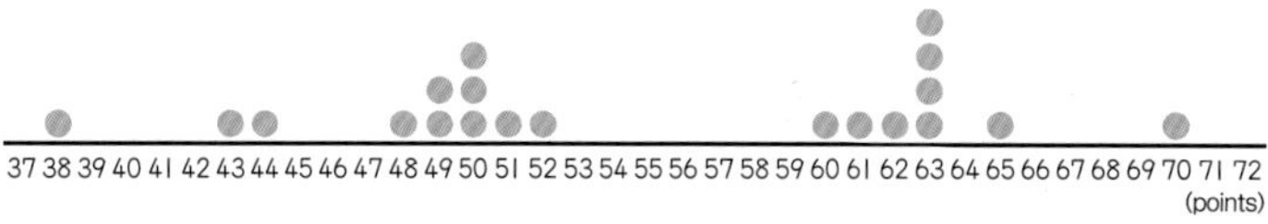

By representing in a table or diagram, we could think about the data without any omissions or overlaps.

→ page 108

1st	2nd	3rd
Ⓨ	Ⓢ	Ⓗ
Ⓨ	Ⓗ	Ⓢ
Ⓢ	Ⓨ	Ⓗ
Ⓢ	Ⓗ	Ⓨ
Ⓗ	Ⓨ	Ⓢ
Ⓗ	Ⓢ	Ⓨ

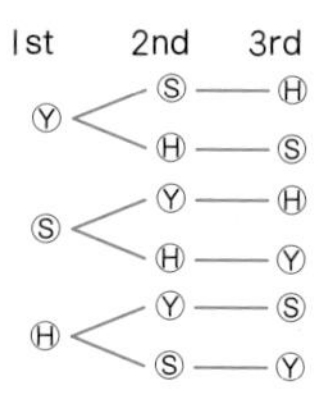

You wonder why?

We could explain what it means to multiply or divide by a fraction.

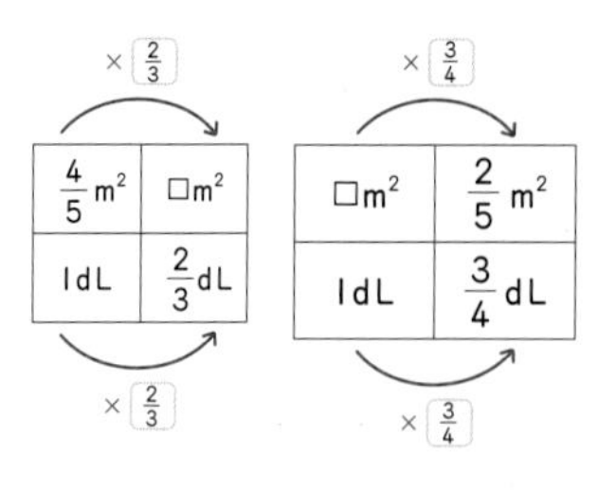

→ pages 61, 75

Based on the properties of line symmetric and point symmetric figures, I could explain why these figures could be drawn.

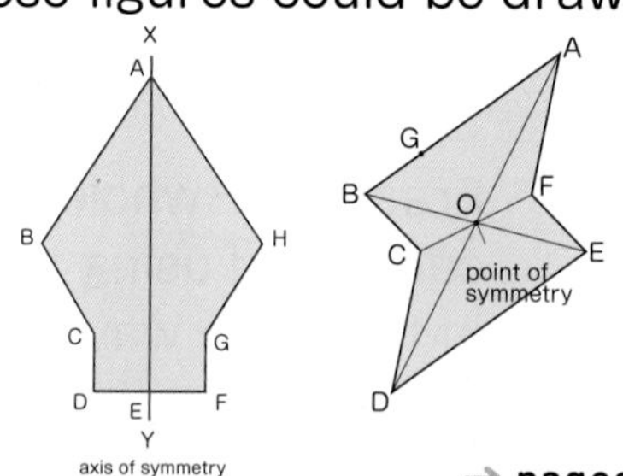

→ pages 15, 21

Symmetry

→ To be used in pages 13, 14, and 19.
Please cut these out for use.

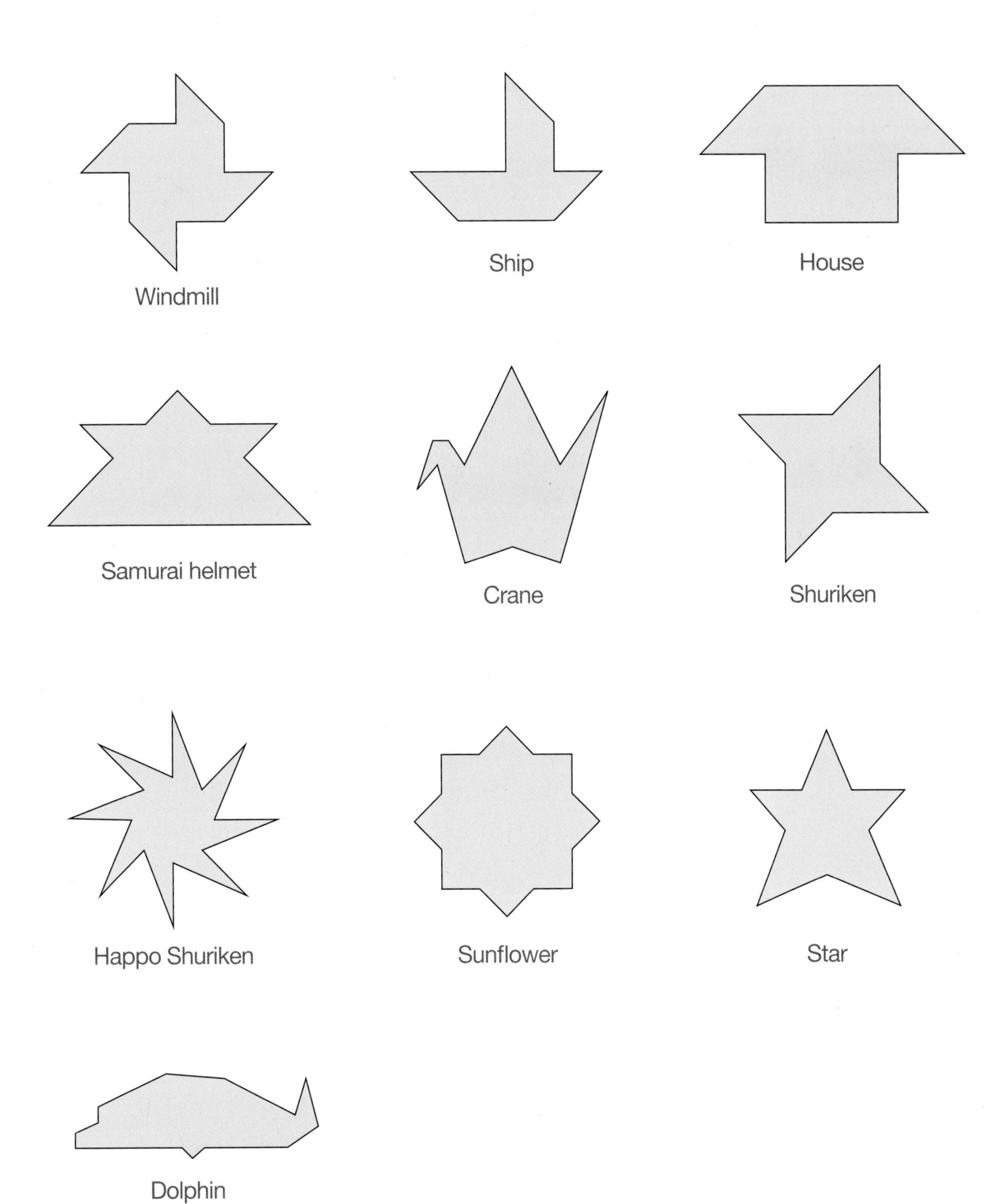

Area of a circle

→ To be used in pages 131 and 132.
Please cut these out for use.

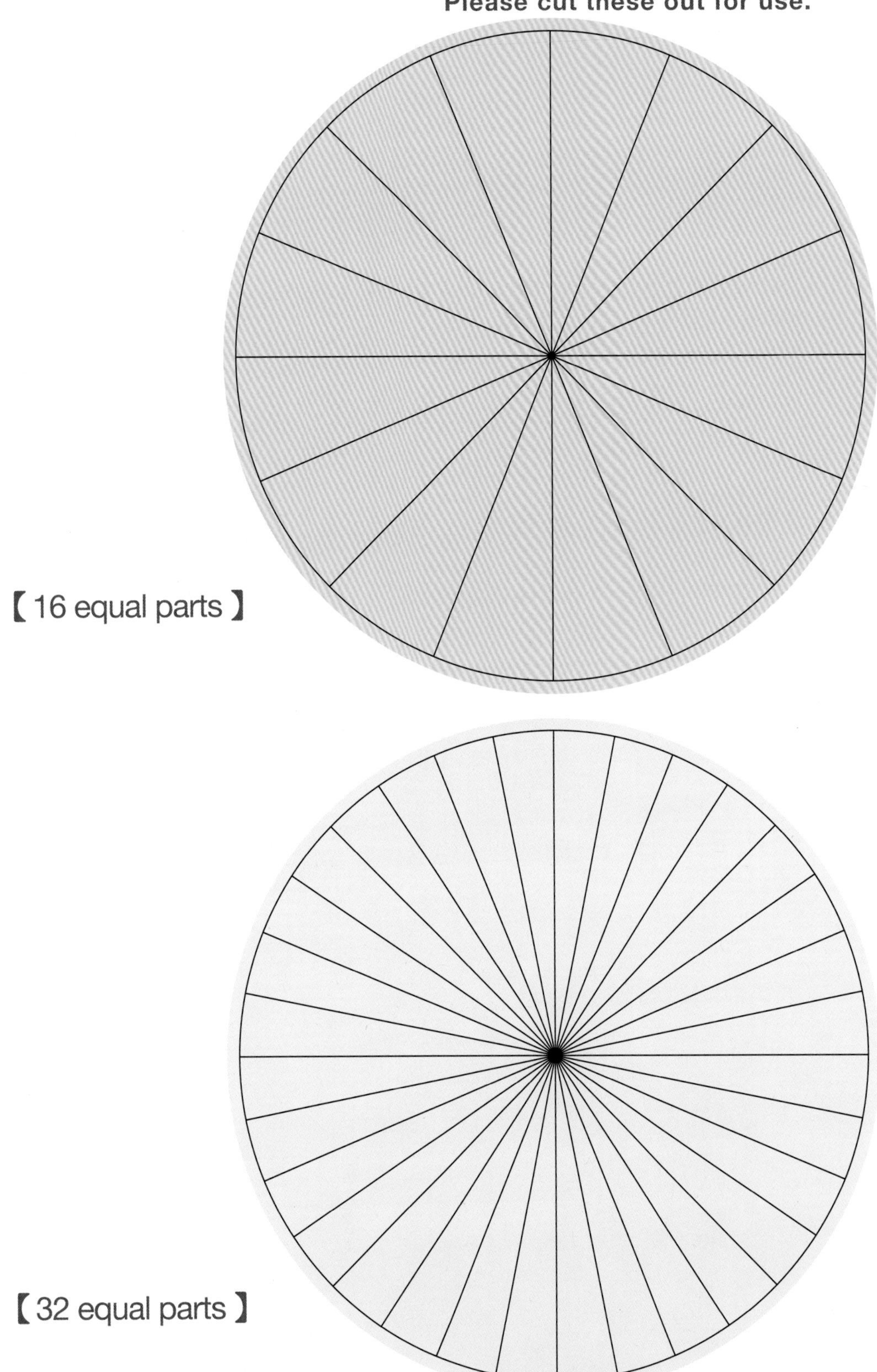

Cooking cards

→ To be used in pages 210 and 211.
Please cut these out for use.

Staple food	One slice of 6-slice bread 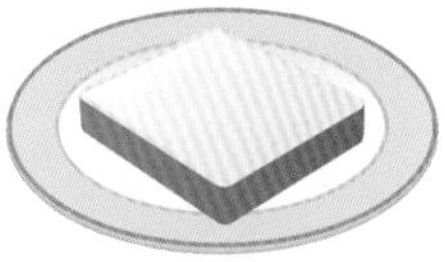	One bowl of rice 	Yakisoba with sauce 	Spaghetti with meat sauce 
Soup	Pork miso soup 	Seaweed soup 	Miso soup 	Corn soup 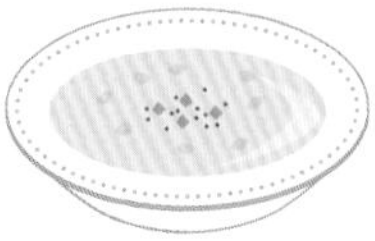
Main dish	Salmon meuniere 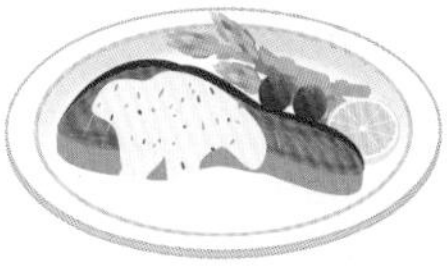	Hamburg steak 	Ginger pork 	Grilled salted horse mackerel 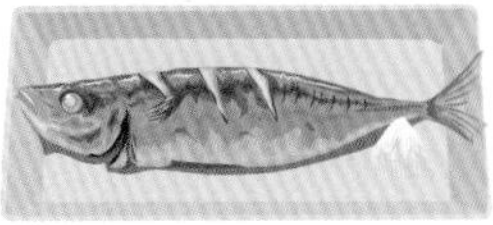
Side dish	Kinpira burdock 	Spinach with sesame dressing 	Potato salad 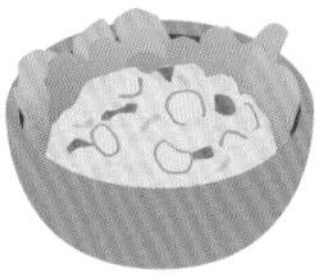	Simmered hijiki 

Spaghetti with meat sauce Ⓒ spaghetti Ⓟ minced beef Ⓥ onion	Yakisoba with sauce Ⓒ chinese noodle Ⓟ pork Ⓥ cabbage Ⓥ onion	One bowl of rice Ⓒ rice	One slice of 6-slice bread Ⓒ bread	Staple food
Corn soup Ⓥ corn [I] milk	Miso soup Ⓟ fried tofu Ⓟ tofu [I] seaweed Ⓟ miso	Seaweed soup [I] seaweed Ⓥ green onion	Pork miso soup Ⓟ pork Ⓥ radish Ⓥ carrot Ⓥ burdock Ⓥ green onion Ⓟ miso	Soup
Grilled salted horse mackerel Ⓟ horse mackerel Ⓥ grated radish	Ginger pork Ⓟ pork Ⓥ cabbage Ⓥ tomato	Hamburg steak Ⓟ minced beef Ⓥ onion	Salmon meuniere Ⓟ salmon [L] butter Ⓥ tomato Ⓥ asparagus	Main dish
Simmered hijiki [I] hijiki Ⓥ carrot Ⓟ fried tofu	Potato salad Ⓒ potato Ⓥ carrot Ⓥ cucumber	Spinach with sesame dressing Ⓥ spinach [L] sesame	Kinpira burdock Ⓥ burdock Ⓥ carrot	Side dish

Memo

Editors of Original Japanese Edition

[Head of Editors]

Shin Hitotsumatsu (Kyoto University), Yoshio Okada (Hiroshima University)

[Supervising Editors]

Toshiyuki Akai (Hyogo University), Toshikazu Ikeda (Yokohama National University), Shunji Kurosawa (The former Rikkyo University), Hiroshi Tanaka (The former Tsukuba Univ. Elementary School), Kosho Masaki (The former Kokugakuin Tochigi Junior College), Yasushi Yanase (Tamagawa University)

[Editors]

Shoji Aoyama (Tsukuba Univ. Elementary School), Kengo Ishihama (Showa Gakuin Elementary School), Hiroshi Imazaki (Hiroshima Bunkyo University), Atsumi Ueda (Hiroshima University), Tetsuro Uemura (Kagoshima University), Yoshihiro Echigo (Tokyo Gakugei Univ. Setagaya Elementary School), Hisao Oikawa (Yamato University), Hironori Osawa (Yamagata University Graduate School), Tomoyoshi Owada (Shizuoka University), Nobuhiro Ozaki (Seikei Elementary School), Masahiko Ozaki (Kansai Univ. Elementary School), Kentaro Ono (Musashino University), Hiroshi Kazama (Fukui University), Michihiro Kawasaki (Oita University), Miho Kawasaki (Shizuoka University), Yoshiko Kambe (Tokai University), Yukio Kinoshita (Kwansei Gakuin Elementary School), Tomoko Kimura (Tamon Elementary School, Setagaya City), Satoshi Kusaka (Naruto University of Education), Kensuke Kubota (Naruo Higashi Elementary School, Nishinomiya City), Itsushi Kuramitsu (The former University of the Ryukyus), Maiko Kochi (Kounan Elementary School, Toshima City), Chihiro Kozuki (Daiyon. Elementary School, Hino City), Goto Manabu (Hakuoh University), Michihiro Goto (Tokyo Gakugei Univ. Oizumi Elementary School), Hidenori Kobayashi (Hiroshima Univ. Shinonome Elementary School), Akira Saito (Shibata Gakuen University Graduate School), Masahiko Sakamoto (The former Tokoha University Graduate School), Junichi Sato (Kunitachigakuen Elementary School), Hisatsugu Shimizu (Keio Yochisha Elementary School), Ryo Shoda (Seikei University), Masaaki Sugihara (University of the Sacred Heart, Tokyo), Jun Suzuki (Gakushuin Primary School), Shigeki Takazawa (Shiga University), Chitoshi Takeo (Nanzan Primary School), Hidemi Tanaka (Tsukuba Univ. Elementary School), Toshiyuki Nakata (Tsukuba Univ. Elementary School), Hirokazu Nagashima (Daisan. Elementary School, Kokubunji City), Minako Nagata (Futaba Primary School), Kiyoto Nagama (Hiyagon Elementary School, Okinawa City), Satoshi Natsusaka (Tsukuba Univ. Elementary School), Izumi Nishitani (Gunma University), Kazuhiko Nunokawa (Joetsu University of Education), Shunichi Nomura (Waseda University), Mantaro Higuchi (Kori Nevers Gakuin Elementary School), Satoshi Hirakawa (Showa Gakuin Elementary School), Kenta Maeda (Keio Yokohama Elementary School), Hiroyuki Masukawa (University of the Sacred Heart, Tokyo), Shoichiro Machida (Saitama University), Keiko Matsui (Hasuike Elementary School, Harima Town), Katsunori Matsuoka (Naragakuen University), Yasunari Matsuoka (Matsushima Elementary School, Naha City), Yoichi Matsuzawa (Joetsu University of Education), Satoshi Matsumura (Fuji Women's University), Takatoshi Matsumura (Tokoha University), Kentaro Maruyama (Yokohama National Univ. Kamakura Elementary School), Kazuhiko Miyagi (Homei Elementary School Affiliated with J.W.U), Aki Murata (University of California, Berkeley), Takafumi Morimoto (Tsukuba Univ. Elementary School), Yoshihiko Moriya (The former Kunitachigakuen Elementary School), Junichi Yamamoto (Oi Elementary School, Oi Town), Yoshikazu Yamamoto (Showa Gakuin Elementary School), Shinya Wada (Kagoshima University), Keiko Watanabe (Shiga University)

[Reviser]

Katsuto Enomoto (The former Harara Elementary School, Kagoshima City)

[Reviser of Special Needs Education and Universal Design]

Yoshihiro Tanaka (Teikyo Heisei University)

[Cover]

Photo : Tatsuya Tanaka (MINIATURE LIFE)

Design : Ai Aso (ADDIX)

[Text]

Design : Ai Aso, Takayuki Ikebe, Hanako Morisako, Miho Kikuma, Sawako Takahashi, Katsuya Imamura, Satoko Okutsu, Kazumi Sakaguchi, Risa Sakemoto (ADDIX), Ayaka Ikebe

[Illustrations]

Lico, Nanami Sumimoto, Ryoko Totsuka, Kinue Naganawa, Yoshie Hayashi, B-rise

[Photo・Observation data]

Ritmo, Aflo, Pixta, Eco-Mo Foundation, Geospatial Information Authority of Japan, Fujiidera City, Masaki Kubo, TOBU RAILWAY CO.,LTD. & TOBU TOWER SKYTREE CO.,LTD., Chubu Civil Engineering Office